MUSHROOMS
AND TOADSTOOLS OF BRITAIN & EUROPE Vol. 1

Puffballs, Earthstars, Stinkhorns, Chanterelles,
Toothed Fungi, Club Fungi, Coral Fungi, Polypores,
Crust Fungi, Boletes & their relatives,
Russula & Milkcaps

Geoffrey Kibby

Over 650 species illustrated in colour

2nd Edition. Published in Great Britain in September, 2017 by Geoffrey Kibby.

Copyright © Geoffrey Kibby. The author asserts his moral right to be identified as the author of this work. All rights reserved. No part of this publication may be reproduced, stored in a retrieval system or transmitted in any form or by any means, electronic, mechanical, photocopying, recording or otherwise, without prior permission of the copyright owner.

ISBN 978-0-9572094-3-5

Printed and bound by Gomer Press, Wales.
www.gomerprinting.co.uk

Using this book

Throughout this book some common terms and symbols are repeated. The species descriptions have a uniform format as shown below:

• the species name is always in ***bold italic***

• two asterisks (**) mean that the species has not yet been recorded in Britain

• author names follow standardised abbreviations (see p. xiv for more details on nomenclature)

• recent or important synonyms are given, preceded by the abbreviation Syn.

• important structures or characters are in **bold**

• specific epithets (the second part of the name) are translated into English as this helps to explain the derivation of the name or an essential character of the species.

• spore sizes are quoted in microns (µm)

• the scientific name is given for tree species in *italics*, the English names are given on page xxv

• notes on habitat, distribution and any further information about similar species are given at the end of the description

For the meanings of technical terms see the Glossary on page xxiii.

Some additional technical terms
• sensu = in the sense of, often abbreviated to ss.

• non sensu = not in the sense of

• sensu auct. = in the sense of many but not the original author.

A note on the spore drawings
Within a genus the spores are to scale with each other but not between different genera.

Notes on the illustrations
Wherever possible, useful chemical tests have been indicated on the paintings of the fruitbodies. If the fungus is exclusively found under a particular tree then the leaf of that tree may be shown with the mushroom. Mushrooms are not necessarily to scale.

Edibility
This book is intended as an identification guide, and although edibility is mentioned for some very popular and well-known species, as a general rule it is not mentioned for the remainder. Equally, well known toxic and possibly deadly species are indicated but be aware that edibility is not known in many cases. You are advised to consult a book specifically on eating mushrooms if this is your main interest—see Further Reading for some recommendations.

Using the pictorial key (page xxvi)
Try to find a fungus group that most closely resembles the fungus in hand. For many groups this will be quite easy, e.g the puffballs, earthstars or stinkhorns. For others it may be a little more difficult. Turning to the species descriptions and reading the description carefully should help to confirm or rule out your initial determination.

Be aware that the bulk of the gilled mushrooms are in volume 2. The gilled mushrooms included in this volume however, have good distinguishing characters which should help you to decide whether you have one of their species.

Second edition
This edition incorporates corrections to the first edition plus three new species: *Clavaria flavostellifera, Russula puellula* and *Russula tinctipes.*

Cantharellus alborufescens ** (Malençon) Papetti & S. Alberti
[alborufescens = white and reddish brown]
Syn. *C. henrici, C. ilicis, C. lilacinopruinatus*
Cap 3–9 cm across, fleshy, at first rounded, inrolled, then expanding and often with irregular, lobed margin, bright yellow-orange, orange to whitish yellow, without a whitish, pruinose coating. **Hymenium** decurrent, resembling wrinkled, frequently forking shallow 'gills', very pale whitish ochre, bruising brownish. **Stem** cylindric, 3–5 x 1–1.5 cm, solid, white to pale whitish yellow, conspicuously staining reddish brown. **Flesh** pale cream-yellow, firm. **Taste** mild. **Odour** pleasant, fruity. **Basidia** 4–5-spored. **Spores** ellipsoid to slightly constricted, 8.4–11.5 x 4.5–6.5 (-7) µm. **Spore deposit** pale yellowish ochre.

With *Quercus ilex* and *Q. rotundifolia*, usually on calcareous soils, probably widespread in the Mediterranean countries, not yet known in Britain but might be found under introduced evergreen oaks.

Contents

Preface i
Introduction ii
What are fungi? iii
Fungal classification vi
Orders, Families and Genera in Vol. 1 viii
The structure of larger fungi x
How to collect and study fungi xii
Chemical tests xiii
Nomenclature - names and the Science of Naming xiv
Preserving fungi xvi
Collecting equipment xvii
Resources xviii
Further reading xx
Glossary xxiii
Checklist of trees and shrubs xxv
Pictorial key to fungi in Vol. 1 xxvi

Illustrations and descriptions of species . . . . pages 1–221

1. Earthstars 2–7
2. Puffballs & Bird's Nest Fungi 8–17
3. Earthballs, Dyeballs & False Truffles 18–21
4. Stinkhorns & Cage Fungi 22–27
5. Chanterelles 28–33
6. Hedgehog, Tooth & Fan Fungi 34–45
7. Club & Coral Fungi 46–59
8. Polypores & Other Bracket Fungi 60–83
9. Resupinate Fungi 84–97
10. Jelly Fungi 98–105
11. Boletes & Their Relatives 106–139
    Paxillus, Hygrophoropsis, Tapinella, Aphroditeola 140–143
    Gomphidius, Chroogomphus 144–145
12. Russulas, Milkcaps & Their Relatives 146–221
    Lactarius 146–169
    Lactifluus 170–171
    Russula 172–219
    Lentinellus 220–221

Genera & Species Index 222

Preface

The writing of any field guide involving a large group of organisms will always be a difficult task and which species to include is very much the author's personal choice. This book and its companion volume, cover one of the larger divisions of fungi, the Basidiomycota, commonly referred to as basidiomycetes. There are probably 5000+ species of larger basidiomycetes in Britain, by which is meant those species easily visible to the naked eye. With so many species to choose from, which does an author pick? There are a large number of common species which will be found in any guide but that still leaves a large pool of species ranging from the occasional to the rare that could be mentioned. I have made a conscious effort to include species rarely or never seen in any recent field guide as well as some species only recently described as new to science. Several distinctive species known from mainland Europe but not yet recorded from Britain are also included.

This volume illustrates 650 species including several varieties, primarily of fungi which are not true agarics (the majority of mushrooms and toadstools with gills under their caps are agarics). Volume 2 includes over a thousand true agarics. Paintings rather than photographs have been used for a variety of reasons: they allowed me to show and highlight important characters which might be difficult to see in a photograph; species could be shown that I had never personally photographed and of course there is the aesthetic pleasure in producing paintings. In every instance drawings are added of spores and any other microscopic character which might be important or useful in the identification of the fungus.

No book of this sort is ever the product of just one person and this one is no exception. Numerous people have helped with advice, criticism, specimens and photographs or in the financing of its printing. First and foremost must be my friend and colleague Mario Tortelli, who was unstinting with his time and his continuing enthusiasm for this project. His advice and suggestions were invaluable, as was his badgering any time I slacked or felt like giving up! Alick Henrici and Martyn Ainsworth, two well known names in British mycology, were generous with their vast knowledge of all things mycological. For access to the wonderful resources in the libraries of the Royal Botanic Gardens, Kew and of the British Mycological Society I am immensely indebted. To the preceding I must add the names of Henry Beker, Trudy Fleming, Simon Gallagher, Eilidh Gardner, Barry Hodgson, David Lunn, Andy Overall, Roger Phillips, Irene Ridge, Sigrid Werner and Sue White, who were all instrumental in bringing this project to fruition. To them and to all who gave me their encouragement over the years my heartfelt thanks.

Geoffrey Kibby

Introduction

At the tender age of thirteen, more years ago than I care to remember, I was convalescing after surgery. By chance, at the same time, my father, who was a builder and plasterer, was asked to work on a house in Dersingham, Norfolk, directly opposite part of the Queen's large estate there. So my whole family moved for several weeks to stay in the house and I was left with little to do except wander around the beautiful countryside there.

Part of the Queen's estate was a large coniferous woodland surrounded by a fence, and this was on the other side of the main road opposite the house in which we were living. There was a gate into the woodland opposite the front door of the house and on the gate was a sign which said, if my memory serves, "Private, property of HM Queen Elizabeth" or words to that effect. As a self-respecting thirteen year old I promptly ignored the sign and climbed over the gate. I would add that now, in my older, wiser years, I would not advocate to any young person that they trespass, but in my defence I was a child of the 50's and had been brought up on stories by Enid Blyton, Boys' Own comics and the like and this is what boys did and indeed were expected to do. Those were happier, safer times and nothing happened. I dare say if that happened today then alarms might be triggered and hoards of SAS might appear from behind every bush, but nothing did and instead I entered a magical kingdom.

The woodland was shaded, humid and dripping with mosses, and just a few paces along the path from the gate I came across a small toadstool. It was small and an almost glowing amethyst-violet; the most amazing fungus I had ever seen. I had found —although I wasn't aware of it at the time—*Laccaria amethystina*, the Amethyst Deceiver. I still remember the wonder and the magic of that moment, at finding something which I had not known existed. Having found one fungus I went on to find more and from that moment on I was hooked. I am not the first to be so entranced; John Ramsbottom in his wonderful book *Mushrooms & Toadstools,* in the New Naturalist series, describes how two of the most famous mycologists of all, C.H. Persoon (1755–1837) and E.M. Fries (1794–1878) relate that it was the beauty of a fungus which first attracted them, the former by the orange cups of *Aleuria aurantia* and the latter by the shining white, spiny growths of *Hericium coralloides*. Ramsbottom goes on to say that for the great Italian mycologist A. Battarra (1714–1789) it was the scarlet cups of *Sarcoscypha coccinea* which entranced him. So I am in good company. I suspect that for many biologists and naturalists, whatever their particular passion or field of study, they will also remember some "eureka!" moment of their own when they knew that this was what they wanted to study.

So, on our first trip into Hunstanton, the nearest big town, I purchased my first of a long line of mushroom books, the *Observer's Book of Common Fungi*. It did not take me long however, to realise that, excellent though that small guide was as an introduction, many of the fungi I was finding were not to be found in its pages. So a week later I persuaded my long-suffering parents to take me to the bookstore again where they purchased for me the much bigger *Collins Guide to Mushrooms and Toadstools* by Lange & Hora. I could not have made a better choice. That guide, first published in 1963, used a selection of the paintings from the classic iconography of fungi by Jacob Lange, *Flora Agaricina Danica* (the five volumes of which I was to buy many years later). The Collins guide was the standard mushroom book for a whole generation of mycologists, including myself. Proof of its importance is that it was reprinted six times, the last being in 1978. Eventually it was superseded by other guides, all of which I purchased and now, looking around my room as I type this introduction, I see shelf after shelf full of books on fungi from all over the world.

From those early days, fungi went on to become a central part of my life. They have taken me to many different countries around the world, introduced me to so many wonderful people, led me to become the senior editor of the mycology journal *Field Mycology*, to become the author of numerous books of my own and finally to this, perhaps the culmination of my life's work and studies. If this book goes on to help others discover their own passion for fungi then I will be content.

What are fungi?

In Britain we tend to call fungi mushrooms or toadstools. The word mushroom is derived from the French *mousseron* from *mousse*, which means moss, a reflection perhaps on the sort of habitat in which they were commonly found. Toadstool on the other hand is rather more obvious, with variations such as paddockstool or puddockstool in Scotland and the north of England. Certainly, toadstools are found on occasion with a small toad or frog sitting on them as seen in the photo below. For the British, mushrooms tend to equate with edible fungi and toadstools with poisonous fungi but in reality the two names are interchangeable and the distinction is not a good one, many toadstools being perfectly edible and some mushrooms being toxic. Most other countries do not have this artificial distinction and in this book they will all be referred to as either mushrooms or just fungi.

Fungal Biology

Fungi are quite distinct from both plants and animals, although genetically and in their biochemistry they are more closely related to animals. Traditionally however, they were often studied along with plants. So what makes a fungus so unique? The fruitbodies known as mushrooms have structural and chemical characteristics akin to both plants and animals. The cell walls of fungi contain chitin which is the basic constituent found in the exoskeleton of arthropods. Plant cell walls on the other hand contain cellulose, a compound absent in fungi. Unlike plants, fungi lack chlorophyll so have developed other means of acquiring the organic carbon and other nutrients they need to grow; this is similar to animals. Unlike animals however, they do not digest these nutrients internally, instead, fungi do it by secreting enzymes that release nutrients from the chosen food source, which they then absorb directly through the cell walls. Fungi have evolved several different life strategies and ways of feeding and they interact with neighbouring organisms in different ways, of which the principal ones are described below.

Mycorrhizae

Just below the surface of the soil or other substrate is the main body of the fungus. This is the mycelium which, to the eye, looks like fine threads or filaments. The mycelium branches out and in many species of fungi may penetrate the fine roots of a tree or shrub binding it in a special sheath of hyphae around the root tip. This association with the plant roots is called a mycorrhiza. The mycorrhiza facilitates a back and forth exchange of nutrients between the fungus and the tree and is therefore a mutualistic symbiosis, a mutually beneficial relationship.

A genuine toadstool, this is an unstaged photograph of a small toad found resting on the cap of a *Lactarius pubescens* in Scotland.

There are two common forms of mycorrhiza: endomycorrhiza and ectomycorrhiza.
- **Endomycorrhiza** is present in some 85% of flowering plant families. In this relationship not only do the fungal hyphae grow inside the root of the plant but they penetrate the root cell walls and become enclosed in the cell membrane as well. The penetrating hyphae create a greater contact surface area between the hyphae of the fungus and the plant. This heightened contact facilitates a greater transfer of nutrients between the two. Endomycorrhizae have further been classified into five major groups: arbuscular, ericoid, arbutoid, monotropoid, and orchid mycorrhizae where the root cells of the host plant have been directly penetrated by the fungus; endomycorrhiza is the form adopted by many ascomycete fungi (the phylum that includes the cup fungi).
- **Ectomycorrhiza** is formed between the roots of around 10% of plant families (mostly woody plants, including many tree and shrub genera, as well as some orchids), and fungi belonging to the *Basidiomycota*, *Ascomycota*, and *Zygomycota*. Some ectomycorrhizal fungi, e.g. many *Leccinum* and *Suillus* species, are symbiotic with only one particular genus of plant, while other fungi, such as *Amanita*, tend to be generalists that form mycorrhizas with many different plants. An individual tree may have 15 or more different fungal partners at one time. Thousands of ectomycorrhizal fungal species exist, hosted in over 200 genera of plants. A recent study has conservatively estimated global ectomycorrhizal species richness at approximately 7750 species, although, on the basis of estimates of knowns and unknowns in macromycete diversity, a final estimate of ectomycorrhizal species richness would probably be between 20,000 and 25,000.

Ectomycorrhizas consist of a sheath of hyphae, or mantle, covering the root tip and a net of hyphae surrounding the plant cells within the root cortex referred to as a Hartig net. In some cases the hyphae may also penetrate the plant cells, in which case the mycorrhiza is called an ectendomycorrhiza. Outside the root, the fungal mycelium forms an extensive network within the soil and leaf litter. Nutrients have been shown to move between different plants through the fungal network.These relationships between plant and fungus are known to be essential components for the health of forests across the globe, many trees for example being unable to reach their full growth potential in the absence of the fungi.

Saprotrophs
This is the feeding strategy adopted by a whole host of fungi that slowly break down dead wood, dung, leaf litter and even animal carcasses. These fungi (along with many bacteria) form an essential component of our ecosystems, as without them all of these areas would resemble a huge refuse tip of dead and dying matter. Examples of saprotrophic fungi include the common *Agaricus* mushroom purchased in your local shop, and the puffballs which are a prominent feature in woodlands and fields.

Parasites
Parasitic fungi obtain their nutrients directly from the living tissues of their host which can include both plants and animals. Almost every plant including most woodland trees (and many animals, including man), will be subject to some sort of parasitic fungus, in some cases leading to the death of the host plant. In woodlands these fungi can be one of the major causes of timber loss. Examples of common parasitic fungi include the Honey Mushroom, *Armillaria mellea* and many of the bracket fungi or polypores which are so common in our woodlands. Many fungi, having once killed their host will then continue to feed on the dead remains in a saprotrophic manner.

Fungal Reproduction
All of the larger fungi reproduce sexually by producing spores. There are two principal methods of spore production, in the **Phylum *Ascomycota*** (the cup fungi) the spores are formed *internally*, inside a special, usually club-shaped cell called an ascus (Fig. 1) from which the spores are expelled when mature, rather like the bullets from a gun. There are commonly eight spores in each ascus although in some cup fungi there may be 16 or even more. These spores are referred to as ascospores.

In the **Phylum *Basidiomycota*** the spores are formed *externally*, on a special cell called a basidium (Fig. 2). The spores are formed at the tips of structures called sterigmata and commonly there will be four spores although 1, 2, 3 or even more may also occur. These are

Order Boletales
Families and genera included
Boletaceae
(Boletes)
- Aureoboletus p.120
- Baorangia p.118
- Boletus p.108
- Buchwaldoboletus p.120
- Butyriboletus p.116
- Caloboletus p.110
- Chalciporus p.122
- Cyanoboletus p.120
- Hemileccinum p.118
- Hortiboletus p.124
- Imleria p.118
- Imperator p.114
- Lanmaoa p.118
- Leccinum p.130
- Neoboletus p.110
- Phylloporus p.128
- Porphyrellus p.128
- Pseudoboletus p.120
- Rheubarbariboletus p.124
- Rubroboletus p.112
- Strobilomyces p.128
- Suillellus p.112
- Tylopilus p.128
- Xerocomus p.122
- Xerocomellus p.126

Astraeaceae
(Barometer Earthstars)
- Astraeus p.4

Gomphidiaceae
- Chroogomphus p.144
- Gomphidius p.144

Gyroporaceae
- Gyroporus p.138

Hygrophoropsidaceae
(False Chanterelles)
- Aphroditeola p.142
- Hygrophoropsis p.140

Paxillaceae
- Gyrodon p.138
- Paxillus p.140

Pisolithaceae
- Pisolithus p.20

Rhizopogonaceae
- Rhizopogon p.20

Sclerodermataceae
(Earthballs)
- Scleroderma p.18

Suillaceae
- Suillus p.134

Tapinellaceae
- Tapinella p.142

Order Russulales
Families and genera included
Albatrellaceae
- Albatrellus p.82
  (incl. with the polypores for convenience because of the pored hymenium)

Auriscalpiaceae
- Artomyces p.56
- Auriscalpium p.42
- Lentinellus p.220

Hericiaceae
- Hericium p.44

Russulaceae
- Lactarius p.150
- Lactifluus p.170
- Russula p.172

Order Agaricales
Families and genera included
Agaricaceae
(Puffballs)
- Battarrea p.10
- Bovista p.14
- Calvatia p.10
- Lycoperdon p.12
- Mycenastrum p.10
- Queletia p.10
- Tulostoma p.10

(Bird's Nest Fungi)
- Crucibulum p.16
- Cyathus p.16
- Nidularia p.16

Clavariaceae
(Club Fungi)
- Typhula p.48
- Clavaria p.50
- Clavulinopsis p.52
- Ramariopsis p.54

Pterulaceae
- Pterula p.48

Mycenaceae
- Favolaschia p.82
  (incl. with the polypores for convenience because of the pored hymenium)

Note:
Some families and genera in this book have been illustrated out of their correct Order so as to put them with other genera of similar shape and/or structure. So, for example, the families and genera in the Order *Agaricales* have been broken up and placed after genera of very similar appearance but in different Orders, as it makes for easier identification. This table shows the correct scientific placement as currently understood.

The Structure of Larger Fungi

In the larger fungi the basic unit of construction is formed of microscopic cotton-like threads called hyphae. The term mycelium is used to describe the complete web of hyphae that makes up each individual fungus. The mycelium forms the basic vegetative structure, each hypha penetrating and branching throughout the growing medium (such as humus, wood, or leaf litter), absorbing the required nutrients over its entire surface area. From this mycelium the fruitbodies will eventually be initiated. Individual hyphae or parts of the mycelium become highly modified in the fruitbodies, but even in the most complex mushroom the basic hyphal structure can still be easily recognized when viewed through a microscope.

Many people believe that when they observe a mass of toadstools in autumn these are the only parts of the fungus that ever form. In fact all year round hyphae are to be found growing in the soil and only at the time of reproduction do the fruitbodies we all recognize develop and become visible. The function of the fruitbodies is to produce spores, which are dispersed and will germinate to produce new mycelial colonies. The form of the fruitbody was always the traditional basis for classifying fungi, and in the larger fungi described in this book the fruitbody characters are still used to help identify fungi right down to the species level.

The fungal hymenium

Fungi have evolved many ways to disperse their spores and in most of the types illustrated in this book the spore-producing basidia are formed on a specific area called the hymenium. In gilled mushrooms (gills is the popular term for what are technically called lamellae) the hymenium is spread over the surface of each gill. In the toothed mushrooms the hymenium coats the spines or teeth and so on. Exceptions occur in such fungi as puffballs, earthstars, , etc where the basidia and spores are formed inside the ball-shaped fruitbody in special tissues called the gleba or spore mass. In the stinkhorns and cage fungi the spore mass later liquifies, releasing a strong odour to attract invertebrates which eat and disperse the spores.

Fungal spores

The spores are often of critical importance when it comes to identifying a fungus down to species level. Although many species can be identified in the field, based purely on their visual appearance (i.e. the Fly Agaric, *Amanita muscaria* or the Cep, *Boletus edulis*), many others can only be identified reliably if examined microscopically. This can be daunting at first for beginners who may never have used a microscope since their school days but with a little practice, some good literature or perhaps by attending a short course, anyone can soon master the techniques involved. Spores have specific shapes, and some have surface structures such as warts and each species will have spores that fall within a certain size range.

Some of the shapes have very precise technical words to describe them and it is important to understand them as they are used in most mushroom books, including this one. So the most common terms are described and illustrated opposite (Fig. 4) as well as in the glossary.

In measuring fungal spores it is very important that this is done under high magnification to minimise error, preferably at x1000 using an oil immersion objective lens. Any basic book on microscopes will explain these terms or there are many very useful resources on the Internet which will guide you in using your microscope. Spores usually have a visible attachment point at one end—the apiculus—where the spore used to be joined to the sterigma of the basidium. It is usual not to include this when measuring the spore length. If any surface ornamentation is present (warts, spines, network, etc) you should make careful note of the structure and if possible measure the height of the ornamentation. This is particularly important in the genera *Russula* and *Lactarius* for example.

Fungal spores may be completely transparent (hyaline) and without any inherent colour but they may also be coloured and it is usual to try and obtain a spore deposit to assess this. In the case of gilled fungi the cap is usually placed gills down on a piece of glass or on white paper and then covered with a plastic cup or other container to retain moisture. After a couple of hours there should be a sufficient deposit of spores to be able to assess their colour. It is important to let the spore print air dry for a few minutes after removing both cup and mushroom and only then carefully scraping the deposit into a small pile. Placing a glass cover slip on this pile to flatten it is helpful to show the colour accurately. Sticky tape may then be used to seal the edges (see Fig. 8). You will often discover that the spores are much darker than they first appeared when still spread out. Spore colour may be white, cream, yellowish, ochre, rust-brown, brown, dark brown, purplish, black or rarely green or red.

For other odd fungi such as the club fungi, jelly fungi, etc, just lay the whole fungus down on the glass or paper and follow the same procedure. Not all fungi will oblige with a nice print and you may be forced to scrape off or cut off a small piece of the hymenium in order to look at spores, although they are then often at various stages of maturity and this will give a wider range of spore measurements. Measuring a good number of spores will help even out any discrepancies (10–20 spores is good).

Cystidia

A great many fungi have other specialised cells in the hymenium or, in the case of many mushrooms, even on the cap or stem surface. The most common are cystidia and these come in a wide range of shapes and sizes and once again there may be technical terms for these. A selection of common cystidia types are illustrated below (Fig. 5). Cystidia on the edge of a gill or lamella (or spines, pores, etc) are called **cheilocystidia** while those on the face of the gill (or spines, pores, etc) are called **pleurocystidia** – terms you will find used again and again in this and other books.

Cystidia on the stem (or stipe as it is technically called) are called **caulocystidia** while those on the cap (or pileus) are referred to here as **pileocystidia**. Some cystidia will have noticeably thickened cell walls and this is an important character as is any inherent colour. Some will change colour when certain chemicals are applied (viewed under the microscope). A list of useful chemicals is provided on p. xiii.

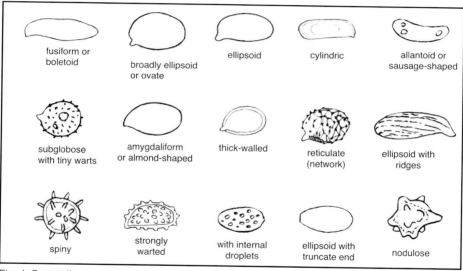

Fig. 4. Spore diagrams showing the variety of shapes and ornamentation.

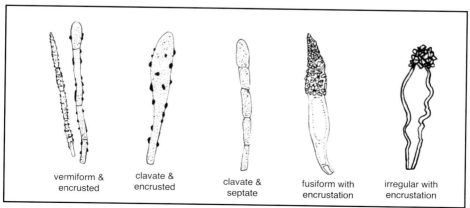

Fig. 5. Cystidia showing shapes and surface ornamentation.

Clamp connections

A character of great importance in separating some genera and even species is the presence or absence of clamp connections. When present these curious structures may be found in various parts of the fruitbody. Sometimes they are confined to the very base of a basidium, at other times just in the cells of the cap cuticle or perhaps throughout the tissues, or they may be absent. Clamp connections are formed by the terminal hypha during elongation. Before the clamp connection is formed this terminal segment contains two nuclei. Once the terminal segment is long enough it begins to form the clamp connection. At the same time, each nucleus undergoes mitotic division to produce two daughter nuclei. As the clamp continues to extend, one of the newly formed nuclei moves into the clamp (red spots). While this is occurring the remaining nuclei begin to move apart from one another to opposite ends of the cell (blue spots). Once all these steps have occurred a septum forms, separating each set of nuclei (Fig. 6). This is to ensure that each cell within the hypha has its own set of nuclei.

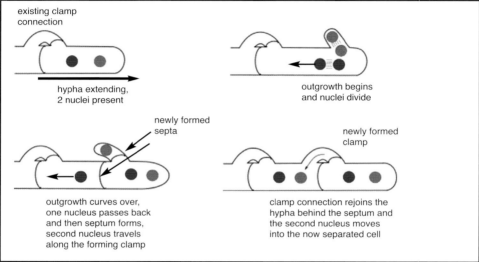

Fig. 6. Diagram showing the process of clamp formation.

Looking for clamps in a microscope preparation can be a thankless task, especially when they are absent! It may take some time to convince yourself that they are indeed missing and not just scarce…Figure 7 shows a fungal hypha with a clamp connection.

Fig. 7. A clamp connection on a fungal hypha.

How to Collect and Study Fungi

To improve your chances of identification there are a number of steps that will help.

• Collect the specimen carefully so that its important features remain intact. Try to handle the specimen as little as possible, delicate features such as hairs, scales, etc may be lost, handling may cause colour changes. Try to collect different stages of growth; young buttons often show characters missing on the mature fruitbody. Only pick where you know it is *allowed*! Many woodlands today have strict regulations regarding mushroom picking.

• Ensure that you collect the whole fruitbody, there may be parts below the soil level that could be left behind if carelessly picked. Do not pick more than required.

• While still in the field make sure to smell the specimen, many fungi have distinctive odours. As smell is a very individual thing it is a good idea to get a second opinion.

- A good, close-up photograph can be enormously helpful, modern smart-phone cameras make this very easy should you not have a real camera to hand. Make sure you photograph from several angles, especially if the fungus has a cap, you need to show the underside also.

- Taste can be very important in some fungi (i.e. *Russula*, *Lactarius*) but should only be assessed if you are certain which genus you have and even then only a tiny piece of tissue should be tasted and then spat out. It is not recommended to taste known toxic fungi AT ALL. For complete beginners this step should be omitted.

- Wrapping the specimen in kitchen foil or greaseproof paper will help keep it intact.

- As soon as possible make detailed notes of the colour, size, shape, and any distinctive features, making particular note of any colour changes after bruising. Cutting the specimen in half may reveal further colour changes and any interesting internal structures.

- Once back home the application of some common chemicals (listed below) to the cap/fungus surface and the flesh may reveal interesting changes.

- Place the specimen down for a spore deposit on glass (microscope slides are useful for this) or on paper as described earlier.

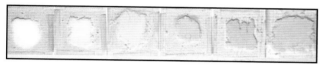

Fig. 8. Examples of *Russula* spore deposits scraped together and sealed under glass microscope cover slips.

Chemical Tests

Listed here are some chemicals which will be useful in the study of fungi. A list of sources for chemicals, microscope supplies, etc is given at the end of this book.

Ammonia. An aqueous solution of 75% ammonia. Common household ammonia is available from many hardware stores and is perfectly suitable for testing fungi. Keep in a tightly sealed dropper bottle. Note that it will become weaker over time as it absorbs moisture.

Congo Red in ammonia. Saturated solution. A universal stain for fungal tissues. Note: a very strong stain, it will stain almost everything, including you!

Cresyl Blue (Brilliant Cresyl Blue). Another very useful stain for fungal tissues, etc.

Ferrous sulphate ($FeSO_4$). 10% solution or crystal (preferred) is a very useful chemical, especially for the genus *Russula* where the reaction on flesh varies from nil to pink, salmon-pink, rust or even dark green.

guaiac. 1 part guaiac to 5 parts 60–70% alcohol. Very useful in the study of *Russula*.

Melzer's Iodine solution. An essential stain for many fungal spores, but especially for *Russula* and *Lactarius* where the ornamentation cannot be seen without this stain. Can only be acquired from specialist suppliers.

PlaqSearch. A commercial dental stain with both red and blue dyes within it, produces very useful staining on fungal cells of all sorts. Becoming very popular as a standard stain. Obtainable from online suppliers (Amazon, etc) in liquid form.

Potassium hydroxide (KOH). 5–10% solution in water (caution - a caustic chemical, will slowly burn holes in clothes and skin). Essential chemical, many fungi turn orange, red, purple, etc when it is applied to outer surface or flesh. A similar solution of caustic soda (sodium hydroxide (NaOH), common drain cleaner) will usually produce identical results.

There are other chemicals in use by mycologists but those listed cover most requirements.

Nomenclature – Names and the Science of Naming

All organisms today, whether fungi, fleas or humans have a scientific name. The common chanterelle mushroom is *Cantharellus cibarius*, the human flea is *Pulex irritans*, and humans are called *Homo sapiens*. The names are in two parts, usually derived from Latin or Greek. The first part of the name indicates the genus to which the organism belongs, so the chanterelle is in genus *Cantharellus*. The second half of the name—the specific epithet—denotes the particular species involved. This system of modern nomenclature (the naming of organisms), was formalised by one important man: Carl Linnaeus.

Carl Linnaeus (1707–1778), later Carl von Linné after being ennobled, published a seminal botanical work, *Species Plantarum* in 1753. He was the first to stabilise the use of just two words as the scientific name of any organism—the binomial system, e.g. *Agaricus campestris*. This is now seen as the beginning of scientific nomenclature. His name is sometimes Latinized into Carolus Linnaeus (Fig. 9).

Before this time a variety of naming systems were in place; often a name was little more than a description of the organism involving many words in a sentence. And of course many common names were in use (and still are today). The problem with common names is that there is no universally agreed set of names; every country and sometimes every district may have its own common name for a fungus. So for *Boletus edulis*, the well known edible fungus, we can call it Cep, Penny Bun, Steinpiltz, Herrenpilz, Porcini, King Bolete and many more. Try communicating a common (vernacular) name, even in English-speaking countries and you will invariably come unstuck; different regions having different common names. A scientific name however is universal, it will be understood wherever you are.

People often feel that scientific names are difficult to learn but in fact we use many scientific names in our daily lives without thinking of it. No one seems to find the garden *Chrysanthemum*, *Clematis* or *Pelargonium* at all unusual or difficult, but these are all scientific names derived from Latin.

The scientific naming of fungi today is bound by a rather rigorous procedure: the *International Code of Botanical Nomenclature*. This is a set of rules, internationally agreed, as to what you can and cannot do when naming fungi. One of the most important rules is that whenever a new species is described the original description must cite an actual collection on which the description and new name is based – this is called the holotype. This specimen must be deposited in a recognised national fungarium (such as the one in the Royal Botanic Gardens, Kew for example). The original description is usually written in Latin (considered a universal scientific language) although this has recently ceased to be a requirement and English is now also accepted. It must be in a published book or journal or online site such as Index Fungorum (www.indexfungorum.org). When very old descriptions have no type material available then an image may be cited as the type, known as an iconotype.

Author citations

If you look at the mushroom names in this book you will see that each name is followed by one or more author's names, often abbreviated, for example *Craterellus cornucopioides* (L.) Pers., the Horn of Plenty mushroom. This is known as the author citation. In this example Linnaeus (abbreviated as L.) was the original author of the epithet 'cornucopioides' in 1753 but he placed it within the genus *Peziza*. The second name, Pers. indicates that the mycologist Persoon later transferred it to the genus *Craterellus* in 1825, where it can be found today. The first author's name is placed in brackets.

Perhaps more confusing are names where the same author appears several times such as *Craterellus lutescens* (Fr.) Fr. In this case Elias Fries described the species as *Cantharellus lutescens* in 1821 but then decided to transfer his own species to the genus *Craterellus* in 1868.

Instances where there is just a single author's name such as *Boletus satanas* Lenz means the fungus in question remains where it was originally placed by the describing author.

Sometimes an author publishes a name but does not fulfil the requirements of the code. For example *Lyophyllum caerulescens* was described by a mycologist called Clémençon but his description was missing one of the points required by the code (he neglected to cite an actual type specimen), so I then validated his description by referencing a type specimen.

The name then became *Lyophyllum caerulescens* Clémençon ex Kibby.

Where a species has been described more than once by different authors (unaware that the other had also described it) the usual rule is that the earliest published name at the rank of species wins, but there are exceptions and deciding which is the correct name to use for any given species can be a difficult and tricky problem. The published rules are extensive and often complicated and not always easy to interpret.

This is one of the reasons that names change over the years (a constant source of annoyance to many…): earlier valid names are sometimes discovered, or someone decides that a mushroom is better placed in a different genus. In the latter case not everyone may agree with the change and you are not forced to use the new name if you disagree. The genus *Boletus* is a good example with very many name changes having taken place in the last 10 years or so, based almost entirely on recent molecular studies involving analysing the fungal DNA.

Fig. 9. Carl Linnaeus, later Carl von Linné, is the founder of our modern nomenclatural system.

One of the many difficult decisions any author of a field guide such as this has to face is which names to use, how 'cutting edge' should one be? I have tried to base my decisions on whether the name changes derive from thorough research, often involving modern DNA techniques and also whether any name change has already been accepted by the wider mycological community; thus I have accepted the recent changes in boletes for example. Such decisions are always something of a balancing act and not everyone will agree with them. To help avoid too much confusion I have provided recent synonyms and you should always be able to find a fungus by its older name in the index.

Which author's names to abbreviate and when to use their initials, is based on the updated list available at Index Fungorum, where you can download a copy of the full list. (www.indexfungorum.org/Names/AuthorsOfFungalNames.asp)

Common abbreviations you will see many times include the following, although there are many hundreds more.

Berk. = M.J. Berkeley
Bres. = Giacomo Bresadola
Bull. = J.B. Bulliard
Fr. = Elias Fries
Huds. = William Hudson
L. = Carolus Linnaeus
Pers. = C.H. Persoon
Quél. = Lucien Quélet
Schwein. = L.D. Schweinitz

An invaluable work for checking what species are officially British, where they have been found and what their current name is (along with the author citation) is the *Checklist of the British & Irish Basidiomycota* by N.W. Legon and A. Henrici (2005) published by the Royal Botanic Gardens, Kew. The printed version is still available but note that this is now updated regularly with new species and name changes, online at: www.basidiochecklist.info and this resource should always be checked for the latest information.

Preserving Fungi

Once you begin studying fungi you may decide that you need to keep a specimen for later study. Perhaps you have been unable to decide on the identity of a collection or want a second opinion. You may not have a microscope and would like someone else who does to check the specimen for you. Keeping a mushroom in the fridge in an airtight container will be fine for a day or two but the fungus will quickly start to spoil and once the tissues start to collapse or rot then it may already be too late. The answer is to air dry the specimen until all of the moisture is removed. There are several ways to do this. If you are really enthusiastic and envisage drying lots of material then a commercial fruit dryer is the quickest and easiest method. They have several stacked trays of plastic mesh and usually have a heated fan which blows warm air up through the stack and is thermostatically controlled; but they can be expensive (Fig. 10).

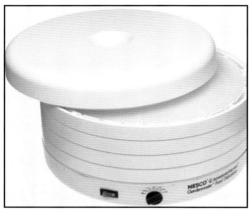

Fig. 10. Commercial fruit dryer.

If it is late autumn or winter and you have the central heating on you can fit some wire mesh over a radiator and place the mushroom (usually cut in half or even in vertical slices if very robust) on the mesh to dry. This is obviously less controllable but actually works very well, although family members may object to having the smell of drying fungi in the living room…

You can make a temporary and indeed portable dryer from a cardboard box and a low wattage light bulb (Fig. 11). Punch two holes wide apart and low down on opposite sides of the box through which you push two wooden or bamboo rods. On these you place a sheet of pre-cut wire mesh. Several layers of mesh may be positioned up the sides of the box if you wish. Near the very bottom of the box, on one side, cut a hole large enough to put through a cable and light socket and then fit the bulb; the bulb should be just off the bottom of the box, not touching the bottom. Finally punch a number of small holes in sides of the lower third of the box on all sides and the top to encourage good air flow. As a safety measure you can line the box first with kitchen foil. When the bulb is switched on the heat generated will rise up and dry the mushroom slices quite well. Note that the bottom of the box may get very warm, so don't rest the box on valuable furniture!

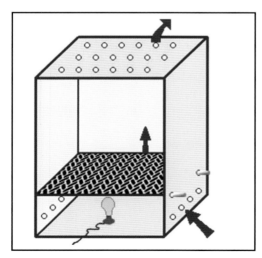

Fig. 11. Home made cardboard box mushroom dryer showing the air flow (blue arrows). Front panel removed for clarity.

Do not overheat your mushrooms otherwise they will simply cook rather than dessicate, hence drying in the oven is not recommended. The most important point is to ensure good warm air flow to ensure rapid extraction of the moisture. When dry the specimen should feel light and crisp but not burnt (very black specimens are likely to be the latter). Once dried these will keep all their important microscopic features almost indefinitely and after suitable resuscitation in water or other more specialised wetting fluids, can be examined under the microscope for cystidia, spores, hyphae, etc. This dried material can also be used for molecular studies.

Collecting Equipment

There are some basic pieces of equipment with which anyone studying fungi should equip themselves, none more important than a hand lens or loupe. These small, usually folding lenses come in a variety of magnifications and qualities and it is worth spending as much as you can afford to get high quality optics. Prices range from a few pounds to several hundred but around £15–40 should get you one of reasonable quality. A magnification of x10–12 will be perfectly adequate and will allow you to see all sorts of details in the field that you might miss with the naked eye (Fig. 12). An important feature is whether the image is flat from edge to edge; avoid lenses where there is a lot of distortion at the edges. Many lenses now have a small LED light incorporated and this is particularly useful in dim woodlands.

Fig. 12. A typical jeweller's loupe with a different lens at each end.

Traditionally, mushrooms were collected in a woven basket and many people still prefer to use one, especially with edible fungi. They are not very practical for tiny fungi however, specimens may get rather bruised if they tumble around in the basket. Wrapping the fungi in a twist of kitchen foil or waxproof paper will help in preserving their delicate characters.

Most mycologists prefer to use a plastic box with internal dividers so each fungus can be kept separately and different size chambers can be used for different sized fungi. Many of these boxes allow the partitions to be moved around or removed, so you can customise the size of the chambers; hardware stores or fishing tackle shops are good places to look. Also, if the lid is transparent you can use a marker pen to write a number on the lid above each chamber. This number can then relate to any notes you might make in the field.

Many collectors use a penknife, garden dibber or some other implement to help remove the fungus from the soil, wood or other substrate, as often the part that is in the ground or wood is as equally important for identification as what is above soil.

A small mirror is a popular accessory, it allows you to see the underside of a mushroom cap without having to pick the fungus. The sort of mirror used by dentists – a small round mirror attached to a thin rod – is particularly useful and is widely available online.

If you are going to photograph fungi in the wild then a small paintbrush to clean dirt and leaves off the specimens will be a great addition to your photographic equipment. If your chosen camera is your smart phone then it will make a huge difference if you purchase one of the several adaptors available that allow you to attach the phone to a standard tripod. These adaptors only cost a few pounds and in a dimly lit forest will ensure you can handle long exposures in low light without the problem of camera shake.

Eventually you may decide you want to purchase a compound microscope to be able to see fungal spores, etc and you don't have to spend a fortune to get a perfectly adequate microscope (although you can). What is important is that it has a selection of different magnification objectives, in particular a x100 oil-immersion objective which will allow you to examine spores at x1000 when used with x10 eyepieces (Fig. 13). If your budget runs to it then getting a microscope with a binocular eyepiece head and what is known as Kohler illumination will definitely improve the quality of the images you can obtain. Contacting any good microscope dealer (several are listed in the resources section on p. xviii) is recommended as they can advise you on different models and pricing and what microscope is best for you.

Stereo microscopes on the other hand work at much lower magnification, usually from around x20–x80, but as they provide a stereoscopic, three-dimensional image and have a much larger working distance between the objectives and the specimen they are invaluable for when you need to work on very tiny specimens or when cutting up a specimen to make a slide to examine under the compound microscope. Models start from around £50.

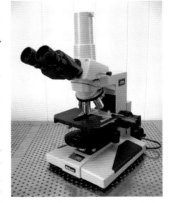

Fig. 13. A compound microscope with binocular head and a third tube for photography.

Resources

Microscope Dealers

Brunel Microscopes Ltd
Unit 2, Vincients Road
Bumpers Farm Industrial Estate
Chippenham
Wiltshire SN14 6NQ
Tel: 01249 462655
Email: mail@brunelmicroscopes.co.uk
Website: www.brunelmicroscopes.co.uk

A long-standing and very popular supplier of microscopes, stains and other chemicals as well as photographic accessories.

Meiji Techno UK Ltd
The Vineyard
Hillside
Axbridge
Somerset BS26 2AN
Tel: 01934 733655
Email: enquiries@meijitechno.co.uk
Website: www.meijitechno.co.uk

Providing microscopes for over 50 years as well as photographic cameras and accessories. A very popular dealer.

Microscience
1556 Menom Rd
Newton Aycliffe
County Durham DL5 6AU
Tel: 01916 451910
Website: https://micro-science.co.uk

A very helpful dealer that supplies microscopes, microscope cameras, loupes and chemicals as well as mycological books.

Mycological Societies

British Mycological Society
Charles Darwin House
12 Roger St
London WC1N 2JU
Tel: 020 76852676
Email: admin@britmycolsoc.info
Website: www.britmycolsoc.org.uk

The Society was founded in 1896 to promote the study of fungi in all their diversity. Since then it has become one of the major mycological societies in the world. They publish a number of journals including *Field Mycology* (of which I am Senior Editor) with articles aimed at all levels from beginners to more advanced mycologists. The BMS organises field meetings for members both in the UK and overseas. They also have an extensive library of mycological books and journals available for members to borrow. Membership details are available on their website.

1. Earthstars

Earthstars represent some of the most intriguing and complex fungi that you may encounter. Many of the species are uncommon to rare, some extremely so and they all have very precise habitat requirements.

They begin as a tough, onion-like ball which may be entirely or partially buried in the soil. The apex of the ball is usually pointed and the outer skin or **exoperidium** (Fig. 1.1) will split open and recurve backwards to form star-like arms or rays. These serve to raise the fungus out of the soil while at the same time pushing back any covering leaves or soil. The opening of these rays will expose the central spore-containing ball which consists of an outer skin called the **endoperidium** and an internal mass of spore-forming tissues called the **gleba**. The apex of the endoperidium forms an opening called the **peristome** (Figs 1.4, 1.5). The spores will be ejected from this aperture by means of raindrops hitting the surface of the ball and thus puffing them out, or by the action of wind passing across the mouth of the peristome and causing a reduction in air pressure at the mouth which essentially 'sucks' the spores out. The spores themselves are produced on **basidia** which may have as many as 8 or even 11 **sterigmata**. These are enmeshed within a mass of thick-walled hyphae called the **capillitium**.

Exoperidium

The number of rays produced as the exoperidium splits, as well as their particular structure, is of considerable importance in the separation of species. Some species may have as few as 5 or 6 rays while others as many as 10 to 13.

The inner surface of the rays is usually smooth while the outer surface has a mycelial layer composed of thin- or thick-walled interwoven hyphae. In some species this mycelial layer binds tightly to the surrounding soil and hence the rays will have an encrusting layer of soil and debris. In other species however the surface remains relatively smooth and this is a useful identification character (Fig. 1.6). The colour of the rays can vary from pale cream to a rich brown or greyish brown.

The thickness of the exoperidium varies between species as does the complexity of its internal structure. In those species with especially thick rays the inner surface frequently cracks and splits as it is unable to flex sufficiently as the rays bend backwards. This is best seen in the common *Geastrum triplex* where the splitting is so extreme as to frequently leave behind a collar surrounding the endoperidial ball.

In two or three species the arms are distinctly hygroscopic, curling inwards if the weather is dry and then reflexing back out when the air is moist.

Three British species are distinguished by the mycelial layer splitting away from the other inner layers. The mycelial layer adheres to the soil so firmly that as the earthstar's rays flex and lift the fungus the mycelial layer remains as a basal cup attached only at the tips of the rays (Fig. 1.2).

Endoperidium

The spore-containing ball exposed when the rays reflex back is usually subglobose but in some species is more ellipsoid or pear-shaped. The surface in some species is distinctly granular from a fine coating of minute calcium oxalate crystals (Fig. 1.4). The size of the crystals is important and fairly constant in the species concerned. The structure of the peristome is of great importance in identification of species. The base of the ball may be directly attached to the centre of the rays (sessile) but in some species is raised up on a distinct **stalk** (stipitate) (Fig. 1.4); this frequently becomes more obvious as the fungus dries out. There may also be a distinct ridge at the base of the ball, just above the stalk, known as the **apophysis** (Fig. 1.4).

Peristome

The aperture through which the spores escape may form a simple, ragged hole while in others this region may be demarcated by a distinct circular ridge or line, often darker or paler in colour. In one group of species an aperture forms a distinct cone and this is ridged or fluted (use a hand lens). This feature is a useful character when identifying species (Fig. 1.5).

Spores

Geastrum spores are more or less globose, ranging from around 3 µm to 8 µm in diameter depending on the species. The size is measured excluding any warts or ridges that may be present. The spores of all species are ornamented with distinct warts and the size and prominence of these warts is again an important feature to note (Fig. 1.3). Spore colour is of less importance than in some genera but varies from medium brown to a dark purplish brown.

Capillitium

Inside the endoperidial ball the tissues consist of interwoven threads of hyphae in a mass called the capillitium, along with spore-forming basidia. These capillitial threads are usually unbranched, thick-walled and either smooth or ornamented with encrusting blobs or warts. Their detailed structure is of importance in the definition of species but requires in-depth microscopy and interpretation and is not included this guide.

Habitat

Earthstars are usually found in drier situations, often on sandy or well-drained soils and good places to look are under hedgerows, or conifer hedges, especially yew (*Taxus*) or cyprus (*Chamaecyparis*) where the soil is often very well sheltered. Many species may also be found in the slacks at the back of coastal sand dunes. Roadside banks where the soil drains quickly and is sheltered by shrubs, etc may also prove worthwhile. A very few species such as *G. triplex* seem less fussy and frequently turn up in woodlands, parks and even urban gardens.

Relationships

Not all earthstars are closely related; the Barometer Earthstar, *Astraeus hygrometricus* is actually related to the boletes while *Geastrum* species are cousins to the stinkhorns.

Fig. 1.1. *Geastrum triplex* with exoperidium just splitting into star-shaped rays. The endoperidium or spore-containing ball is in the centre. The peristome has opened to release the spores.

Fig. 1.2. *Geastrum britannicum* showing the basal, mycelial cup with soil bound to the outside.

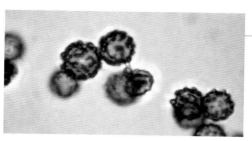

Fig. 1.3. *Geastrum fornicatum* spores showing the large, blunt warts that ornament the surface.

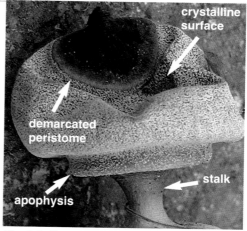

Fig. 1.4. *Geastrum britannicum* with strongly demarcated plicate peristome, crystalline surface, ridged apophysis and basal stalk.

Fig. 1.5. *Geastrum* peristomes: above is simple, ragged or fimbriate, below plicate.

Fig. 1.6. *Geastrum lageniforme* underside showing the smooth surface almost free of soil.

Family *Diplocystidiaceae*

Genus *Astraeus*

Astraeus hygrometricus (Pers.) Morgan
[hygrometricus = like a hygrometer]
Barometer Earthstar
Fruitbody 5–9 cm across when fully opened. **Exoperidium** splitting and forming an irregular star of 7–13 rays; outer surface felty, pale buff, inner surface reddish brown to black, with a whitish layer breaking apart like a mosaic. As the humidity changes the arms will open and close. **Endoperidium** or spore-containing ball pale buff, felty, without a distinct peristome but with a simple, torn apical hole. **Gleba** or spore-mass brown when mature. **Spores** globose, 7.5–10 µm excluding the isolated spines which are 0.3–0.7 µm high.

Usually on dry sandy soils, under mixed broadleaved trees. Uncommon, known from a few southern counties.

Family *Geastraceae*

Genus *Myriostoma*

Myriostoma coliforme (With.: Pers.) Corda
[coliforme = like a colander]
Pepper Pot
Fruitbody subglobose or flattened, 7–15 cm when opened, with 10–18 rays, which recurve back at the tips. **Rays** yellowish brown, smooth or cracking; outer mycelial layer pale brown. **Endoperidial body** subglobose to depressed, 3–5.5 cm across, pale greyish brown, base with up to 17 small stalks; with numerous small stomata or openings. **Stomata** slightly raised with ragged edges. **Gleba** dark brown. **Spores** globose, 3.8–4.5 µm excluding ornamentation, with ridges to 1.2 µm high, some fusing.

On dry, sandy soils on roadsides, hedgerows, etc, often coastal. Very rare in Britain, only recently re-found in Suffolk after a century's absence.

Genus *Geastrum*

Geastrum triplex Jungh.
[triplex = three-fold]
Collared Earthstar
Fruitbody onion-like, then 4–10(-12) cm when open, splitting into 5–7 rays. **Rays** thick-fleshed, inner surface pale greyish buff, usually cracking and leaving a circular collar; outer surface brownish, without encrusting debris. **Endoperidial body** 2–4 cm across, sessile, smooth, pale clay-buff. **Peristome** fibrillose, with a ragged aperture, delimited by a paler, circular depression. **Gleba** dark brown. **Spores** globose, 3.5–4.2 µm excluding ornamentation, with warts to 0.4–0.6 µm high.

On leaf litter, compost, etc, in woodlands, parks and gardens. The commonest earthstar in Britain.

Geastrum fimbriatum Fr.
[fibriatum = fringed]
Sessile Earthstar
Fruitbody 2–6 cm when opened, with 5–9 rays, reflexing. **Rays** with inner surface pale cream-buff, smooth; outer surface pale brown, encrusted with soil and debris. **Endoperidial body** 1–2.5 cm across, sessile or with a very short stalk, grey-buff to brown, minutely roughened. **Peristome** fibrillose-ragged. **Gleba** pale to medium brown. **Spores** globose, 3–3.7 µm excluding ornamentation, with warts 0.1–0.3 µm high.

On calcareous soils, in mixed woods, parks and gardens. Widely distributed in southern England.

Geastrum marginatum Vittad.
[marginatum = with margin, i.e. delimited peristome]
Tiny Earthstar
Syn. *G. minimum sensu auct.*
Fruitbody subglobose then 1.5–3 cm across when opened with 6–10 rays which quickly reflex back. **Rays** fleshy, inner surface cream-buff, often cracking; outer mycelial layer usually encrusted with soil. **Endoperidial body** 0.7–1 cm across, grey-brown, often white granular-pruinose; with a short stalk exposed as the fungus dries. **Peristome** delimited by a circular groove, mouth fimbriate. **Gleba** brown. **Spores** globose, 5–6 µm excluding ornamentation, with warts to 0.7 µm high.

On calcareous soils, often in sand dunes, very rare in Britain, apparently confined to Norfolk.

Geastrum floriforme Vittad.
[floriforme = flower-shaped]
Daisy Earthstar
Fruitbody 2–5 cm across when open, splitting into 6–11 rays, recurving strongly. **Rays** with inner surface greyish brown, buff brown; outer surface with soil adhering; rays strongly hygroscopic. **Endoperidial body** 0.5–1.5 cm across, whitish to pale brown, surface minutely scurfy. **Peristome** ragged-fibrillose, not clearly delimited. **Gleba** dark brown. **Spores** globose, 5.5–7 µm excluding ornamentation, warts to 0.5 µm high.

On dry soils, under various trees, rare in Britain, mainly southern and western counties of England.

Geastrum corollinum (Batsch) Hollós
[corollinum = corolla-like, as in the petals of a flower]
Weather Earthstar
Easily confused with **G. floriforme** (see above) with similarly hygroscopic rays but differing in its delimited peristome, browner endoperidium, and rays without soil on the underside. **Spores** 3.5–4.5 µm excluding ornamentation, with warts to 0.5 µm high.

On well-drained soils, in woods, hedgerows, gardens under shrubs, rare in Britain.

Geastrum rufescens Pers.
[rufescens = reddening]
Rosy Earthstar
Fruitbody 3–9 cm across when opened with 5–8 rays, rather similar to **G. fimbriatum** but differing in its rays developing a distinct reddish pink flush as they age. **Rays** pale cream to buff flushing reddish brown; outer surface with mycelial layer encrusted with soil. **Endoperidial body** 1–3 cm across, clay-buff, sometimes with a slight basal apophysis or ridge. **Gleba** purple-brown. **Spores** 3.8–4.3 µm excluding ornamentation, with warts to 0.4 µm high.

In woods, heaths and dunes, usually on calcareous soils. Uncommon but widespread in Britain.

Geastrum coronatum Pers.
[coronatum = like a crown]
Crowned Earthstar
Fruitbody subglobose then 5–8 (10) cm across when opened, with 7–12 rays which reflex back. **Rays** fleshy, inner surface greyish buff; outer surface with mycelial layer usually encrusted with soil. **Endoperidial body** 2–3 cm across, dark grey-brown-blackish, surface ± smooth, with an apophysis and short stalk. **Peristome** well delimited with a circular groove, mouth fimbriate. **Gleba** dark brown to purple-brown. **Spores** globose, 4.5–5.5 µm excluding ornamentation, with warts to 0.6 µm high.

On calcareous or sandy soils, often in hedgerows or roadsides. Rare in Britain although widely distributed.

Astraeus, Myriostoma, Geastrum

Astraeus hygrometricus

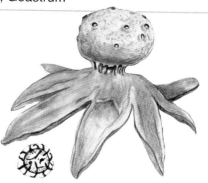
Myriostoma coliforme

Geastrum species with fimbriate, non-plicate peristomes

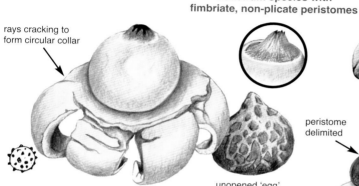

rays cracking to form circular collar

unopened 'egg'

Geastrum triplex

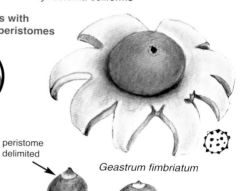

peristome delimited

Geastrum fimbriatum

Geastrum marginatum

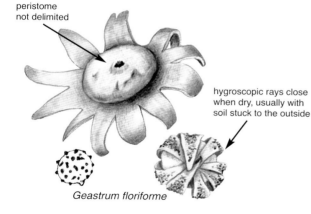

peristome not delimited

hygroscopic rays close when dry, usually with soil stuck to the outside

Geastrum floriforme

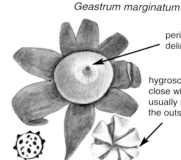

peristome delimited

hygroscopic rays close when dry, usually smooth on the outside

Geastrum corollinum

Geastrum rufescens

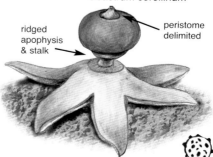

ridged apophysis & stalk

peristome delimited

Geastrum coronatum

Geastrum schmidelii Vittad.
[schmidelii = after Prof. C.C. Schmidel, 1718–1792]
Dwarf Earthstar
Fruitbody subglobose-ovate, then 1.5–4 cm across when opened, with 5–8 rays, thick-fleshed, from horizontal to deeply recurved. **Rays** fleshy, inner surface pale buff, darker with age, smooth; outer surface with mycelial layer encrusted with soil. **Endoperidial body** 0.5–1.2 cm across, grey-brown to clay, with distinct apophysis and a short stalk (more easily seen as it dries). **Peristome** well delimited with circular groove and strikingly plicate. **Gleba** dark brown. **Spores** globose, 5–5.6 µm excluding ornamentation, with warts to 0.7 µm high, often in clumps.

In coastal dunes amongst moss and grass in the dune slacks. Widespread in Britain, mainly in England and Wales, rare in Scotland.

Geastrum campestre Morgan
[campestre = from fields]
Field Earthstar
Fruitbody subglobose-ovate, 1–2 cm across, then 2–5 cm across when opened, with 7–10 rays, hygroscopic with rays closing up when dry, re-opening when moist. **Rays** pale to dark brown when mature, smooth, outer surface with mycelial layer encrusted with soil. **Endoperidial body** 1.2–1.5 cm across, with a very short stalk, surface pale grey, minutely granular. **Peristome** conical, strongly plicate, clearly delimited with a circular ridge. **Gleba** dark brown. **Spores** globose, 4.5–5.7 µm excluding ornamentation, with warts to 0.6 µm high.

On calcareous, well-drained soils in parks, gardens and hedgerows. Very rare in Britain, known from only two collections in Surrey and Kent.

Geastrum berkeleyi Massee
[berkeleyi = after Rev. Berkeley, 1803–1889, mycologist]
Berkeley's Earthstar
Syn. *G. pseudostriatum*
Fruitbody subglobose-ovate, 1–5 cm across, then 4–10 cm across when opened, with 5–9 rays, thick-fleshed, not hygroscopic, usually arched downwards. **Rays** dark reddish brown when mature, often cracking or fragmenting; outer surface with mycelial layer encrusted with soil. **Endoperidial body** globose-pyriform, 1.3–3.5 cm across, with a short stalk and usually with a distinct apophysis below, pale grey-brown with minutely granular, crystalline texture. **Peristome** conical, slender, strongly plicate, clearly delimited with a circular groove. **Gleba** dark purple-brown. **Spores** globose, 4.7–6 µm excluding ornamentation, with warts to 0.5 µm high.

On calcareous soils in mixed woodlands, hedgerows and gardens. Very rare in Britain, possibly extinct with very few authentic records.

Geastrum striatum DC.
[striatum = for its striated or grooved peristome]
Striated Earthstar
Syn. *G. bryantii*
Fruitbody subglobose-ovate to onion-like, 2–3 cm across, then 3–6.5 cm across when opened, with 6–9 rays, not hygroscopic, usually arched downwards. **Rays** pale buff, cream-fawn then browner with age, ± smooth, often splitting; outer surface with mycelial layer strongly encrusted with soil. **Endoperidial body** rather rounded-flattened, pale whitish grey to pale fawn, with a distinct apophysis and basal collar, raised up on a stalk 3–6 mm high. **Peristome** conical, strongly striated and clearly delimited. **Gleba** purple-brown. **Spores** globose, 3.8–4.8 µm excluding ornamentation, with warts to 0.7 µm high.

In woods along pathsides, gardens and hedgerows, often on calcareous soils, widely distributed in Britain and one of our commoner species.

Geastrum pectinatum Pers.
[pectinatum = toothed like a comb]
Beaked Earthstar
Fruitbody subglobose-ovate, 3–7.5 cm across when opened, with 6–9 rays, not hygroscopic, horizontal to somewhat recurved. **Rays** pale yellowish cream internally, outer surface with mycelial layer strongly encrusting soil. **Endoperidial body** subglobose to flattened, 1.5–2.5 cm across, pale to dark grey, mealy-pruinose, with stalk to 0.6 cm high. **Peristome** conical, plicate, clearly delimited with a circular ridge. **Gleba** dark brown. **Spores** globose, 4.2–4.8 µm excluding ornamentation, with warts to 1.2 µm high.

Usually under conifers in woods and parks, less often with broadleaved trees, widespread and frequent in England and Wales, less common further north.

Geastrum fornicatum (Huds.) Hook.
[fornicatum = arched]
Arched Earthstar
Fruitbody subglobose-ovate, 4–8 cm when expanded, with 4–5 rays which are extremely reflexed and connected to a basal cup (actually the outer mycelial layer). **Rays** smooth, pale to dark reddish brown, breaking free of the outer mycelial layer which remains firmly bonded to the soil. **Endoperidial body** 1.5–2.5 cm across, grey-brown, finely granular to ± smooth, with an apophysis. **Peristome** fibrillose-ragged, hardly or not at all delimited. **Gleba** dark brown. **Spores** globose, 3.5–4.2 µm excluding ornamentation, with warts to 0.2 µm high.

On soil under broadleaved trees, rather uncommon but widespread in Britain, not recorded from Scotland.

Geastrum britannicum J.C. Zamora
[britannicum = from Britain]
British Earthstar
Fruitbody subglobose-ovate, 3–5 cm across, with 5–7 rays which are extremely reflexed and connected to a basal cup (actually the outer mycelial layer). **Rays** smooth, pale cream breaking free of the outer mycelial layer which remains firmly bonded to the soil, sometimes breaking free of the arms in old specimens. **Endoperidial body** 1–1.5 cm across, pale grey-brown, strongly roughened, crystalline, very conspicuous when dry. **Peristome** conical, slightly plicate and well delimited, dark brown. **Gleba** dark brown. **Spores** globose, 3–3.8 µm excluding ornamentation, with warts to 0.5 µm high.

Recently described (2015) this species is widespread in southern England, on soils under a variety of trees. Distinguished by its strongly granular endoperidium, delimited peristome, more numerous rays and very small spores.

Geastrum quadrifidum Pers.: Pers.
[quadrifidum = split into four]
Four-rayed Earthstar
Fruitbody subglobose-ovate, 1.3–3.7 cm across, with 4–5 rays, deeply reflexed and joined to a basal mycelial cup. **Rays** pale pinkish buff. **Endoperidial body** taller than wide, 0.7–1.3 cm across, pale grey, finely pruinose. **Peristome** dark grey, ragged, well delimited with circular ridge. **Gleba** dark brown. **Spores** 4.5–5.5 µm excluding ornamentation, with warts to 0.5 µm high.

Under *Fagus* on calcareous soils in Britain, mainly southern England, but also under conifers elsewhere in Europe, rather uncommon.

Geastrum

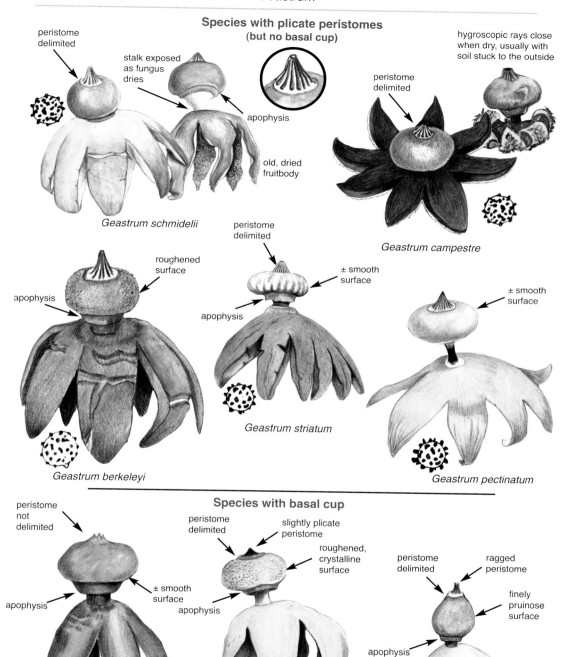

2. Puffballs & Bird's Nest Fungi

Traditionally, fungi that produced their spores on basidia contained within a fruitbody, as opposed to the familiar agaric type whose basidia are formed externally on lamellae, tubes, spines, etc, were lumped together into a large group called the gasteromycetes. However, it was long felt by most mycologists that this group was an artificial assemblage of sometimes unrelated fungi and subsequent molecular studies have supported this view. The earthstars (*Geastraceae*) for example are now in their own order, the *Geastrales*, while the earthballs (*Sclerodermataceae*) are one of several families now contained within the order *Boletales*. The stinkhorns in their turn are separated off into the order *Phallales*. One of the commonest groups however — the puffballs — are actually much more closely related to the gilled fungi in the order *Agaricales*.

In the British Isles the largest genus is *Lycoperdon* with approximately 23 species, along with a number of much smaller genera represented by just a few species or even just one species each. *Lycoperdon* represents one of the rather rare instances where DNA studies have shown that several smaller genera are best subsumed within this one genus, thus the genera *Vascellum*, *Bovistella*, and *Handkea* are now considered as not being different enough as to support generic separation. They have therefore been returned to *Lycoperdon* where they were all placed in the past.

Habitat is an important feature to note, some species for example being strictly found in grasslands, others in sand dunes or dry sandy banks and some being less fussy, occurring in a wide range of habitats. Some of the genera such as *Battarrea* and *Mycenastrum* are much commoner in the warmer, more Mediterranean climes of southern Europe and are considered a rare prize when found in Britain, usually in the drier, coastal regions of the country. Others prefer the colder regions of the north or the high slopes of mountains.

The basic structure of all these varied genera consists of a rounded sac within which the basidia form their spores but this sac may be simple and emerge directly from the ground (*Lycoperdon*, *Bovista*, etc) or be raised up on a more complex stem (*Tulostoma*, *Battarrea*, etc, Fig. 2.1). This stem and rounded spore-sac may in turn be formed within an outer sac which remains at the stem base as it erupts and extends upwards and thus forms a volva-like structure (*Battarea*). Most species are found on soil or in leaf litter but a few grow from fallen logs (e.g. *Lycoperdon pyriforme*). Some genera (e.g. *Battarrea*) begin deep in the soil, emerging only as they mature. In the bird's nest fungi particularly complex fruitbodies have evolved, resembling tiny nests of 'eggs' or **peridioles**, which are dispersed by the splashing of raindrops (Fig. 2.3). Each egg-like spore-sac has a long, coiled, sticky thread trailing behind and this both aids in its projectile flight and also helps anchor the mass to grass stems, etc, awaiting the grazing of some passing animal to ingest the spores and pass them out unharmed in a parcel of dung.

The shape and structure of the spore-sac and in particular the structure of the outer surface layer (**exoperidium**) is a useful identificatory character. The surface is frequently formed of warts or spines, especially in the genus *Lycoperdon* and the size of these warts and their arrangement (often in small groups) is unique to each species (Fig. 4). These frequently fall or are rubbed off leaving a distinctive pattern on the inner layer (**endoperidium**). These walls will rupture, either via an apical aperture which may form a distinct small funnel (**peristome**) or sometimes by the entire wall breaking down and flaking away as happens in *Calvatia* or *Mycenastrum*.

Inside the spore-sac is found the spore producing tissue (**gleba**), consisting of a mass of interwoven hyphae (the **capillitium**) and the basidia on which the spores are formed. Unlike in a typical gilled fungus the spores are not actively projected off the basidium but are released by passive detachment. The structure of the capillitial threads is important; they often have very thick walls and in some genera may have large spikes or recurved hooks (Fig. 2.2). The colour of the gleba is important, olive-brown, purple-brown and grey-brown being common, this colour developing as the spores mature, the immature gleba often being white. In some species of *Lycoperdon* a distinct subgleba may be present, forming the 'flesh' of the stem, sometimes separated from the gleba by a membrane. In some genera, such as *Calvatia*, *Bovista* and *Lycoperdon*, once the peridial wall breaks down the gleba may slowly release its spores over many weeks, the entire fruitbody sometimes tumbling around and dispersing spores, in much the same way that the famous tumbleweed of North America also rolls around dispersing its seeds.

Spores vary from completely smooth to ornamented with warts and/or ridges and the size of the spores (length and width) combined with the structure of any ornamentation is important in the identification of species (Fig. 2.5).

A few species of puffballs are considered good edibles, in particular the Giant Puffball, *Calvatia gigantea*, before the spore mass ripens and while the gleba is still white and firm. A few species in other parts of the world are reported as toxic, so as with all fungi, eating species with which you are not familiar or where edibility is not recorded is best avoided.

Lycoperdon, Bovista

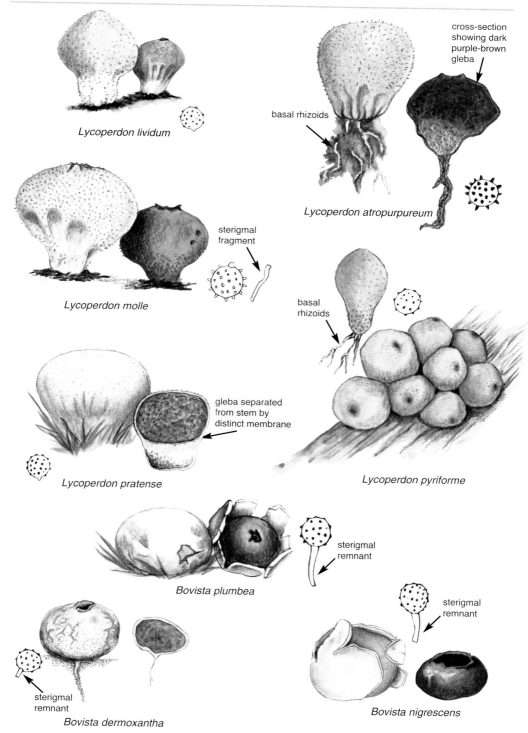

Genus *Cyathus*

Cyathus striatus (Huds.) Pers.
[striatus = finely striped]
Fluted Bird's Nest
Fruitbody in the form of a deep goblet, 8–15 mm tall, 6–8 mm wide. **Peridium** rust-brown to dark brown, densely shaggy-hairy, at first entirely enclosing the contents then gradually expanding to reveal a thin white veil or **epiphragm** which splits to reveal the inner chamber. **Inner surface** silvery grey, vertically ridged or fluted. **Peridioles** or spore balls lens-shaped, pale grey, 12–16 in number, each attached to the inner wall by a thin, coiled thread called the funicular cord. **Spores** 12–21 x 7–12 µm, oblong-ellipsoid, hyaline, smooth and thick-walled. **Hyphae** from peridium surface simple, thick-walled.

The peridioles are forcefully ejected as rain drops splash into the open cups, the long, coiled funicular cord stretching out and often wrapping around grass stems, etc. The basal mass of hyphae called the hapteron is sticky and aids in attaching the ejected peridiole to any grass stem. The peridiole is then ingested by grazing animals along with the vegetation and the spores pass unharmed through the animal to be deposited in their dung elsewhere.

C. striatus is widespread and fairly common throughout Britain on fallen twigs, woody debris, etc.

Cyathus olla (Batsch: Pers.) Pers.
[olla = resembling a pot or cup]
Field Bird's Nest
Fruitbody in the form of a shallow cup with a widely flared margin, 8–15 mm high, 8–15 mm wide. **Peridium** yellow buff to greyish yellow, silky-tomentose, then smooth, totally enclosed at first then opening to reveal a whitish veil or **epiphragm** which splits to reveal the inner chamber. **Inner surface** silver-grey, smooth to faintly fluted. **Peridioles** lens-shaped, dark grey, 2–3 mm across, 8–10 in number, each attached to the chamber wall by a thin thread-like cord or funiculus. **Spores** ellipsoid, 9–12 (-14) x 6.5–8.5 µm, hyaline, thick-walled. **Hyphae** from peridium surface simple, thick-walled.

On fallen twigs, rotted humus, especially common on woodchip mulching, also in dune slacks.

Widespread and common throughout Britain and the most common species of *Cyathus*. In recent years it has undergone a population explosion owing to the widespread use of woodchip mulch in gardens and parks.

Cyathus stercoreus (Schwein.) De Toni in Sacc.
[stercoreus = stinking (of dung)]
Dung Bird's Nest
Fruitbody in the form of a small cup or goblet, 6–10 mm tall, 4–8 mm across. **Peridium** covering entire fruitbody, ochre-brown to dark brown, shaggy-tomentose, breaking open at the apex. **Epiphragm** or veil pale greyish ochre, rupturing to reveal the inner cup and peridioles. **Inner surface** smooth, dark grey. **Peridioles** 1–2 mm diameter, black, each attached by a fine, thread-like funicular cord. **Spores** subglobose, massive, 23–33 x 17–30 µm, hyaline, thick-walled (up to 5 µm). **Hyphae** from peridium surface simple, thick-walled.

Usually on rabbit pellets, in sand dunes, often among marram grass. Elsewhere in Europe on dung and manured soil or even bonfire sites. Very rare in Britain, mainly from Wales and Galloway.

May be mistaken for *Cyathus olla* which can also grow in dunes but that species has much smaller spores and is usually on woody debris, twigs, etc.

Genus *Crucibulum*

Crucibulum laeve (Huds.) Kambly
[laeve = smooth]
Syn. *C. vulgare*
White-egg Bird's Nest
Fruitbody in the form of small cylindric cups, 5–8 mm high, 7–8 mm across, at first almost globose then becoming more cylindric. **Peridium** thick, felty-shaggy, pale greyish to cinnamon-brown, darkening with age, breaking up into small scales. Opening of the cup covered by a fragile, membranous, ochraceous **epiphragm** which breaks apart to reveal the inner cup and peridioles. **Inner surface** smooth and shiny, pale cream. **Peridioles** 1.5–2 mm in diameter, 10–15 (-20) in number, whitish, each attached to the cup wall by a thin, threadlike funiculus. **Spores** ellipsoid-oblong, 6–10 x 3.5–5 µm. **Hyphae** from the surface of the peridium yellow-brown, with spiny outgrowths.

On twigs, woody debris and rotting herbaceous stems and compost, often in large numbers. Widespread and fairly common throughout Britain, especially late autumn and winter.

Genus *Nidularia*

Nidularia deformis (Willd.: Pers.) Fr.
[deformis = deformed]
Syn. *N. confluens, N. farcta*
Pea-shaped Bird's Nest
Fruitbody in the form of a small, globose to slightly flattened ball, 3–10 mm across, broadly attached at the base. **Peridium** pale cream to pale cinnamon-brown, minutely shaggy-tomentose or floccose, completely enclosing peridioles then tearing apart and finally falling away. **Inner surface** smooth. **Peridioles** 1–2 mm across, lens-shaped, numerous, chestnut-brown, embedded in sticky mucilage, without a funicular cord. **Spores** 5–9.5 x 4–6 µm, ovoid to broadly ellipsoid, hyaline, smooth. **Hyphae** from peridium surface with numerous spiny outgrowths.

Often in large numbers on dead, fallen twigs and branches, wood-chips and woody debris.

Widespread throughout Britain but rather uncommon, mainly fruiting during summer and autumn. The only species in Northern Europe.

Genus *Sphaerobolus*

Sphaerobolus stellatus (Tode) Pers.
[stellatus = star-like]
Fruitbody at first spherical, 1–2 mm across, whitish to pale straw or orange-yellow. **Exoperidium** multilayered, rupturing to a star-shaped opening. Osmotic pressure within the body propels the brownish glebal spore mass out. **Endoperidium** white, everting outwards after the gleba has been expelled. **Spores** ellipsoid, smooth, thick-walled, 7.5–10 x 4–5.5 µm.

In groups on rotting wood and decomposing plant remains. Widespread in parks, gardens especially on flower beds mulched with woodchips. Common and widespread.

Cyathus, Crucibulum, Nidularia, Sphaerobolus

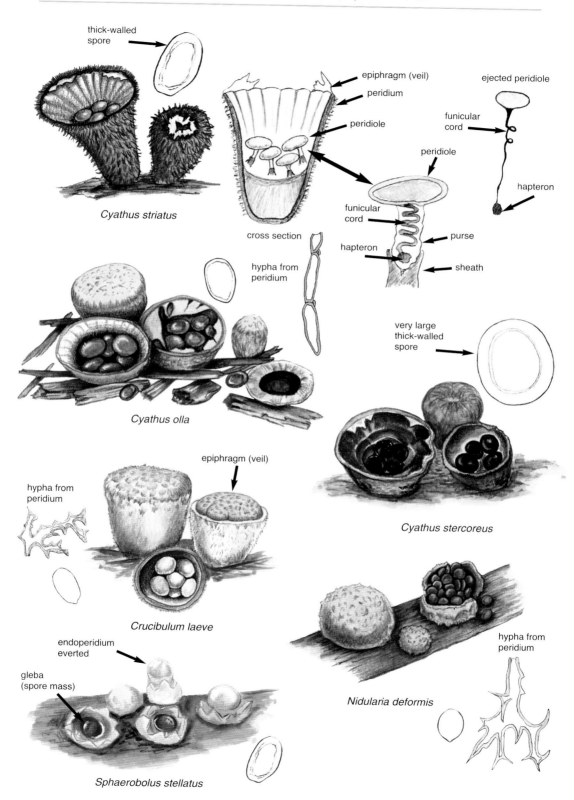

3. Earthballs, Dyeballs & False Truffles

> **Family *Sclerodermataceae***
> Fruitbody globose, tough and fleshy, sometimes with a small to well-developed pseudostem, outer skin (peridium) well developed. Spore mass or gleba at first white then soon purplish and finally dark brown. Spores ± globose with ornamentation of spines and/or ridges. Six species in Britain.

Scleroderma citrinum Pers.: Pers.
[citrinum = of a citrine-yellow colour]
Common Earthball
Fruitbody 5–15 cm across, subglobose to ovoid, often a little flattened, with a very short, often almost absent pseudostem. **Peridium** very thick, 2–5 mm, varying from almost smooth with tiny scales to broken up into thick polygonal scales, pale whitish ochre to bright orange-yellow; whitish in cross-section. **Gleba** white and tough at first then soon purplish black and finally dark olive-brown and powdery. **Spores** globose, 10–13 µm, with spines and crests up to 1.5 µm high, forming a partial to almost complete reticulum. **Clamp connections** present.

On acid soils in mixed woodlands and heaths from early summer until late autumn. This is the commonest species of earthball found throughout Britain. It is also host for the bolete, ***Pseudoboletus parasiticus***.

Scleroderma bovista Fr.
[bovista = puffball like]
Potato Earthball
Fruitbody 2–5 cm across, subglobose, ellipsoid, often slightly flattened, usually with a very short pseudostem. **Peridium** thin, 1–1.5 mm thick, dull yellowish to grey-brown, soon breaking up into small, tessellated patches, purplish red when bruised. **Gleba** white at first then soon purplish black, marbled, finally powdery and blackish olive. **Spores** globose, 11–14.5 µm (excluding spines), with a complete reticulum of spines and ridges, up to 1–2 µm tall. **Clamp connections** present.

Widespread and common throughout Britain, favouring loamy, sandy soils, often along paths, and trackways, roadsides, and under mixed broadleaved trees.

Scleroderma cepa Fr.
[cepa = like an onion]
Onion Earthball
Fruitbody 2–6 cm across, subglobose, ovoid, usually without a pseudostem. **Peridium** thick, 1–1.5 mm thick, dull yellowish to grey-brown, soon breaking up into small, tessellated patches, yellowish when bruised. **Gleba** white at first then soon purplish black, marbled, finally powdery and blackish olive. **Spores** globose, 9–14 µm (excluding spines), with isolated coarse spines, up to 2.5 µm tall. **Clamp connections** absent.

Widespread but uncommon in Britain, favouring sandy soils, often along paths and roadsides, under *Quercus*.

Scleroderma verrucosum (Bull.: Pers.) Pers.
[verrucosum = with warts]
Scaly Earthball
Fruitbody 2–8 cm across, up to 12 cm tall, subglobose, ovoid, usually with a pronounced, well-developed pseudostem which spreads out into white mycelial cords. **Peridium** thin, 0.5–1 mm thick, dull yellowish to grey-brown, soon breaking up into small, innate squamules, yellowish when bruised. **Gleba** white at first then soon purplish black, marbled, finally powdery and blackish olive. **Spores** globose, 9–11 µm (excluding spines), with isolated slender spines, up to 1.5 µm tall. **Clamp connections** absent.

Widespread and common in Britain, especially on sandy soils, woodlands and open heaths.

Scleroderma areolatum Ehrenb.
[areolatum = with areolae]
Leopard-spotted Earthball
Fruitbody 2–5 cm, subglobose, ovoid, usually with a rather short pseudostem ending in a few mycelial cords. **Peridium** thin, 0.5–1 mm thick, dull yellowish to grey-brown, soon breaking up into small, tessellated patches, each patch often surrounded by a faint, darker line or areola; pinkish violet when bruised. **Gleba** white at first then soon purplish black, marbled, finally powdery and blackish olive. **Spores** globose, 11–15 µm (excluding spines), with isolated coarse spines, up to 2.5 µm tall. **Clamp connections** absent.

Widespread but uncommon in Britain, favouring pathsides, woodland edges, etc, possibly associated with *Quercus*.

Scleroderma polyrhizum (J.F. Gmel.) Pers.
[polyrhizum = with many roots]
Syn. *S. geaster*
Fruitbody 6–18 cm across, tough, fleshy, with a compact mycelial base but rarely a true pseudostipe. Buried in the soil and then erupting out. **Peridium** 3–5 mm thick, pale yellow-ochre to greyish yellow with age, surface dry, smooth to finely squamulose, splitting into unequal star-like lobes which reflex back rather like an earthstar (*Geastrum* spp.) to expose the powdery gleba (sporemass). **Gleba** olive-brown to sepia or purplish umber. **Spores** globose 8–14 µm (excluding warts), with irregular ridges and small warts to 0.8 µm high, without a true reticulum. **Clamp connections** present.

A Mediterranean species found in sandy soils it has been reported a few times from southern England although there is some doubt as to the accuracy of these records. Good modern collections are needed to confirm its presence here.

Scleroderma meridionale Demoulin & Malençon
[meridionale = southern]
Fruitbody 4–10 cm across, subglobose, ovoid, usually with a very pronounced, well-developed pseudostem, up to 12 cm long, buried in the sand and often covered with sand grains. **Peridium** rather thick, 2–4 mm thick, ochre-yellow to grey-brown, roughened to almost smooth, pale vinaceous when bruised. **Gleba** white at first then soon purplish black, marbled, finally powdery and blackish olive. **Spores** globose, 9–15 µm (excluding spines), with a complete reticulum or network of spines and ridges, with spines up to 1.5 µm tall, often curved. **Clamp connections** present.

Very rare in mainland Britain, with only one record known from the Somerset coast in pure sand under *Pinus pinaster*, and another from under pines on the island of Jersey. Widely distributed in the Mediterranean and also known from the USA.

Distinguished by its rather smooth skin, prominent mycelial base, very reticulate large spores and its habitat under pines in dry sandy soils, including sand dunes in coastal areas.

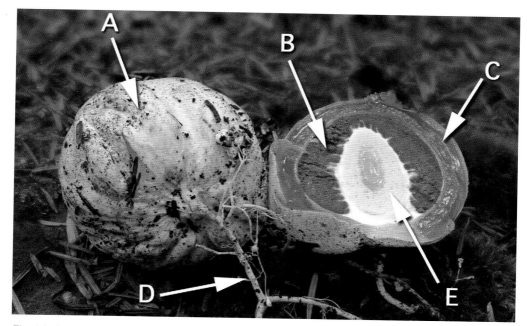

Fig. 4.1. An 'egg' of *Phallus impudicus* sectioned to show the internal structures.
A. The outer skin or peridium. B. Spore mass or gleba. C. Mucilage layer. D. Mycelial cords.
E. Compressed receptacle which expands to form the stem. Photograph © Geoffrey Kibby.

Fig. 4.2. *Phallus impudicus*, the Common Stinkhorn, showing the spore mass spread over the reticulate cap, being devoured by flies. The peridium at the base of the spongy receptacle is hidden in the soil. Photo © Birger Fricke.

Fig. 4.3. Devil's Fingers, *Clathrus archeri* is one of the more colourful species in the UK. An introduced alien it is well established in some southern counties. Photograph © Geoffrey Kibby.

> **Family *Clathraceae***
> Contains those genera in which the spore mass or gleba is spread within or inside the structures of the receptacle, rather than on the outside. Fruitbody cage-like or with long tentacle-like arms.

Genus *Clathrus*

Clathrus ruber Mich.: Pers.
[ruber = red]
Red Cage Fungus
Fruitbody at first an egg-like peridium enclosing the spore-producing receptacle, egg 3–6 cm across. **Peridium** thin, membranous, white to greyish ochre, with a thick, basal, mycelial cord, mucilaginous on inner surface. **Receptacle** at first compressed, soon expanding into a complex, cage-like structure, 10–12 cm across, fragile, spongy, pale pink to bright scarlet, surface wrinkled. **Gleba** olivaceous-brown then almost blackish, liquifying, extremely odorous, foetid, spread over the inner surface of the lattice-like arms. **Spores** ellipsoid-cylindric, 4–6 x 1.5–3.2 μm, hyaline to pale greenish yellow. **Basidia** 6-spored.
    Amongst leaf litter in gardens, hedgerows, woods and parklands. Rare in Britain although possibly spreading, mainly in southern England but with a few records in southern Scotland and Dublin.

Clathrus archeri (Berk.) Dring
[archeri = after William Archer, 1820–1874, naturalist]
Devil's Fingers
Fruitbody at first an egg-like peridium enclosing the spore-producing receptacle, egg 2–4 cm across, globose-ovoid. **Peridium** thin, membranous, cream to greyish buff, with furrows marking the position of the arms as they expand, mucilaginous on inner surface. **Receptacle** a short, hollow, spongy stem 3–5 (-7) cm tall, 1–2.5 cm diameter, whitish to pinkish red, with 4–7 tapering, spongy arms, united at their tips at first then soon flexing back to form a starfish shape, from pinkish to deep scarlet. **Gleba** olive-brown at first then darkening to blackish olive, spread over the inner surface of the arms, soon liquifying and attracting flies by its foul, foetid odour. **Spores** ellipsoid to elongate-ovoid, 4.5–8.5 x 2.3–3.5 μm. **Basidia** 6-spored.
    On soil, in decaying woodchips, under bamboo in the leaf litter, in parks and gardens.
    An introduced species originally from Australasia and the Pacific, very uncommon in Britain but seems to be spreading in the southern counties.

Genus *Aseröe*

Aseröe rubra Labill.
[rubra = deep red]
Starfish Fungus
Fruitbody at first an egg-like peridium enclosing the spore-producing receptacle, egg 2–4 cm, beginning partially below ground, with basal mycelial cord. **Peridium** dirty white to greyish buff, thin, membranous, gelatinous on inner surface. **Receptacle** compressed inside the egg, then rapidly expanding to form a hollow, spongy stem 5–9 cm tall, 1–2.5 cm diameter, pale pink, apex forming a flattened disc bearing 5–10 bifurcating arms, tapering, bright scarlet, 3–5 cm long. **Gleba** olive-brown to blackish, spread over the disc, soon liquifying to produce a foul but weak odour which attracts flies. **Spores** ellipsoid to elongate-ovoid, 4.5–7 x 1.7–2.5 μm, pale brown. **Basidia** 6-spored?
    In leaf litter and grass in open woodlands on rich soils. Extremely rare, known only from the Royal Botanic Gardens, Kew and from a small population in Oxshott Heath, Surrey in a birch woodland.
    An introduced species distributed throughout the tropics.

Genus *Ileodictyon*

Ileodictyon cibarium Tul.
[cibarium = relating to food, supposedly edible]
Basket Fungus
Fruitbody at first an egg-like peridium enclosing the spore-producing receptacle, egg 4–7 cm diameter, bursting open to release the receptacle. **Peridium** whitish, with reticulate grooves. **Receptacle** forming an open lattice-work cage, pure white, sometimes expanding almost explosively from the egg, each arm of the lattice wrinkled, concertina-like. **Gleba** olive-brown, spread over the inside of the arms, odour foetid. **Spores** ellipsoid to elongate-ovoid, 4.5–7 x 2–2.8 μm, hyaline to slightly greenish brown.
    In woodland clearings or on disturbed soils, very rare, known from only a few southern England localities.
    An introduced species from New Zealand and Australasia.

Ileodictyon gracile Berk.
[gracile = slender]
Differing from *I. cibarium* by its smaller fruitbody with very slender tubular arms which do not show concertina-like wrinkles. In gardens, woodland edges, etc, very rare, only recently added to the British list with a collection from Suffolk.

> **Family *Lysuraceae***
> More closely related to the *Phallaceae* than to the *Clathraceae* the European species form a columnar stem surmounted by short arms, grooved on the outside and with the gleba on the inner surface.

Genus *Lysurus*

Lysurus cruciatus (Lepr. & Mont.) Lloyd
[cruciatus = in the form of a cross]
Lizard's Claw
Fruitbody at first an egg-like peridium enclosing the spore-producing receptacle, egg 3–5 cm across. **Peridium** white, membranous, bursting at the apex. **Receptacle** forming an off-white, hollow, cylindric stem 4–10 cm tall, 1.5–2 cm diameter, sometimes flushed pink, surmounted by 4–7 short, transversely wrinkled arms, fused at first then opening up, white to bright reddish, grooved on the outside, with the gleba spread on the inside. **Gleba** olive-brown to blackish brown, slowly liquifying with a weakly foetid odour. **Spores** ellipsoid to elongate-ovoid, 4.5–6 x 1.5–2.3 μm, hyaline.
    On manured soils, stable refuse, greenhouse mulching, very rare in Britain.

Lysurus mokusin ** (L.) Fr.
[mokusin = possibly after native American pointed shoe or moccasin]
Ribbed Lizard's Claw
Very similar to *L. cruciatus* but the stem with four to six longitudinal ribs or buttresses and with the arms fused. Not yet recorded in Britain but with some mainland European records (Italy, Germany, etc).

Clathrus, Aseröe, Ileodictyon, Lysurus

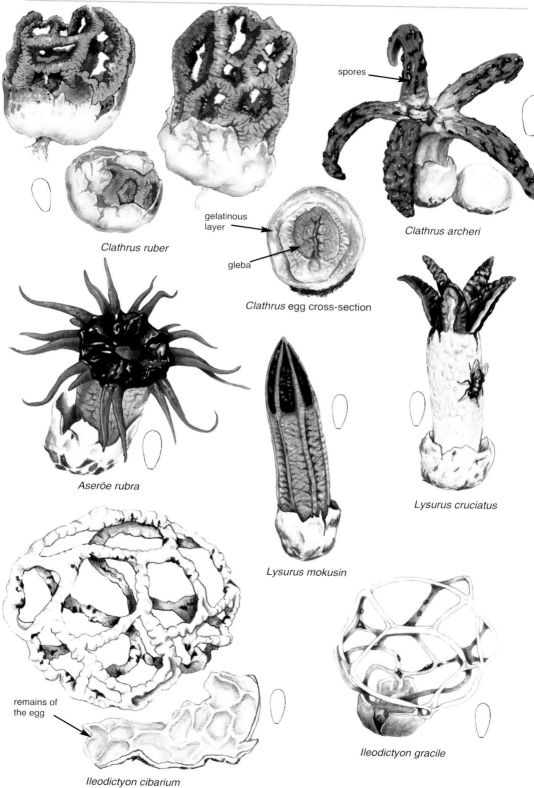

> **Family *Phallaceae***
> Forming rather phallic fruitbodies with a sterile stem and the gleba spread over a separated cap or head region, sometimes with a netlike veil or indusium.

Genus *Phallus*

Phallus impudicus L.: Pers.
[impudicus = shameless or rude]
Common Stinkhorn
Fruitbody an egg-like peridium, globose to ovoid, 4–6 cm across, with basal mycelial cords. **Peridium** white to pale cream, membranous with a thick gelatinous layer internally, splitting apically to release the receptacle, remaining as a basal volva. **Receptacle** forming a tall, sterile, spongy, hollow white stem which expands in a few hours. **Cap** 3–5 cm in diameter, attached by an apical disc to the top of the stem, with a reticulate-ribbed surface over which the gleba is spread. **Gleba** olive-brown, soon liquifying and becoming blackish olive with a strong, foetid odour, soon completely removed by flies and other invertebrates. **Spores** narrowly ovoid, 4–5.6 x 1.8–2.8 µm.

Associated with dead wood, stumps, etc in woodlands, both broadleaved and coniferous, widespread and common everywhere.

The rare variety *togatus* is distinguished by the short, ragged, net-like veil or indusium hanging below the cap.

Phallus duplicatus Bosc.
[duplicatus = twice as large, i.e. the indusium]
Fruitbody an egg-like peridium, globose to ovoid, 4–7 cm across, with basal mycelial cords. **Peridium** white to pale cream, membranous with a thick gelatinous layer internally, splitting apically to release the receptacle, remaining as a basal volva. **Receptacle** forming a tall, sterile, spongy, hollow white stem which expands in a few hours. **Cap** 3–5 cm in diameter, attached by an apical disc to the top of the stem, with a reticulate-ribbed surface over which the gleba is spread. **Indusium** forming a large, white net or lace skirt hanging from beneath the cap margin, the holes of the net become smaller towards the margin which has a distinct, even margin (unlike the var. *togatus* of *P. impudicus* which has a ragged margin). **Gleba** olive-brown, soon liquifying and becoming blackish olive with a strong, foetid odour, soon completely removed by flies and other invertebrates. **Spores** narrowly ovoid to bacilliform, 3–4.5 x 1.5–2.1 µm.

Associated with dead wood, stumps, etc in woodlands, extremely rare in the UK and Europe, only one confirmed record in England. Although reported occasionally most records seem to refer to *P. impudicus* var. *togatus*.

Phallus hadriani Vent.: Pers.
[hadriani = associated with the Adriatic region]
Dune Stinkhorn
Fruitbody an egg-like peridium, globose to ovoid, 3–7 cm across, with basal mycelial cords. **Peridium** white to pale cream, soon flushing pinkish magenta, membranous with a thick gelatinous layer internally, splitting apically to release the receptacle, remaining as a basal volva. **Receptacle** forming a tall, sterile, spongy, hollow, white stem which expands in a few hours. **Cap** 3–5 cm in diameter, attached by an apical disc to the top of the stem, with a reticulate-ribbed surface over which the gleba is spread. The disc at the top usually has a dentate or crenulate margin. **Gleba** olive-brown, soon liquifying and becoming blackish olive with a weak, foetid to pleasant odour, soon completely removed by flies and other invertebrates. **Spores** narrowly ovoid to bacilliform, 3–6 x 2–3.2 µm.

Occasional in coastal dunes, widespread, especially in the south, more common in southern Europe. The common stinkhorn *P. impudicus* can also grow in sand dunes so you have to be sure that the eggs are flushed pinkish and that the disc is dentate. Several authors have remarked on the less foetid, almost pleasant odour of *P. hadriani*, resembling liquorice or hyacinths.

Phallus rubicundus (Bosc.) Fr.**
[rubicundus = red-flushed]
Fruitbody an egg-like peridium, globose to ovoid, 3–4 cm across, with basal mycelial cords. **Peridium** white to pale cream, membranous with a thin gelatinous layer internally, splitting apically to release the receptacle, remaining as a basal volva. **Receptacle** forming a tall, sterile, spongy, hollow pinkish to bright red or orange stem which expands in a few hours. **Cap** 3–4 cm high, thimble-shaped to bell-shaped, developing a central perforation, with a finely granular-pitted surface over which the gleba is spread. **Gleba** olive-brown, soon liquifying and becoming blackish olive with a strong, foetid odour. **Spores** long-elliptical; 3.5-5 x 1.5-2.5 µm.

On rich humus, woodchips, woody debris, a tropical species now widespread around the world in parks and gardens on garden mulch. Not yet recorded in Britain but reported from Italy and perhaps spreading.

Genus *Mutinus*

Mutinus caninus (Huds.: Pers.) Fr.
[caninus = pertaining to dogs]
Dog Stinkhorn
Fruitbody an egg-like to pear-shaped peridium, 2–4 cm high, with basal mycelial cords. **Peridium** white to pale cream, membranous with a thin gelatinous layer internally, splitting apically to release the receptacle, remaining as a basal volva. **Receptacle** forming a tall, sterile, spongy, hollow whitish to pale orange stem, tapered above with a bright reddish orange head, separated from the stem by a constriction, initially covered in the smooth grey-brown gleba. **Head** with a finely bumpy (not sponge-like) surface. **Gleba** greyish to olive-brown, then liquifying to dark blackish olive and with a slightly foetid odour. **Spores** narrow, ellipsoid, 4.5–6.5 x 1.8–3 µm.

On soil in leaf litter, often near dead wood or stumps in broadleaved woodland. Frequent to common throughout Britain from summer to late autumn.

Mutinus ravenelii (Berk. & Curtis) E. Fisch.
[ravenelii = after H.W. Ravenel, 1814–1887, mycologist]
Red Stinkhorn
Fruitbody an egg-like to pear-shaped peridium, 2–3 cm high, with basal mycelial cords. **Peridium** white to pale cream, membranous with a thin gelatinous layer internally, splitting apically to release the receptacle, remaining as a basal volva. **Receptacle** forming a tall, spongy, hollow, pink to deep carmine-red stem, tapered above and merging into the bright carmine-red head initially covered in the smooth gleba. **Head** with a finely bumpy (not sponge-like) surface. **Gleba** greyish to olive-brown, then dark blackish olive, with a very strong, foetid odour. **Spores** narrow, ellipsoid, 5–7 x 1.8–2.5 µm.

On soil in leaf litter, bamboo beds, or rotten grass. Rare but with several records in southern England, probably introduced from North America.

Phallus, Mutinus

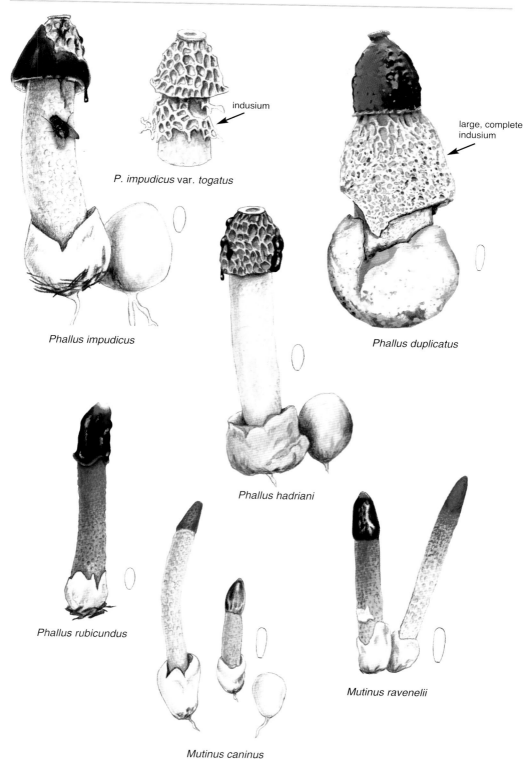

5. Chanterelles

The *Cantharellales* are an order of fungi in the class *Agaricomycetes*. The order includes not only the highly esteemed chanterelles (*Cantharellaceae*), but also some of the tooth fungi (*Hydnaceae*), as well as the club fungi (*Clavulinaceae*) which are treated elsewhere in this book.

Species within the order are variously ectomycorrhizal, saprotrophic, associated with orchids, or facultative plant pathogens. Those of economic importance include edible and commercially collected *Cantharellus*, *Craterellus*, and *Hydnum* species.

Family *Cantharellaceae*

The genus *Cantharellus* is characterised by fleshy fruitbodies with the hymenium developed into irregular folds or more complex lamellae-like forked wrinkles (Fig. 1) and in particular a solid, fleshy cap and stem. Their spores are smooth, ellipsoid and non-amyloid; spore deposits range from cream to pinkish yellow or pale ochre. The species often exhibit yellow, orange or red colours (although other colours such as grey are possible), contain carotenoid pigments and many are rich in vitamin D. The species may also exhibit a noticeable odour of apricots when fresh which adds to their appeal.

The most famous species, *Cantharellus cibarius* was formerly considered to have an almost world-wide distribution but more detailed investigations using molecular techniques have shown this not to be the case. A number of species have now been described in North America, Asia and the Mediterranean countries, which look superficially similar but actually have distinctive morphological characters. The majority of these species, along with the true *C. cibarius* are highly esteemed edibles.

Craterellus species on the other hand have thinner, less fleshy fruitbodies and the hymenium varies in different species from shallow wrinkles (Fig. 5.1) to complex forked wrinkles (Fig. 5.2) or even almost smooth (Fig. 5.3). The fruitbody and in particular the stem is usually partially or entirely hollow and tube-like and for this reason they have been given the common name of trumpet fungi (Fig. 5.4). *Craterellus* was formerly also considered to differ by the absence of clamp connections, but as now defined some *Craterellus* species do have clamps.

Pseudocraterellus, often treated as a separate genus, is also included in *Craterellus* by many authors, based on both recent molecular work and morphology and that placement is followed here.

Cantharellus and *Craterellus* basidia can bear as many as 6 spores with 4–6 being usual, although as few as 2 have also been recorded.

It has been suggested that *Cantharellus* species are all mycorrhizal whereas *Craterellus* are partially or entirely saprotrophic, which would certainly agree with some field observations on the common species that I have made.

Species can for the most part be identified macroscopically, but spore measurements and other microscopic details are also important as always. It has been shown that many species can produce on occasion all-white fruitbodies and this has led to much confusion and the publishing of various species which are just forms of existing species.

In Britain we have just five species of *Cantharellus* and six of *Craterellus* while several more are found in the Mediterranean region, some of which may well be present here also, albeit very rarely.

Family *Gomphaceae*

Not closely related to the chanterelles but similar in form, the family *Gomphaceae* is represented by the one genus *Gomphus* in Britain, but that is now considered as probably extinct, the species *G. clavatus* not having been recorded here for over 80 years. If you find this species you have a prize indeed and should definitely contact one of the mycological institutes to report it. Because of the similarity in shape it is included here with the chanterelles.

Fig. 5.1. *Craterellus sinuosus* is an uncommon species, sometimes included in a separate genus *Pseudocraterellus*. Photo © Gerhard Schuster.

Fig. 5.2. The blunt, multiply forked, gill-like hymenium of *Cantharellus cibarius* showing the prominent cross-veins.

Fig. 5.3. The almost smooth hymenium of *Craterellus cornucopioides* forms the outer surface of the funnel-shaped fruitbody.

Fig. 5.4. Some chanterelles like the Trumpet Chanterelle, *Craterellus tubaeformis* can grow in very large numbers. The colour and form of the hymenium are important identificatory features.

> **Family *Cantharellaceae***
> Two British genera: *Cantharellus* and *Craterellus*, with hymenium as either shallow, wrinkled, pseudogills or a ± smooth or wrinkled surface. *Craterellus* species have hollow stems. Basidia are generally 5–6-spored.

Genus *Cantharellus*

Cantharellus cibarius Fr.
[cibarius = relating to food, edible]
Chanterelle
Cap 4–12 cm across, at first rounded with inrolled margin, then expanding, often with irregular, wavy margin, bright egg-yellow, apricot-yellow, bruising brownish, surface smooth to suede-like. **Hymenium** resembling wrinkled, frequently forking blunt 'gills', bright egg-yellow to slightly pinkish apricot. **Stem** 3–10 x 0.8–3 cm, cylindric-tapered, concolorous with cap, paler below. **Flesh** whitish. **Taste** mild. **Odour** of apricots. **Basidia** 4–5-spored. **Spores** are ellipsoid, 7.5–9 x 4–6.5 μm. **Spore deposit** pale pinkish yellow.

C. cibarius is especially abundant in the wetter, more humid woodlands of Scotland, northern England and Wales. It occurs with a wide range of host trees including *Betula*, *Fagus* and *Quercus* among others.

Cantharellus amethysteus (Quél.) Sacc.
[amethysteus = of amethyst colour]
Cap 3–6 cm across, at first rounded, inrolled, then expanding and often with irregular, lobed margin, surface dull yellow to egg-yellow with a greyish violet felty tomentum which breaks up into small squamules or silky zones, especially at the margin. **Hymenium** decurrent, resembling wrinkled, frequently forking blunt 'gills', pale ochre-yellow to slightly pinkish apricot. **Stem** cylindric-tapered, 2–4 x 1.5–2.5 cm, solid, pale ochre, often flushed violaceous, bruising tawny yellow. **Flesh** pale cream-yellow, firm. $FeSO_4$ on flesh of stem base reddish grey, all other species just grey. **Taste** mild. **Odour** pleasant, fruity. **Basidia** 4–5-spored. **Spores** ellipsoid to slightly constricted, 8–10 (-12) x 4.5–6 μm. **Spore deposit** white.

Usually in grass under *Quercus*, *Fagus*, *Betula*. Uncommon but widespread in Britain.

Very variable, old specimens are hardly violaceous at all but usually some trace of tiny scales will remain.

Cantharellus pallens Pilát
[pallens = whitish or becoming pale]
Syn. *C. subpruinosus, C. pallidus?*
Cap 3–6 cm across, at first rounded, inrolled, then expanding and often with irregular, lobed margin, pale lemon-yellow to whitish beige or pinkish with a pruinose white 'frosting', bruising strongly rust-orange. **Hymenium** decurrent, resembling wrinkled, frequently forking blunt 'gills', pale lemon-yellow to slightly pinkish white, often more yellow-orange at margin. **Stem** cylindric-tapered, 2–4 x 1.5–2.5 cm, solid, whitish to pale yellow, bruising tawny rust. **Flesh** pale cream-yellow, firm. **Taste** mild. **Odour** pleasant, fruity. **Basidia** 4–5-spored. **Spores** ellipsoid to slightly constricted, 7.5–10 x 4.5–6.5 μm. **Spore deposit** pale yellowish orange.

On calcareous soils under broadleaved trees. A rare species, known from southern England, much commoner in the Mediterranean countries.

Cantharellus ferruginascens P.D. Orton
[ferruginascens = staining rust-brown]
Cap 2–5 (-7) cm across, at first rounded, inrolled, then expanding and often with irregular, lobed margin, pale lemon-yellow to greenish yellow with orange tints, finely velvety, bruising strongly rust-orange. **Hymenium** decurrent, resembling wrinkled, frequently forking blunt 'gills', whitish to pale yellow to slightly pinkish white, sometimes with reddish patches. **Stem** cylindric-tapered, 2–4 x 0.6–2 cm, solid, whitish to pale yellow, bruising strongly orange-brown to reddish brown. **Flesh** pale cream-yellow, firm. **Taste** mild. **Odour** pleasant, fruity. **Basidia** 4–5-spored. **Spores** ellipsoid to slightly constricted, 7.5–10.6 x 4.2–5.6 μm. **Spore deposit** pale yellowish.

With *Quercus* or *Fagus*, rare, mainly in southern England but distribution uncertain owing to confusion with other species, especially *C. pallens*.

Cantharellus friesii Quél.
[friesii = after Elias Fries, 1794–1878, mycologist]
Cap 3–6 cm across, at first rounded, inrolled, then expanding and often with irregular, lobed margin, pale lemon-yellow to pink-yellow or pinkish with a pruinose white 'frosting', bruising strongly rust-orange. **Hymenium** decurrent, resembling wrinkled, frequently forking blunt 'gills', pale lemon-yellow to slightly pinkish white. **Stem** cylindric-tapered, 2–4 x 1.5–2.5 cm, solid, whitish to pale yellow, bruising tawny rust. **Flesh** pale cream-yellow, firm. **Taste** mild. **Odour** pleasant, fruity. **Basidia** 2–5-spored. **Spores** ellipsoid to slightly constricted, 7.5–12 x 4.5–6.5 μm. **Spore deposit** pale yellowish orange.

Usually found singly or in scattered groups, under *Fagus* or occasionally *Betula*. Our smallest British *Cantharellus* species, rare, mainly in northerly counties.

Cantharellus romagnesianus* Eyssart. & Buyck
[romagnesianus = after H. Romagnesi, 1912–1999, mycologist]
Cap 1–3 (-4) cm across, thin-fleshed, at first rounded, inrolled, then expanding and often with irregular, lobed margin, bright yellow-orange, lemon-yellow to slightly greenish yellow. **Hymenium** decurrent, resembling wrinkled, frequently forking shallow 'gills', pale lemon-yellow to slightly ochraceous. **Stem** long, thin, cylindric, 1–3 x 0.2–0.6 cm, solid, pale to bright yellow, often reddish at the base. **Flesh** pale cream-yellow, firm. **Taste** mild. **Odour** pleasant, fruity. **Basidia** 4–5-spored. **Spores** ellipsoid to slightly constricted, 7.5–12 (-14) x 3.8–5.8 (-6.4) μm. **Spore deposit** pale yellowish orange.

With broadleaved trees and conifers. Recorded from France and Spain, not yet known in Britain but might be found in our Atlantic woods on the west coast.

Cantharellus alborufescens* (Malençon) Papetti & S. Alberti
[alborufescens = white and reddish brown]
Syn. *C. henrici, C. ilicis, C. lilacinopruinatus*
Cap 3–9 cm across, fleshy, at first rounded, inrolled, then expanding and often with irregular, lobed margin, bright yellow-orange, orange to whitish yellow, without a whitish, pruinose coating. **Hymenium** decurrent, resembling wrinkled, frequently forking shallow 'gills', very pale whitish ochre, bruising brownish. **Stem** cylindric, 3–5 x 1–1.5 cm, solid, white to pale whitish yellow, conspicuously staining reddish brown. **Flesh** pale cream-yellow, firm. **Taste** mild. **Odour** pleasant, fruity. **Basidia** 4–5-spored. **Spores** ellipsoid to slightly constricted, 8.4–11.5 x 4.5–6.5 (-7) μm. **Spore deposit** pale yellowish ochre.

With *Quercus ilex* and *Q. rotundifolia*, usually on calcareous soils, probably widespread in the Mediterranean countries, not yet known in Britain but might be found under introduced evergreen oaks.

Cantharellus

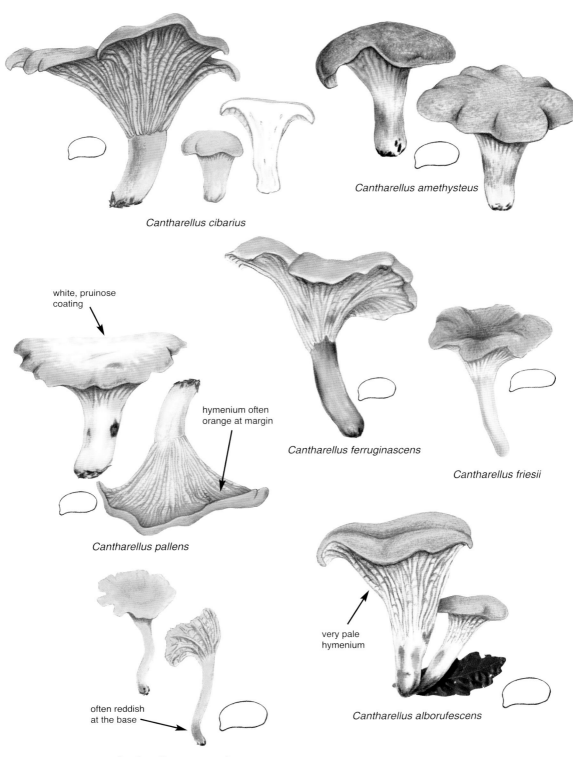

Genus *Craterellus*

Craterellus ianthinoxanthus* (Maire) Pérez-De-Greg.
[ianthinoxanthus = with lilac and yellow colours]
Syn. *Cantharellus ianthinoxanthus* (Maire) Kühner,
Cap 1–4 (-6) cm across, thin-fleshed, at first rounded, inrolled, soon depressed and then funnel-shaped, with irregular, lobed margin, ochre-yellow to pale ochre-brown. **Hymenium** decurrent, resembling wrinkled, frequently forking shallow veins, pale violaceous grey. **Stem** cylindric, 2–4 x 0.5–1.3 cm, hollow with age, pale to bright ochre-yellow. **Flesh** pale cream-yellow, firm, greying after 24 hours. **Taste** mild. **Odour** pleasant, fruity. **Basidia** 4–5-spored. **Spores** ellipsoid to slightly constricted, 9.5–12 (-13) x 6.3–7.6 (-8) μm. **Spore deposit** pinkish ochre.

With *Quercus* and *Fagus* on acidic or calcareous soils, France, Italy and Switzerland. Not yet British but might occur here.

Craterellus melanoxeros (Desm.: Fr.) Pérez-De-Greg.
[melanoxeros = dark and dry]
Syn. *Cantharellus melanoxeros* Desm.
Cap 1–4 (-6) cm across, thin-fleshed, at first rounded, inrolled, soon depressed and then funnel-shaped, with irregular, lobed margin, ochre-yellow to pale ochre-brown, bruising blackish. **Hymenium** decurrent, resembling wrinkled, frequently forking shallow veins, ochre then pale violaceous grey. **Stem** cylindric, 4–5 x 0.5–1.5 cm, solid to pithy, pale to bright ochre-yellow, bruising blackish. **Flesh** pale cream-yellow, firm, quickly greying then blackening. **Taste** mild to peppery. **Odour** pleasant, fruity. **Basidia** 4–5-spored. **Spores** ellipsoid to slightly constricted, 8–11 x 5.5–7.5. **Spore deposit** cream.

In clusters in leaf litter under *Fagus* and *Quercus*. Rare in Britain, known from a few southern counties.

Craterellus cinereus (L.: Fr.) Pers.
[cinereus = ash-grey]
Fruitbody 1–6 cm across, 2–6 cm high, thin-fleshed, deeply funnel-shaped, with irregular, lobed margin, dark brown to blackish, surface dry with tiny erect scales or tufts. **Hymenium** deeply decurrent, smooth to finely wrinkled, pale grey to ochre-grey. **Flesh** pale greyish black. **Taste** mild to slightly bitter. **Odour** strikingly fruity. **Basidia** usually 5-spored. **Spores** ellipsoid to slightly constricted, 7.7–10 x 4.6–7 μm. **Spore deposit** pale ochre. This species lacks clamp connections in its tissues.

With *Quercus* and *Fagus* and sometimes under conifers, usually on calcareous soils, often in moss, possibly indicative of ancient woodlands. Widespread throughout Britain but very uncommon. Widely regarded as an edible fungus.

Craterellus cornucopioides (L.: Fr.) Pers.
[cornucopioides = funnel-shaped]
Horn of Plenty
Fruitbody 1–6 cm across, 2–6 cm high, thin-fleshed, deeply funnel-shaped, with irregular, lobed margin, dark brown to blackish, surface dry with tiny erect scales. **Hymenium** deeply decurrent, smooth to finely wrinkled, pale grey to ochre-grey. **Flesh** pale greyish black. **Taste** mild to slightly bitter. **Odour** pleasant, fruity. **Basidia** 2-spored. **Spores** ellipsoid to slightly constricted, 9.5–13 x 6.4–9 μm. **Spore deposit** white. This species lacks clamp connections in its tissues.

With *Quercus* and *Fagus* and sometimes under conifers, often in moss. Widespread and frequent throughout Britain. Widely regarded as an edible fungus.

Craterellus sinuosus (L.: Fr.) Pers.
[sinuosus = sinuate, wavy]
Syn. *Pseudocraterellus sinuosus, P. undulatus*
Fruitbody 1–2.5 cm across, 2–5 cm high, thin-fleshed, deeply funnel-shaped, with irregular, lobed margin, dark brown to blackish, surface dry, smooth to finely radially fibrillose. **Hymenium** deeply decurrent, smooth to finely wrinkled with age, pale grey to ochre-grey. **Flesh** pale greyish black. **Taste** mild to slightly bitter. **Odour** rather weak. **Basidia** 4–5-spored. **Spores** ellipsoid to slightly constricted, 8.8–11.5 x 5–7.5 μm. **Spore deposit** yellowish ochre. This species lacks clamp connections in its tissues.

With *Quercus* and *Fagus*, often in moss. Widespread but distinctly rare to uncommon throughout Britain.

Craterellus tubaeformis (L.: Fr.) Pers.
[tubaeformis = tube-shaped]
Syn. *C. infundibuliformis*
Fruitbody 1–6 cm across, 2–6 cm high, thin-fleshed, deeply funnel-shaped, with irregular, lobed margin, dark brown to blackish, surface dry with tiny erect scales. **Hymenium** deeply decurrent, smooth to finely wrinkled, pale grey to ochre-grey. **Flesh** pale greyish black. **Taste** mild to slightly bitter. **Odour** pleasant, fruity. **Basidia** usually 4-spored. **Spores** ellipsoid to slightly constricted, 9.5–13 x 6.4–9 μm. **Spore deposit** white.

In mixed woods, often in large drifts in wet gulleys, stream sides and damp moss. Widespread and frequent throughout Britain. Some collections are much yellower than others.

Craterellus lutescens (Fr.: Fr.) Fr.
[lutescens = yellowish]
Syn. *Cantharellus aurora, C. xanthopus*
Fruitbody 1–4 cm across, 2–6 cm high, thin-fleshed, deeply funnel-shaped, with irregular, lobed margin, chestnut-brown to ochraceous, surface dry or with tiny erect scales or tufts. **Hymenium** decurrent, smooth to finely wrinkled, pale yellow to yellow-orange **Flesh** pale yellow. **Taste** mild to slightly peppery. **Odour** pleasant, aromatic-fruity. **Basidia** usually 4–5-spored. **Spores** ellipsoid to slightly constricted, 9.5–12 x 6.5–9 μm. **Spore deposit** cream.

In mossy conifer woods (*Pinus* and *Picea*), often at the edge of streams, swamps, etc, frequently in *Sphagnum* moss, often in large numbers. Usually on calcareous soils. Rare in Britain, most commonly found in Scotland and Ireland, especially in mountainous areas.

Family *Gomphaceae*
One chanterelle-like genus in Britain, possibly extinct.

Genus *Gomphus*

Gomphus clavatus S.F. Gray
[clavatus = club-shaped]
Fruitbody 5–12 cm high, at first flattened on top, then expanding, becoming irregular, lobed and depressed at centre, rich violet when young then top fading to tan-ochre, smooth to wrinkled. **Hymenium** strongly decurrent, formed of numerous irregular wrinkles and folds, usually with cross-veins, bright violet then fading to greyish violet, lilac-pink and finally ochraceous. **Flesh** thick, white, watery. **Taste** mild to slightly bitter. **Odour** weak. **Spores** ellipsoid, with blunt warts, 10–14 x 4.5–5.5 μm. **Spore deposit** pale cream-ochre.

In mossy conifer or *Fagus* woods, very rare in Britain, known from only a handful of southern records and not recorded for over 80 years, possibly extinct.

Craterellus, Gomphus

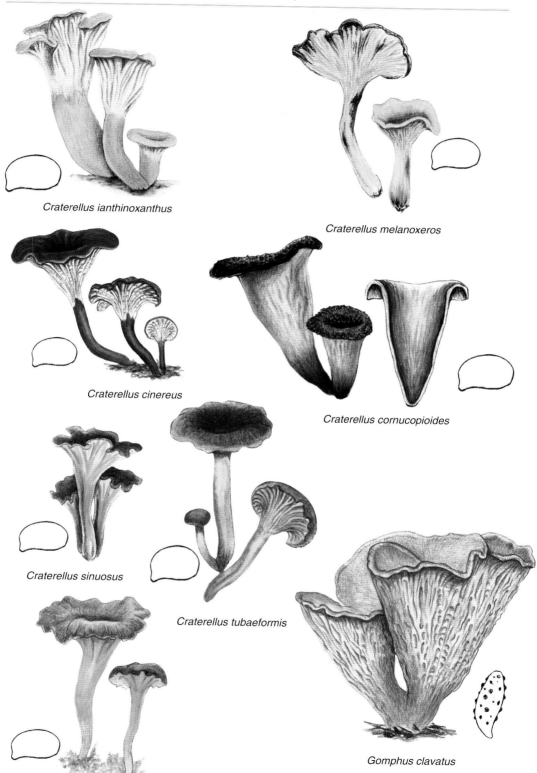

Craterellus ianthinoxanthus

Craterellus melanoxeros

Craterellus cinereus

Craterellus cornucopioides

Craterellus sinuosus

Craterellus tubaeformis

Craterellus lutescens

Gomphus clavatus

6. Hedgehog, Tooth & Fan Fungi

At one time all the genera with spines on the underside of the cap were included in the family *Hydnaceae*. Recent DNA studies have shown that many of the genera, including *Hydnellum*, *Bankera*, *Sarcodon*, *Phellodon*, etc are actually not closely related to *Hydnum* but are instead placed in the order *Thelephorales*. The large, white species in the genus *Hericium*, with pendent spines and growing from the trunks of trees, are actually members of the order *Russulales*. These genera all demonstrate how body form and structure is not always a good indicator of relationships, such structures having evolved independantly several times in different unrelated groups of fungi.

Family *Hydnaceae*
A popular group of fungi collected around the world as highly prized edible fungi. In Britain we have just one principal genus: *Hydnum*. Traditionally we have had just two or three species but that has increased recently following molecular studies and studies are still ongoing so there may well be more to come. The spores are ellipsoid to globose and smooth and the spore deposit is white to cream.

Family *Bankeraceae*
Four genera are dealt with here: *Bankera*, *Hydnellum*, *Phellodon* and *Sarcodon*, a fifth genus *Boletopsis* is poroid and placed with the polypores.

As with many other fungi, recent molecular studies have shown that there are more species still to be described, often very difficult to distinguish from each other morphologically.

Their fruitbodies are often very irregular in outline and many exhibit the phenomenon of indeterminate growth, whereby the growing margin can meet and flow around an obstruction. Hence you frequently see blades of grass or pine needles appearing to grow though a cap (Fig. 6.1). This contrasts with gilled agarics for example, which have determinate growth, where the cap shape and structure is preformed in the tiny primordial fruitbody and what appears to be growth is mostly expansion. If such a cap meets a blade of grass it can only push it aside, not flow around it. Indeterminate growth is also seen in other orders such as the *Polyporales*. The cap surface may be soft and tomentose, smooth, eroded and pitted (scrobiculate) or even coarsely scaly (Fig. 6.2).

The spores of species in the *Bankeraceae* range from white in the genera *Bankera* and *Phellodon* to brown in *Hydnellum* and *Sarcodon*. The spores may be spiny to coarsely warted or lumpy (tuberculate).

The flesh of these toothed genera is usually very tough and leathery and may exhibit striking colours and/or colour changes, especially in the stem and may be distinctly zonate. The odour in some species can be very strong of fenugreek, often referred to as like stock cubes (especially when drying), spice, aniseed or even fruity. The taste varies from almost mild to bitter or strongly acrid.

The colour of the spines and whether their attachment to the stem is decurrent or adnate can be important diagnostic characters (Fig. 6.1).

A common feature in many species is the formation of amber to blood-red droplets on the cap or oozing from the spines.

A number of species are used around the world, in areas where they are still much more common than in Britain, as a source of dyes for woollens and fabrics. They yield rich colours which vary depending upon the particular mordant used in the dyeing process.

Most of the species are uncommon to rare in Britain and considered vulnerable or endangered due to habitat change or loss and possibly also pollution. Some species are only found in coniferous forests, others may occur with broadleaved trees.

Family *Auriscalpiaceae*
This family of fungi is related to mushrooms of the genus *Russula*, difficult to believe when looking just at their body forms. This family encompasses types with spines, coral-like branches and gills; each treated separately here in those sections. *Auriscalpium vulgare*, the Earpick Fungus is very easy to identify but often difficult to spot, being small and well camouflaged against the forest floor where it grows.

Family *Hericiaceae*
This family is perhaps most well known for the genus *Hericium*, all species of which form large, white to cream growths on wood with pendent spines. They are considered endangered in Britain due to habitat loss and should not be picked.

Fan Fungi
Family *Thelephoraceae*
Often referred to as earthfans, their common name gives a clue to the general shape and habitat of many of the species (Fig. 6.3). Not all the species are terrestrial however, some being common resupinate or corticioid forms on wood.

Only one genus is dealt with here: *Thelephora*, with just six species recorded in Britain, some very rarely. They have brown, warted or spiny spores (Fig. 6.4), fan-like or coral-like fruitbodies and can be found in both coniferous and broadleaved woodlands. They are usually very dull brown or greyish in colour. Some have a very strong odour of rotting garlic. The flesh may have the rather unusual reaction of turning blue with potassium hydroxide. The hyphae usually have clamp connections.

Fig. 6.1. The beautiful *Hydnellum caeruleum* with sky-blue fruitbody, including the spines. Note the pine needles trapped in the flesh. Photo © Walt Sturgeon.

Fig. 6.2. The cap of *Sarcodon squamosus* is extraordinarily scaly; it is one of our largest species in this order.

Fig. 6.3. *Thelephora palmata* is one of the less common species of Earth Fan and has a repulsive odour of rotting garlic. Photo © H. Krisp.

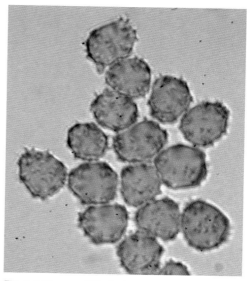

Fig. 6.4. Spores of *Thelephora palmata* showing the irregular outline with small spines. Photo © Ron Pastorino.

> **Family *Hydnaceae***
> One British genus: *Hydnum*, with hymenium as tiny, densely packed spines or pegs. *Hydnum* species are still under investigation via their DNA to determine the number of species present and their distribution in Europe. Like the chanterelles they often have similar orange-yellow to apricot colours.
>
> Recent DNA-based studies have shown that more species exist in Europe than traditionally included in field guides; some of these new species are included here but there may well be more yet to be described.

Genus *Hydnum*

Hydnum repandum L.: Fr.
[repandum = with uneven or wavy margin]
Cap 3–15 cm across, thick-fleshed, rounded, with irregular, sometimes lobed, inrolled margin, pale pinkish cream, apricot, pinkish orange, more rarely almost white. **Spines** deeply decurrent, 4–8 mm long, pointed, rounded to flattened in section, pale whitish pink. **Stem** cylindric-clavate, 3–7 x 1–3 cm, whitish bruising reddish. **Flesh** pale cream. **Taste** mild to slightly bitter. **Odour** pleasant, fruity. **Basidia** usually 4-spored. **Spores** subglobose, 6.5–8.5 x 5.5–7 µm. **Spore deposit** pale cream.

In broadleaved woods of *Fagus*, *Quercus* or *Betula* but also known in mossy conifer woods (*Pinus* and *Picea*), often in large numbers. Widespread in Britain and locally common although perhaps becoming rarer in some parts because of pollution, etc. Widely collected as an edible species.

Hydnum albidum Peck**
[albidum = of a whitish appearance]
Cap 3–10 cm across, thick-fleshed, rounded, with irregular, sometimes lobed, inrolled margin, pure white to slightly yellowish with age, bruising dull ochre. **Spines** deeply decurrent, 4–8 mm long, pointed, rounded to flattened in section, orange-pink. **Stem** cylindric-clavate, 2–5 x 1–3 cm. **Flesh** pale cream. **Taste** strongly bitter. **Odour** pleasant, fruity. **Basidia** usually 4-spored. **Spores** subglobose, 4–5 x 3.5–4 µm. **Spore deposit** pale cream.

In broadleaved woods of *Fagus*, *Quercus* or *Betula* but also known in conifer woods (*Pinus* and *Picea*), usually on calcareous soils.

Not recorded in Britain but widespread in Europe and might well occur here. The contrasting white cap and pinkish spines are very striking and the rather small spores easily distinguish it from pale forms of the common **H. repandum**.

Hydnum rufescens Fr.
[rufescens = of a reddish appearance]
Cap 3–7 cm across, thin-fleshed, rounded, with irregular, inrolled margin, bright reddish orange, brick-red to yellowish orange. **Spines** free to adnate, not decurrent, 4–5 mm long, pointed, rounded to flattened in section, pale yellow-orange to orange-pink. **Stem** cylindric-clavate, 2–7 x 0.5–1 cm. **Flesh** pale cream. **Taste** mild. **Odour** weak, slightly fruity. **Basidia** usually 4-spored. **Spores** subglobose, 6.5–9 x 5.5–7.0 µm. **Spore deposit** pale cream.

In rather wet, broadleaved woods of *Fagus*, *Quercus* or *Betula* but also more rarely with *Corylus*.

Widespread in Britain and fairly common in some localities. Although the cap may be depressed at the centre it rarely has the distinctive hole down the centre of **H. umbilicatum**.

Hydnum ellipsosporum Ostrow & Beenken.
[ellipsosporum = with ellipsoid spores]
Cap 3–5 cm across, thin-fleshed, rounded, with an inrolled margin before expanding, bright yellow-orange, to brick-red. **Spines** free to adnate, not decurrent, 4–5 mm long, pointed, rounded to flattened in section, pale yellow-orange to orange-pink. **Stem** cylindric-clavate, 2–5 x 0.5–1 cm. **Flesh** pale cream. **Taste** mild. **Odour** weak, slightly fruity. **Basidia** usually 4-spored. **Spores** subglobose, 9–11 (-12) x 6–7.5 µm. **Spore deposit** pale cream.

In mixed woods, both broadleaved and coniferous, its distribution in Britain is uncertain owing to its only being recently described. Collections are known from Scotland, Wales and England. Its ellipsoid spores allow its easy separation from the common **H. rufescens**.

Hydnum umbilicatum Peck
[umbilicatum = with a navel]
Cap 2.5–5 cm across, thin-fleshed, rounded, with an inrolled margin before expanding, with a prominent depression at the centre forming a tube down into the stem, cream-buff to pale brick-red. **Spines** free to adnate, not decurrent, 2–4 mm long, pointed, rounded to flattened in section, pale yellow-orange to orange-pink. **Stem** cylindric-clavate, 2–5 x 0.5–1 cm. **Flesh** pale cream. **Taste** mild. **Odour** weak, slightly fruity. **Basidia** usually 4-spored. **Spores** subglobose, 8–10 x 7.5–8.5 µm. **Spore deposit** white.

In mixed woods, both broadleaved and coniferous, its distribution in Britain is uncertain owing to confusion with other, similar species. It is uncertain whether the species in Britain is the same as the original *H. umbilicatum* as described by Peck in North America and it is likely that a new name will be needed for European material.

Hydnum vesterholtii Olariaga *et al*.**
[vesterholtii = after J. Vesterholt, mycologist]
Cap 2–4 cm across, thin-fleshed, rounded, with an inrolled margin before expanding, with a slight depression at the centre, sometimes slightly umbilicate, cream to pale buff. **Spines** free to adnate, sometimes decurrent, 1–2 mm long, pointed, rounded to flattened in section, pale ochre. **Stem** cylindric-clavate, 2–5 x 0.3–0.7 cm, staining ochre when handled. **Flesh** pale cream. **Taste** mild to bitter. **Odour** weak, slightly fruity. **Basidia** 3–5-spored. **Spores** ovoid to broadly ellipsoid, 8–9 x 6–7.5 µm. **Spore deposit** cream.

In mixed woods, especially *Fagus* and *Abies*, not yet recorded in Britain but described from France and also recorded from Spain and Italy, so might possibly be present here.

This species is characterised by its slender, very pale and sometimes umbilicate fruitbodies, often with decurrent spines and possessing ovoid spores. In similar habitats and with the same distribution is **H. ovoideisporum**, which shares very similar ovoid spores but differs in its rich, orange-tawny cap colours.

Hydnum

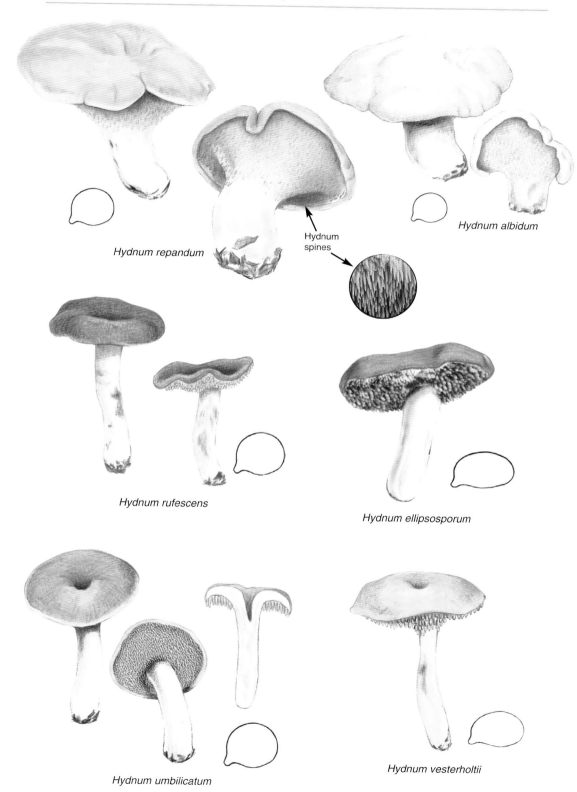

Hydnum repandum

Hydnum spines

Hydnum albidum

Hydnum rufescens

Hydnum ellipsosporum

Hydnum umbilicatum

Hydnum vesterholtii

> **Family *Bankeraceae***
> The five genera treated in this volume produce their hymenium on pendent spines on the underside of the cap, with the exception of the genus *Boletopsis* which has a poroid hymenium (see the polypores p.82). The spores vary from spiny to angular-warted and are whitish to brown in deposit.

Genus *Phellodon*

Phellodon melaleucus (Schwein.: Fr.) P. Karst.
[melaleucus = very dark and light]
Fruitbody with irregularly lobed 'cap' 3–10 cm across, but often with many caps fused together, depressed at centre with margins undulating, surface with fine, radiating wrinkles, finely tomentose to silky, blackish grey, or sometimes faintly lilac, fading to dull brown, margins whitish. **Spines** 1–3 mm long, whitish to dull grey-brown. **Stem** 1–3 x 0.5–2 cm, blackish, smooth to finely fibrillose, several stems often arising together. **Flesh** zoned, brownish to grey-brown, tough and fibrous, turning green with KOH. **Odour** when drying strongly spicy-aromatic (bouillon stock cubes). **Taste** mild to bitter. **Spores** subglobose, 3.5–4.5 x 3–4 µm, with small but distinct spines, non-amyloid. **Spore deposit** white.

In both coniferous and broadleaved forests, widespread in Britain. The similar *P. niger* differs in its tomentose stem and generally darker, blackish flesh.

Phellodon tomentosus (L.: Fr.) Banker
[tomentosus = with felty surface]
Fruitbody with rounded or oval caps, often fused to neighbouring caps to form a short chain, funnel or goblet-shaped, yellowish brown to grey-brown or reddish, with concentric zones, often with darker grey-brown central spot, surface minutely tomentose to radially wrinkled, much paler at the margin. **Spines** 1–2 mm, whitish to pale pinkish grey. **Stem** cylindric, 3–4 x 0.5–1.5 cm, minutely tomentose, concolorous with the cap to darker grey-brown. **Flesh** pale buff to yellowish brown, slightly zonate, **not** turning green with KOH. **Odour** strong of fenugreek as it dries. **Taste** mild to bitter. **Spores** subglobose, 3–4.5 x 3–4 µm, with small warts, non-amyloid. **Spore deposit** white.

May be confused with *P. confluens* which is not concentrically zoned and is usually found with broadleaved trees.

Genus *Bankera*

Bankera fuligineoalba (J.C. Schmidt: Fr.) Pouzar
[fuligineoalba = sooty-brown and white]
Fruitbody with a regularly shaped (determinate) cap, 4–15 cm across, rounded then slightly depressed, smooth to tomentose-squamulose, whitish then darker reddish brown, pinkish brown, not zoned. **Spines** 2–5 mm long whitish then soon grey. **Stem** cylindric, 2–5 x 2–4 cm, finely tomentose, dull greyish brown, usually whitish at the apex. **Flesh** rather soft, white turning slightly pink, not zoned. **Odour** weak when fresh, intense, spicy of bouillon cubes when dry. **Taste** mild. **Spores** subglobose to broadly ellipsoid, 4.5–5.5 x 2.5–2.5 µm, with pointed spines. **Spore deposit** white.

Mycorrhizal with *Pinus*. Uncommon in native pine forests in Scotland, possibly extinct in England.

Bankera violascens (Alb. & Schwein.: Fr.) Pouzar
Grows with *Picea*, and is pale purple-brown with pale greyish white spines and rounder spores. There are a few British sites, all in Scotland, almost certainly an introduced species.

Genus *Hydnellum*

Hydnellum caeruleum (Hornem.) P. Karst.
[caeruleum = sky blue]
Cap 2–10 cm across, rounded then expanding to slightly depressed, velvety or roughened-scrobiculate, often with concentric bands or grooves, pale to bright blue-grey when young, browner with age, sometimes exuding amber droplets. **Spines** decurrent, 3–6 mm long, bluish then brownish. **Stem** cylindric, 2–5 x 2–3 cm, velvety, orange then orange-brown. **Flesh** zonate in the cap, tough, corky, pale orange zoned with bluish, deeper orange-brown in the stem. **Odour** mealy. **Taste** mild. **Spores** ellipsoid but irregular, knobbly-tuberculate in outline, 4.5–6 x 4–4.5 µm. **Spore deposit** brown.

Mainly in coniferous woods but occasionally with *Fagus*, very rare in Britain, known only from native Scottish pine woods.

H. suaveolens, not yet British, is similarly bluish but has blue stem flesh and a strong smell of anise.

Hydnellum peckii Banker in Peck
[peckii = after C.H. Peck, 1833–1913, American mycologist]
Cap 3–12 cm across, at first rounded, cushion-like, then expanding into a flattened and finally dish-shaped, irregular funnel, surface velvety, felty becoming scrobiculate, whitish then flushing pinkish red, finally dark reddish brown at the centre. Usually exuding copious blood-red drops in damp conditions. **Spines** decurrent, 3–5 mm long, whitish then pinkish. **Stem** 2–4 x 2–3 cm, velvety, reddish brown. **Flesh** pinkish brown, zonate. **Odour** slightly sour. **Taste** very acrid. **Spores** ellipsoid to subglobose with coarse tubercles, 5–6.5 x 3.5–5 µm. **Spore deposit** brown.

In coniferous woods of *Pinus* or *Picea*, very uncommon to rare in Britain, in native pine woods of the Scottish Highlands.

Hydnellum aurantiacum (Alb. & Schwein.) P. Karst.
[aurantiacum = orange]
Cap 3–9 cm across, rounded then soon flattened, lumpy, radially ridged, tomentose, pale to dark orange-brown, whitish at the margin. **Spines** 2–4 mm, decurrent, white at first then pale brownish. **Stem** 2–5 x 1–2.5 cm, cylindric-rooting, orange to orange-brown, tomentose. **Flesh** pale, whitish buff in the cap, darker orange below, zonate. **Odour** weak. **Taste** mild. **Spores** ellipsoid with coarse tubercles, 5.5–7 x 4–5 µm. **Spore deposit** brown.

Associated with *Pinus sylvestris* in Britain on sandy soils, rare, known only from native pine woods in Scotland.

The hyphae of the flesh are without clamp connections which is diagnostic among British species.

Phellodon, Bankera, Hydnellum

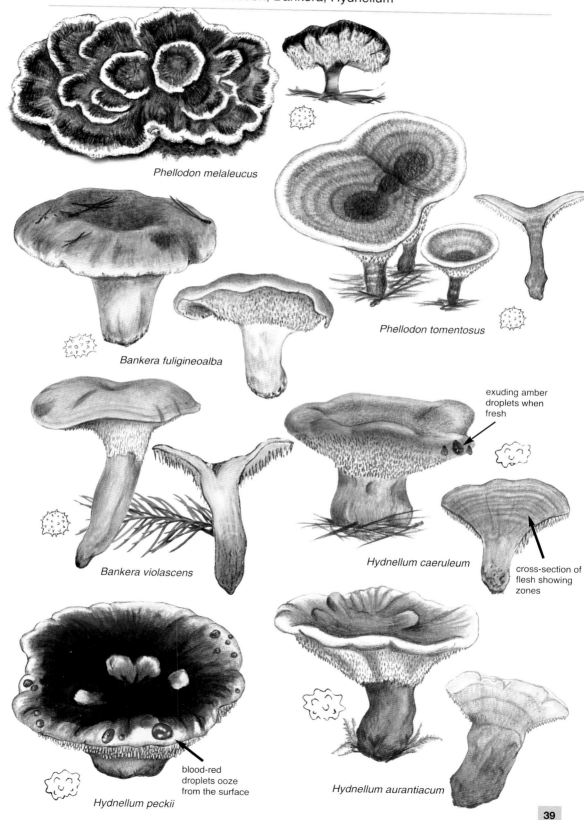

Hydnellum concrescens (Pers.) Banker
[concrescens = becoming grown together]
Cap 3–9 cm across, rounded then soon flattened, depressed, irregularly lobed and usually with many caps fused together, upper surface lumpy, concentrically zoned, tomentose then smooth, pale to dark reddish brown, darker, almost black at centre, whitish at the margin. **Spines** more or less central, 2–4 mm, decurrent, white at first then pale brownish. **Stem** 2–5 x 1–1.5 cm, cylindric-rooting, often bulbous, orange to orange-brown, tomentose. **Flesh** pale, whitish buff in the cap, darker brown below, zonate, rapidly dark green with strong alkalis. **Odour** weak of meal. **Taste** slightly bitter. **Spores** ellipsoid with coarse tubercles, 4.5–5.5 x 3–3.5 µm. **Spore deposit** brown.

In leaf litter of both broadleaved and coniferous woods, widespread in Britain.

Both this species and the rather similar ***H. scrobiculatum*** are being investigated molecularly as they probably represent a species complex and the British collections may represent as yet undescribed species.

Hydnellum scrobiculatum (Fr.) P. Karst.
[scrobiculatum = pitted]
Cap 3–5 cm across, solitary to caespitose or even several caps fused together, thin-fleshed, depressed, upper surface finely tomentose, very irregular with coarse lumps and lacerated excrescences, radially ridged, reddish brown to grey-brown darker at the centre and with paler growing margin. **Spines** decurrent, 2–3 mm long, brown to purplish brown. **Stem** more or less central, tomentose, cylindric to bulbous, reddish brown. **Flesh** rather thin, red-brown to purplish brown. **Odour** of new ground meal. **Taste** mild to spicy. **Spores** subglobose with coarse tubercles, 5–6.5 x 4–5.5 µm. **Spore deposit** brown.

In coniferous and mixed woodlands, especially with *Pinus*; widely distributed but uncommon. Like ***H. concrescens*** this is possibly a species complex and the exact identity of British collections remains to be determined.

Hydnellum ferrugineum (Fr.: Fr.) P. Karst.
[ferrugineum = rust-coloured]
Cap 3–10 cm across, rounded, cushion-like then flattened to depressed, surface velvety-hairy, ragged to scrobiculate, reddish brown to dark brown, often oozing red droplets in wet weather. **Spines** decurrent, to 5 mm long, whitish to purple-brown with age. **Stem** more or less central, 2–5 x 1–2 cm, cylindric, reddish brown, surface spongy-velutinous. **Flesh** tough, buff-brown, pinkish brown, darker in base, with irregular zones, exuding a pale brown stain in strong alkali solution. **Odour** mealy. **Taste** mild. **Spores** ellipsoid with large, coarse warts or tubercles, 5.5–7.5 x 4.5–5.5 µm. **Spore deposit** brown.

In coniferous woods, usually with *Pinus*, occasionally with *Picea*, usually on poor, sandy soils. Widely distributed in Europe but rare in Britain, although known from Scotland and down to the southeast of England as well as Wales and Ireland.

Hydnellum spongiosipes (Peck) Pouzar
[spongiosipes = spongy stem]
Cap 3–10 cm, convex to cushion-shaped, lumpy, gradually flattening, often with concentric grooves, surface tomentose or velvety, clay-brown to reddish brown, then purple-brown when old or wet. **Spines** 2–5 mm, decurrent, pinkish buff to red-brown. **Stem** 2–5 x 2–5 cm, swollen, clavate, brown and densely tomentose, rather soft and spongy with age. **Flesh** dark reddish brown, rather obscurely zonate. **Odour** farinaceous. **Taste** mild, mealy. **Spores** subglobose, warted-tuberculate, 6–7 x 5–5.5 µm. **Spore deposit** brown.

In broadleaved woods of *Quercus*, *Fagus* and *Castanea*, usually on sandy soils, mainly in the southern English counties, uncommon.

Very similar is ***H. ferrugineum*** but that species occurs in conifer woods and has paler flesh. Both species may exude red droplets when wet.

Hydnellum cumulatum K.A. Harrison
[cumulatum = mounded or heaped up]
Cap 3–9 cm, rounded or irregularly domed-lumpy, not funnel shaped and not zonate but may have radial ridges, surface finely tomentose, ochre-brown, nut-brown to deep reddish brown when wet with noticeably paler, whitish margin. **Spines** 2–3 mm, greyish ochre to pinkish brown, adnate-decurrent. **Stem** 3–6 x 2–4 cm, concolorous with the cap or a little darker, tomentose. **Flesh** firm, tough and corky, brownish. **Odour** musty, earthy, unpleasant, even slightly farinaceous. **Taste** mealy-musty. **Spores** rounded with large flattened to bifurcated warts, 3.5–5 x 3–4 µm. **Spore deposit** brown.

Associated with *Pinus* this is a rare species found in a very few locations in Scotland but often mistaken for the similar ***H. spongiosipes*** or ***H. ferrugineum*** which have larger spores and often funnel-shaped caps which may have concentric colour zones.

Hydnellum gracilipes (P. Karst.) P. Karst.
[gracilipes = with slender stem]
Cap 0.5–4 cm, soft, thin-fleshed, surface rather woolly-hairy, becoming flattened, funnel-shaped to almost resupinate, frequently stuck to the covering vegetation or woody debris, pinkish to pinkish ochre, darker when wet. **Spines** 1–2 mm, pinkish white, decurrent, often all the way down the stem. **Stem** very gracile, 1–3 x 0.3–0.6 cm, concolorous with the cap, fragile. **Flesh** very thin, pinkish brown. **Odour** musty. **Taste** musty. **Spores** rounded, 4–4.5 x 3–3.5 µm, with blunt, flattened or often two-pronged warts or tubercles. **Spore deposit** brown.

Only added to the British list in 2011 this small, delicate species is found in Scotland, in pine woods usually along tracksides or near banks of watercourses, but always under overhanging vegetation of common heather, *Calluna vulgaris* and pine litter so very difficult to spot. The woody stems of the heather are used by the fungus to support its fragile and delicate fruitbodies.

In Scandinavia the species is often found under fallen pine logs with the cap adhering to the underside of the log, almost like a resupinate species. So although rare it may be under-recorded as its ecology makes it difficult to find. The blunt, often two-pronged spore warts are a good confirmatory character.

Fig. 7.1. *Clavulina coralloides* along with the other species in the genus is related to the familiar chanterelle in the order *Cantharellales*.

Fig. 7.2. *Clavaria* species are often found in open grasslands, heaths and lawns. This is *C. laeticolor*.

Fig. 7.3 *Ramaria* species such as *R. abietina* are species of woodland, being mycorrhizal with specific tree species.

> **Family *Typhulaceae***
> Fruitbody forming simple clubs producing smooth, ellipsoid, hyaline spores. Usually on dead or decaying leaves, stems, etc.

Genus *Typhula*

Typhula fistulosa (Holmsk.) Olariaga
[fistulosa = hollow stemmed]
Syn. *Macrotyphula fistulosa*
Fruitbody a long, slender club 3–20 cm tall, 0.5–1 cm wide. **Club** stiff, narrowly clavate with a blunt tip, smooth to slightly irregular, ochre-yellow to reddish ochre, minutely pubescent with stiff hairs at base. **Hymenium** on the upper 2/3 of the club. **Flesh** hollow yellowish. **Odour** nil. **Taste** mild. **Spores** ellipsoid, smooth, 10–18 x 5.5–8 µm. **Spore deposit** white.
On soil, leaf litter and twigs in broadleaved woodlands. Common and widespread.

Typhula contorta (Holmsk.) Olariaga
[contorta = twisted or contorted]
Syn. *Macrotyphula fistulosa* var. *contorta*
Fruitbody often caespitose, simple or often forked, 2–3 cm tall. **Club** usually contorted, bent, sometimes blunt-ended, greyish ochre, ochraceous brown, base usually without stiff hairs. **Flesh** hollow, yellowish. **Odour** nil. **Taste** mild. **Spores** ellipsoid-subfusiform, smooth, 14.5–19 (20.8) x 6.3-9.5 µm. **Spore deposit** white.
On dead, usually attached, branches or twigs of *Corylus* and *Alnus*. Widespread but occasional. Often treated as a variety of ***T. fistulosa*** but its restricted host range, shape and broader spores support its separation.

Typhula juncea (Alb. & Schwein : Fr.) P. Karst.
[juncea = pertaining to rushes]
Syn. *Macrotyphula juncea*
Fruitbody a very slender, long club, 3–12 cm tall, 0.5–1.5 mm wide, curved and flexuose. **Club** pointed, smooth, ochre-yellow, ochre to orange-brown, minutely pubescent at base. **Hymenium** spread over the upper 2/3 of the club. **Flesh** solid, later hollow, pale yellowish. **Odour** rather sour. **Taste** unpleasant, sour. **Spores** ellipsoid-amygdaliform, smooth, 7–10.5 x 4–5.4 µm. **Spore deposit** white.
On decaying stems of herbs, leaves, etc, often on the fallen petioles of *Fraxinus*, *Acer* and *Fagus*, in damp broadleaved woods. Widespread and quite common.

Typhula phacorrhiza (Reichard) Fr.
[phacorrhiza = with a tough root]
Fruitbody very slender, filiform clubs arising from a small, flattened disc-shaped sclerotium. **Club** 5–10 cm tall, 0.5–1 mm thick, ochre-yellow to greyish ochre, stem a little darker than upper portion, base often curved, running over leaf surface to attach to the sclerotium. **Flesh** thin, yellowish. **Odour** nil. **Taste** nil. **Sclerotia** flattened, disc-like, white then reddish brown, with pale mycelial strands running over the leaf. **Spores** cylindric-ellipsoid, smooth, 10–12.2 (13.6) x 4.6–5.5 µm. **Spore deposit** whitish.
On decaying leaves, especially *Fraxinus* but also on *Alnus* and *Corylus*. Widespread and locally common.

Typhula erythropus (Pers.) Fr.
[erythropus = red foot]
Fruitbody a slender club with a distinct head, entire club 1–3 cm tall. **Head** cylindric-clavate, white, 0.5–1 mm thick. **Stem** 0.1–0.3 mm thick, blackish at base, reddish above fading to ochre-white. **Odour** nil. **Sclerotia** tiny, rounded, reddish brown, embedded in the substrate on which it is growing. **Spores** ellipsoid, smooth, 5–7 x 2.5–3 µm. **Spore deposit** whitish.
On rotting debris of herbs, twigs, leaves in broadleaved woods. Common and widespread.

Typhula micans (Pers.) Berthier
[micans = with a metallic lustre]
Fruitbody tiny, entire club 1–3 mm tall with fertile head distinct from the short stem. **Head** rose-pink when fresh, fading to whitish when dry, 0.8–2 mm tall, pruinose, with shining granules. **Stem** cylindric, 0.1–1 mm, pinkish hyaline, smooth. **Spores** broadly ellipsoid, smooth, 9–12 x 4.5–6 µm. **Sclerotia** absent.
On dead leaves and stems of herbaceous plants. Widespread but not often reported as easily overlooked.

Typhula incarnata Fr.
[incarnata = flesh coloured]
Fruitbody forming tiny, slender clubs on long stems, 0.5–3 cm tall, arising from sclerotia. **Head** up to 4 x 0.3 mm, rose-pink. **Stem** narrower than the head, 10–30 mm long, white to pale grey. **Spores** ellipsoid, 9–11 x 3.5–5 µm. **Sclerotia** 0.5–4 mm across, reddish brown.
On dead and decaying grasses and some cereal crops where it causes Grey Snow Mould disease. Widespread but not often reported.

Typhula quisquiliaris (Fr.) Corner
[quisquiliaris = pertaining to rubbish or debris]
Fruitbody tiny, frequently in rows, entire club up to 7 mm tall. **Head** clavate-cylindric, 1.5–4 x 1–2.5 mm, white. **Stem** 2–3 x 0.3–0.4 mm, white, finely hairy. **Spores** narrowly ellipsoid, smooth, 9–14 x 4–5.5 µm. **Sclerotia** embedded in the substrate, pale yellow, 1–3 mm across.
On dead stalks of bracken, *Pteridium aquilinum*, more rarely on dead culms of *Juncus* or even leaves of deciduous trees. Common and widespread.

Typhula corallina Quél. & Pat.
[corallina = coral-like]
Fruitbody an elongate club, 2–10 mm tall, with an elongated head and short stem. **Head** cylindric-pointed, white, 0.2–0.5 mm thick. **Stem** slightly narrower than the head, short, greyish white. **Spores** ellipsoid, smooth 6.4–9 (9.6) x 3.6–5 µm. **Sclerotia** rounded to irregular, yellowish to ochre-brown, up to 2 mm across, often with multiple fruitbodies emerging from each sclerotium.
On leaves and stems of dead herbaceous plants, fern fronds, etc. Widespread but not often recorded.

> **Family *Pterulaceae***
> Delicate, feathery corals to simple clubs. 2- or 4-spored. Cystidia present in some species.

Pterula multifida (Chevall.) Fr.
[multifida = many parts]
Fruitbody a delicate, coral-like tuft of slender clubs, up to 6 cm tall, branching from a basal stem, pale whitish ochre, greyish to ochre-brown. **Basidia** 4-spored. **Spores** ellipsoid, smooth, with many droplets, 5–6 x 2.5–3.5 µm. **Cystidia** absent.
Widespread on leaf litter, conifer needles, fern fronds, etc, but uncommon.
P. debilis forms small groups of solitary clubs, loosely branched. Its spores are ellipsoid, 7–9.5 x 3.5–4 µm. Cystidia present. Rare on *Juncus* stems in bogs.

Typhula, Pterula

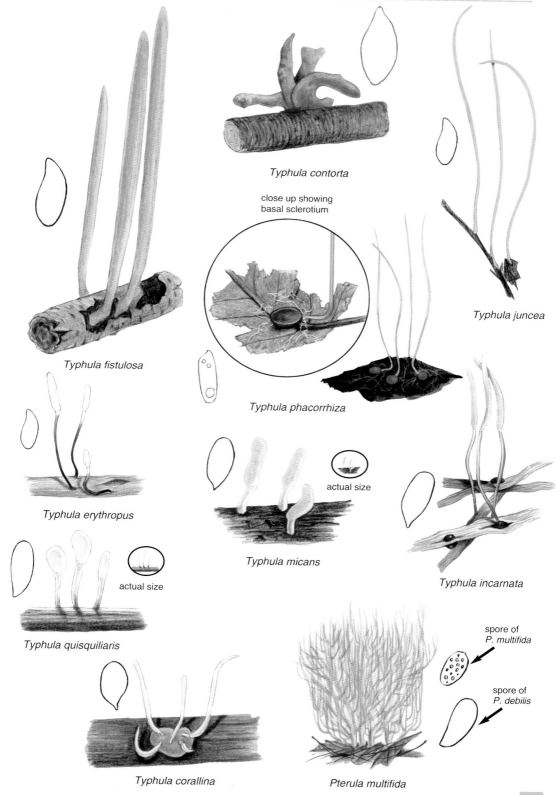

> **Family *Clavariaceae***
> Fruitbody forming simple to more complex clubs producing smooth to spiny, mainly hyaline spores. All may have clamp connections at the base of the basidia but only *Clavulinopsis* and *Ramariopsis* possess clamps in the hyphae of the flesh.

Genus *Clavaria*

Clavaria asperulispora G.F. Atk.
[asperulispora = prickly spore]
Fruitbody a simple club with an indistinct stem, up to 6 x 0.5 cm. **Club** is fusiform to clavate at the tip, smooth to irregularly ridged or grooved, blackish throughout with a slight olivaceous tint. **Flesh** rather soft, brittle. **Spores** globose to subglobose, thick-walled, hyaline, 3.5–4 x 3–4 µm, with fine warts. **Clamp connections** absent.

British collections on mossy soil under *Taxus baccata*. This appears to be a genuinely very rare species with only a handful of records. The colour combined with round, warty spores allow easy identification.

Clavaria greletii Boud.
[greletii = after Louis-Joseph Grelet, mycologist]
Fruitbody a simple club, up to 10 x 0.5 cm. **Club** is mostly rounded-clavate at the tip, smooth to somewhat rugulose, blackish grey, often with a blue-grey pruinose appearance. **Flesh** soft, brittle to slightly elastic. **Spores** subglobose to slightly ellipsoid, thin-walled, smooth, hyaline, 7–10.5 x 6–9.5 µm. A few spores have been recorded with scattered large warts or pegs; these perhaps only occur after spores fall in a deposit rather directly than off the fruitbody. **Clamp connections** hoop-like at base of basidia.

In mossy grassland, also grass in sand dunes, widespread but rare with very few British records.

Clavaria atroumbrina Corner
[atroumbrina = blackish umber]
Syn. *C. pullei*?
Fruitbody simple, usually fasciculate, tubular, up to 10 x 0.5 cm. **Club** longitudinally ridged, dark sepia brown, reddish brown to blackish with age, paler at base. **Flesh** soft, brittle. **Spores** oblong-ellipsoid, thin-walled, smooth, hyaline, 5–6.5 x 2.5–3.5 µm. **Clamp connections** absent.

In upland pastures, mossy lawns, etc, very uncommon but widespread in Britain. Originally described from North America, there is a recent molecular study which suggests that the British collections may represent a different, perhaps undescribed species.

Clavaria atrofusca Velen. **
[atrofusca = blackish reddish brown]
Fruitbody simple, solitary or in small clusters, cylindric to tapered at the tip, up to 5 x 0.3 cm. **Club** often flattened, smooth to irregularly ridged or grooved, matt, fuscous-black. **Flesh** elastic, soft. **Spores** ellipsoid-oblong, thick-walled, 6–8 x 3.6–4.6 µm, black to brownish black, with prominent warts. **Clamp connections** absent.

On mossy soil. No authentic British records but should occur here. The elongate, thick-walled, warty spores make it easy to identify.

Clavaria fumosa Pers.
[fumosa = smoky-grey]
Fruitbody simple, usually in dense, fasciculate tufts. **Club** 2–12 x 0.3–1 cm, fusiform, often slightly compressed or twisted, pale grey-brown, pinkish brown, pale ochre-brown, smooth. **Flesh** firm, brittle. **Spores** ellipsoid, smooth, 5.5–8 x 3.5–4 µm, packed with small granules or oil droplets. **Clamp connections** absent.

In grass in meadows, lawns, etc. Widespread and common everywhere.

Clavaria tenuipes Berk. & Broome
[tenuipes = tapered, slender foot]
Syn. *C. krieglsteineri*
Fruitbody simple, solitary or in small groups, occasionally subfasciculate, forming a swollen head with a slender stem. **Club** cylindric to broadly clavate, often compressed laterally, 15–60 x 2–10 mm, pale greyish clay to cream or dull buff. **Flesh** soft, elastic, hollow when old. **Stem** 5–15 x 1 mm, often very distinctly separated from the head. **Spores** broadly ellipsoid, 6–9.5 (-12) x 4–5.5 µm. **Clamp connections** like loops at base of basidia.

Amongst short grass or on bare soil, heaths, grasslands and woods. Widespread in Britain but rare.

Clavaria crosslandii Cotton
[crosslandii = after Charles Crossland, 1844–1916, British mycologist]
Syn. *C. tenuipes ss. Schild, C. guilleminii ss. Brit. Authors*
Fruitbody simple, solitary or in small groups, up to 3.5 cm high. **Club** often flattened or compressed laterally, clavate, broad at the apex, pale grey to pale brownish. **Stem** narrower than the club. **Flesh** soft, brittle. **Spores** ellipsoid, without oil droplets, 4–6 x 2.5–3.5 µm. **Clamp connections** absent.

In short grass and mosses in woods or often on fire sites with *Funaria hygrometrica*. Widespread and probably common but little recorded. Frequently confused in the past with *C. tenuipes* but that has larger spores, or sometimes recorded as *C. guilleminii* (not authentically British).

Clavaria acuta Sowerby
[acuta = pointed]
Syn. *C. asterospora, C. falcata?*
Fruitbody simple, solitary to small groups, sometimes subfasciculate, usually with a smooth and glossy stem distinct from the club. **Club** cylindric-fusoid, pure white to pale greyish cream, 15–60 x 5 mm. **Stem** 0.5–15 x 1–3 mm. **Flesh** soft, hollow when mature. **Spores** subglobose to tear-shaped, 7–9 x 5–7 µm, sometimes mixed with thick-walled, star-shaped spores or even entirely so. **Clamp connections** like loops at base of basidia.

Widespread and common everywhere. The spores may develop prominent spines as they mature and this gave rise to the name *C. asterospora*.

Clavaria fragilis Holmsk.
[fragilis = fragile]
Syn. *C. vermicularis*
Fruitbody simple, in dense, fasciculate clumps, club not very distinct from the stem. **Clubs** to 12 cm tall, fusiform to narrowly clavate, or sinuous, rounded to slightly compressed, surface smooth, pure white to pale cream, apex often yellowish. **Stem** indistinct, white. **Flesh** soft, brittle, hollow when mature. **Spores** ellipsoid to subglobose, 4.5–6 (-7) x 3–4 µm with numerous internal droplets. **Clamp connections** absent.

In unimproved grasslands, more rarely in woodlands. Widespread and common everywhere.

Clavaria

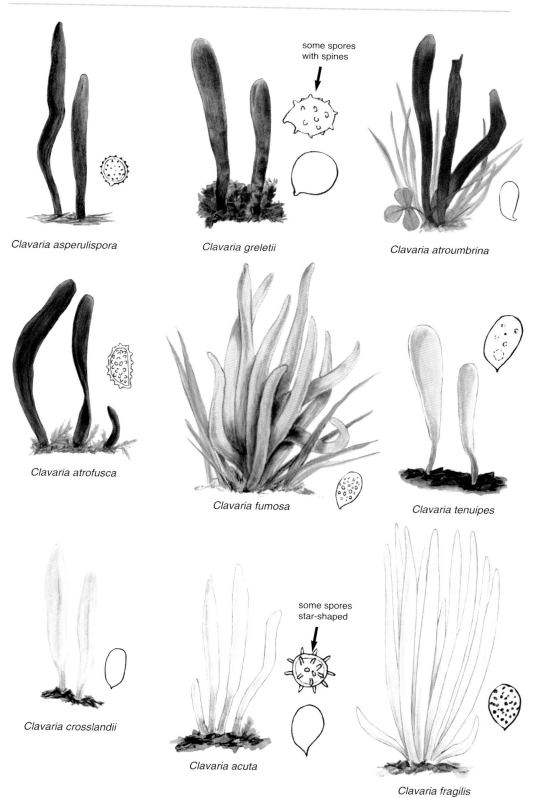

Clavaria argillacea Pers.
[argillacea = clay coloured]
Fruitbody simple, usually in small tufts, more rarely solitary, up to 8 cm tall. **Club** broadly clavate, sometimes attenuated at the tip, pale to dark clay-ochre, straw yellow to yellow. **Stem** up to 1.5 cm long, darker yellow than the club, translucent and shining. **Flesh**, soft, brittle, hollow when mature. **Spores** oblong-ellipsoid, 9–13 x 4.5–6 μm. **Clamp connections** hoop-like at base of basidia.
On acid soils in heathlands, moorlands, often in short turf and thought to be mycorrhizal with *Ericaceae* such as *Calluna*.
C. argillacea var. *sphagnicola* differs in its more slender clubs, longer stem up to 3 cm, broader spores (6–7 μm) and habitat in *Sphagnum* bogs.

Clavaria flavipes Pers.
[flavipes = yellow foot]
Syn. *C. straminea*
Fruitbody simple, slender, cylindric, solitary or in small groups, up to 10 cm tall. **Club** often narrower at apex, straw-yellow to pale greyish yellow, smooth to grooved, often twisted or flattened. **Stem** distinct from the club, yellowish brown, smooth, translucent. **Flesh** soft, brittle, hollow when mature. **Spores** subglobose, 6–8 μm in diameter. **Clamp connections** hoop-like at base of basidia.
On acidic or calcareous soils in short grass or in woodlands. Widespread but rarely recorded.

Clavaria amoenoides Corner, Thind & Anand
[amoenoides = beautiful]
Fruitbody simple, densely fasciculate, up to 14 cm tall by 1 cm wide. **Club** fusiform with narrow, pointed tip, often flattened and twisted, pale yellow with tips browning when old. **Stem** short, indistinct. **Flesh** soft, brittle, hollow when old. **Spores** ellipsoid-oblong to bean-shaped, 6–8 x 3.5–4.5 μm. **Clamp connections** hoop-like at base of basidia.
In mossy grass. Collections are known from Wales and Scotland, probably genuinely rare.

Clavaria incarnata Weinm.
[incarnata = flesh coloured]
Fruitbody simple, solitary or in small clusters, up to 7 cm high. **Club** cylindric, slender, up to 5 mm wide, with an obtuse or pointed apex, smooth, salmon-pink, flesh coloured, pale purplish pink fading to cream. **Stem** fairly distinct, usually darker. **Flesh** soft, brittle. **Spores** ellipsoid, 6.5–10 x 3.5–6 μm, sometimes mixed with thick-walled, star-shaped spores. **Clamp connections** hoop-like at base of basidia.
In unimproved grasslands, lawns, sometimes in woodlands with records near *Taxus*. Widespread but rarely reported.

Clavaria flavostellifera Olariaga et al.
[flavostellifera = yellowish with ornamented spores]
Recently found in Britain this is close to *C. incarnata* with yellow to pink fruitbodies but with shorter spores 5–7.5(8) x 4.5–6 μm. Although originally described from Spain with yellow clubs the British collection was strongly pink.
In grass and woodlands on calcareous soil.

Clavaria rosea Fr.
[rosea = rose-coloured]
Fruitbody simple, solitary or in small groups, up to 7 cm high. **Club** cylindric to broadly clavate with rounded apex, smooth to rugulose, bright rose-pink to lilaceous pink. **Stem** indistinct. **Flesh** soft, fragile. **Spores** ellipsoid, 5–8 x 2.5–3.5 μm. **Clamp connections** absent.
On calcareous soils in grass in woodland clearings and open meadows. Widespread in Britain but very uncommon and not often reported.

Clavaria zollingeri Lév.
[zollingeri = after Heinrich Zollinger, Swiss mycologist]
Fruitbody complex, with multiple branches reaching up to 8 cm high, 6 cm across. **Clubs** dividing at the tips into even smaller points, united at a central base, vivid violet, lavender to pinkish. **Flesh** firm, brittle, violet. **Odour** pleasant. **Taste** mild. **Spores** subglobose-ellipsoid, 5.5–7 x 4.5–5.5 μm. **Clamp connections** absent.
In unimproved grasslands and sometimes woodlands, widespread but rather rare. Its vivid colour and multiple branches, combined with smooth spores make it unmistakable.

Clavaria salentina Agnello & Baglivo **
[salentina = after Salento, Italy]
Fruitbody simple, in small, fasciculate clusters up to 4.5 cm high. **Clubs** fusiform to clavate or spatulate, bright salmon-orange, reddish orange to yellow-rust, paler at the tips. **Stem** hardly distinct, slightly paler. **Flesh** soft, brittle. **Odour** mouldy. **Taste** nil. **Spores** ellipsoid, 4.8–7 x 3.0–4.6 μm. **Clamp connections** absent.
Pastures, unimproved grasslands. Described from Italy, possibly widespread in the Mediterranean regions.

Genus *Clavulinopsis*

Clavulinopsis corniculata (Fr.) Corner
[corniculata = with horn-like appendage]
Fruitbody densely fasciculate, branching and twisting, up to 6 cm tall. **Clubs** vary from simple, fusiform or clavate to flattened, bifurcated or with multiple rounded 'horns', often very tangled, yellow to yellow-orange or ochre-yellow. **Flesh** solid, fragile, yellowish. **Odour** sour. **Taste** bitter. **Spores** subglobose, thick-walled, with prominent apiculus. **Basidia** vary from 2 to 4-spored.
In meadows, lawns, rarely woodlands. Widespread and common everywhere. The forma *bispora* appears to be much less common although probably under-recorded.

Clavulinopsis fusiformis (Sowerby) Corner
[fusiformis = spindle-shaped]
Fruitbody forming dense clumps of tall clubs up to 10 cm tall. **Clubs** simple, fusiform, cylindric to compressed, bright golden-yellow, browner as they wither, especially at the tips. **Flesh** soft, brittle, hollow with age, yellowish. **Odour** weak but pleasant. **Taste** bitter. **Spores** subglobose, thick-walled, with prominent apiculus.
In grasslands, lawns, heathlands. Widespread and common everywhere.

Clavulinopsis helvola (Fr.) Corner
[helvola = pale yellow]
Fruitbody simple, usually solitary or in small groups, up to 6 cm tall. **Club** slender, cylindric-fusiform, smooth to compressed-furrowed longitudinally, egg-yellow, orange with whitish base. **Stem** usually distinct, white to pale yellow. **Flesh** soft, fragile, yellowish. **Odour** nil. **Taste** slightly bitter. **Spores** ellipsoid with large, blunt warts and tubercles.
On acidic or basic soils, in grasslands, lawns, heaths, sometimes in woodlands on calcareous soils. Common and widespread. Very similar macroscopically to *C. laeticolor* or *C. luteoalba* but easily separated by its warted spores

Ramaria, Artomyces

> **Family *Clavulinaceae***
> One British genus: *Clavulina*, with hymenium spread over the external surface of coral-like fruitbodies. Basidia mostly 2-spored, smooth spores.

Genus *Clavulina*

Clavulina coralloides (L.) J. Schröt.
[coralloides = like a coral in form]
Syn. *C. cristata*
Fruitbody 2–6 (-8) cm high, consisting of multiple branches each ending in sharp, many-forked tips, with the branches fusing at the base into a trunk, the whole resembling an undersea coral. Very variable in shape and colour, the branches vary from short and squat to elongated and slender, from almost pure white to cream or smoky greyish lilac. **Flesh** from soft to fairly firm, brittle, whitish. **Taste** mild. **Odour** earthy or mouldy. **Basidia** usually 2-spored. **Spores** broadly ellipsoid to subglobose, 7–9 x 6–7.5 μm. **Spore deposit** cream.
Common and widespread everywhere, in mixed woods, sometimes in pastures.
Extremely variable, particularly as it is frequently parasitised by an ascomycete fungus, **Helminthosphaeria clavariorum**, which causes the branches to become grey to lilac or blackish grey, thus resembling *C. cinerea*. That species is best distinguished by its blunt branches and larger spores.

Clavulina incarnata (Corner) Olariaga
[incarnata = flushed pink]
Syn. *C. cristata* var. *incarnata*
Differs from *C. coralloides* in its more slender, pink-hued branches and, uniquely in *Clavulina*, in possessing cystidia.

Clavulina cinerea (Bull.) J. Schröt.
[cinerea = with grey tones]
Fruitbody 4–8 (-11) cm high, with numerous branches fused into a short trunk, whitish to ochre at the base, greyish lilac, lavender-grey, often with paler, more yellowish tips, branches with blunt tips. **Flesh** rather soft, whitish. **Taste** mild. **Odour** earthy-mouldy. **Basidia** usually 2-spored. **Spores** broadly ellipsoid to subglobose, 8–10 x 7–8 μm. **Spore deposit** cream.
Very common and widespread in mixed woodlands everywhere. Often infected with the ascomycete fungus **Helminthosphaeria clavariorum** which turns the bases of the branches dark grey. May be confused with the greyish lilac, parasitised forms of *C. cristata* which however has sharper-tipped branches and smaller spores.

Clavulina reae Olariaga
(reae = after Carlton Rea, mycologist]
Syn. *C. cinerea* var. *gracilis*
Differs from *C. cinerea* by its long, slender stem and the numerous thin, tapering, acute branches.
On bare soil in woods. Possibly frequent but distribution uncertain as formerly included under *C. cinerea*.

Clavulina rugosa (Bull.) J. Schröt.
[rugosa = roughened or wrinkled]
Fruitbody 5–8 (-10) cm high, with few to numerous distorted, club-like branches fused into a short trunk, dingy white to pale ochre, branches usually wrinkled-lumpy with broad, blunt tips. **Flesh** rather soft, whitish. **Taste** mild. **Odour** earthy-mouldy. **Basidia** usually 2-spored. **Spores** broadly ellipsoid to subglobose, 9–13.5 x 7.5–10 μm. **Spore deposit** cream.
Common and widespread everywhere, in mossy woods, along tracksides, ditches, etc.
Might be confused with the similarly white *C. coralloides* but with larger spores and broader, more flattened and blunt branches.

> **Family *Clavariadelphaceae***
> Large, club-like forms, all rare or uncommon.

Genus *Clavariadelphus*

Clavariadelphus pistillaris (L.) Donk
[pistillaris = resembling a pestle]
Fruitbody 5–20 cm high, at first narrowly fusiform then swelling to become broadly club-shaped. Hymenium spread over the external surface, becoming finely longitudinally wrinkled, bright ochre-yellow when young, suffused with lilac below, fading to brownish ochre with age, often with a whitish bloom on the surface, saffron yellow with KOH, dark green with $FeSO_4$. **Flesh** firm to spongy, white, discolouring violaceous brown. **Taste** bitter. **Odour** weak but pleasant. **Spores** ellipsoid, smooth, 11–13 x 6–7 μm. **Spore deposit** pale ochre.
Widespread in Britain but uncommon to rare and possibly declining, in *Fagus* woods.

Clavariadelphus truncatus Donk**
[truncatus = with a flattened or truncated head]
Fruitbody 5–12 cm high, at first broadly club-shaped, soon progressively more flattened with a distinct, wrinkled edge. **Hymenium** spread over the external surface, becoming finely longitudinally wrinkled, dull ochre when young, suffused with reddish brown below, fading to brownish ochre with age, often with a whitish bloom on the surface, colouring bright red with KOH, dark green with $FeSO_4$. **Flesh** firm to spongy, greyish white, discolouring violaceous brown. **Taste** mild. **Odour** weak but pleasant. **Spores** broadly ellipsoid, smooth, 9–12 x 5.3–6.8 μm. **Spore deposit** pale ochre.
In coniferous woods, absent in Britain with previous records shown to be in error.

Clavariadelphus ligula (Schaeff.: Fr.) Donk
[ligula = resembling a small tongue]
Fruitbody 3–10 cm high, 0.5–15 cm wide, cylindric-clavate, usually a little compressed, grooved-wrinkled. **Hymenium** spread over the external surface, becoming finely longitudinally wrinkled, dull ochre when young, ochre-yellow to brownish yellow with age. **Flesh** firm to spongy, white, unchanging. **Taste** mild. **Odour** weak but pleasant. **Spores** broadly ellipsoid, smooth, 10–14 x 3–4.5 μm. **Spore deposit** pale ochre.
Looking like a slender, smaller *C. pistillaris* this very rare species in Britain is found in Scottish conifer woods and is easily distinguished by its long, narrow spores.

Clavulina, Clavariadelphus

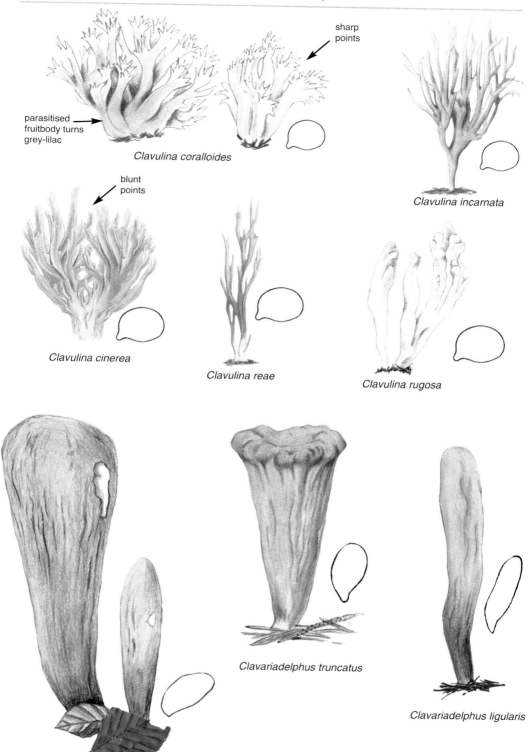

8. Polypores & Other Bracket Fungi

The polypores or bracket fungi may look very similar in body form but are actually an assemblage of often unrelated fungi with a tubular, pored hymenium as well as many resupinate fungi and a few gilled fungi (mainly in the genus *Lentinus*, dealt with in a forthcoming companion volume) and even some with spines or teeth. The principal orders include the *Polyporales* and the *Hymenochaetales* but many other orders also form bracket-like fruitbodies. The 'polypores' are therefore a grouping of convenience. Species are generally saprotrophic, most of them wood-rotters, and some are parasites. Those of economic importance include several pathogens of forestry plantations and amenity trees. A few species cause damage by rotting structural timber. Some polypores are commercially cultivated and marketed for use as food items or in traditional medicine.

The polypores include some of our largest and longest lived fruitbodies, some capable of reaching a metre across. During any walk in the woods you will almost certainly come across one or more of the commoner species shown here. Not all of the families and genera that form what would typically be called a polypore are represented in this volume. Many are beyond the scope of a field guide, requiring specialist monographs for their identification.

As with all other fungi, recent molecular work is providing new insights into the relationships in this group, with the result that some species have shifted into different genera, with some very surprising changes. In the end it will produce a clearer and more 'natural' arrangement but some well-established names will have to change.

The group includes a variety of spore shapes and spore colours, usually ellipsoid or cylindric and white- or brown-spored. The spores vary from thin-walled and very simple (*Polyporus*) to very thick-walled with complex wall structures (*Ganoderma*).

The structure of the hyphae

In bracket fungi they usually consist of one, two or three kinds: monomitic, dimitic and trimitic respectively (Fig. 8.1).

Monomitic species have **generative hyphae** which are simple, smooth, thin- to thick-walled. They frequently branch and, depending on the species concerned, may have clamp connections at the septa. Generative hyphae give rise to all the other types, usually with a clamp connection separating the different types.

Dimitic species have the simple hyphae plus more complex thick-walled, usually (but not always) unbranched hyphae (**skeletal hyphae**). The latter help to give the fruitbody a more robust, tougher texture. **Trimitic** species have additional **binding hyphae**. These are branched, very thick-walled or sometimes even solid and non-septate. They intertwine themselves between and around the other hyphae.

Another, less common type of hyphae are **gloeoplerous hyphae**, these are generally wide, thin-walled and with highly refractive contents which make them appear different to the surrounding generative hyphae.

Cystidia, setae and setal hyphae

The hymenium of polypores, and sometimes the tissues of the flesh, may produce distinctive cells referred to as cystidia or setae. In some species the hyphae may end in seta-like structures. The shape, thickness and overall dimensions of these structures are important identification features (Fig. 8.2).

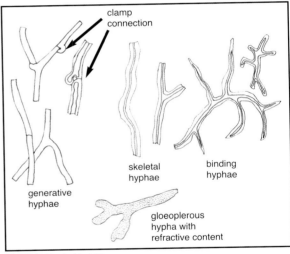

Fig. 8.1. Hyphal types.

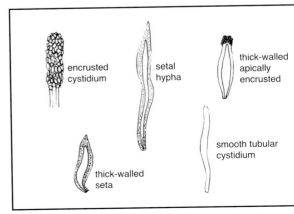

Fig. 8.2. Cystidia, setae and setal hyphae.

Fig. 8.3. The genus *Polyporus* gives its name to the order *Polyporales*. Consisting of relatively few species in Europe, they often have a stem as seen here in the common Dryad's Saddle, *P. squamosus*.

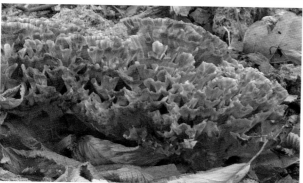

Fig. 8.4. The polypores include genera and species with a wide variety of body shapes and hymenial surfaces including fan-shaped structures without tubes or pores and with spores produced on their more or less smooth outer surface. Illustrated here is *Podoscypha multizonata*.

Fig. 8.5. Many polypores are parasitic on trees such as this Dyer's Mazegill, *Phaeolus schweinitzii* which usually forms its fruitbodies at the base of conifers. The common name reflects its use in dyeing woollens and fabrics.

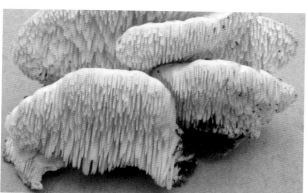

Fig. 8.6. A few genera in the *Polyporales* form their basidia and spores on pointed or flattened spines or teeth, closely resembling species in the unrelated genus *Hericium* in the order *Thelephorales*. This is *Spongipellis pachyodon*, a species in the family *Meruliaceae*.

Family *Ganodermataceae*

Genus *Ganoderma*

Ganoderma australe (Fr.) Pat.
[australe = southern]
Southern Bracket
Fruitbody perennial, starting as a hard, rounded knob, expanding to form a semicircular bracket, often lasting many years as new pore layers are added. Surface very hard, woody, very difficult to depress with a finger, smooth and shiny becoming tuberculate, wrinkled or knobbly with age, greyish brown, ochre-brown to rust, often covered with rust-brown spores. **Pores** white, 4–6 per mm, bruising instantly brown where touched. **Tubes** up to 7 cm deep, dark reddish brown, with only very thin tissue between each layer. **Flesh** tough, woody, reddish brown. **Spores** (measuring the dark, inner spore) 9–10 x 6.2–7.1 µm plus a transparent, outer exospore. **Spore deposit** bright rust-brown.

Widespread and common on a variety of trees but especially *Fagus*, particularly common in the southern counties.

Ganoderma applanatum (Pers.) Pat.
[applanatum = flattened or horizontally expanded]
Artist's Bracket
Macroscopically this species looks essentially identical to *G. australe* but there are some important and consistent differences which will help to distinguish the two species. The upper crust of *G. applanatum* is thinner, less than 0.75 mm thick compared to that of *G. australe* which is over 0.75 mm. The result is that the crust can be easily depressed with the pressure of a finger; not possible with *G. australe* which is very hard.

A cross-section will reveal the layers of tubes laid down in previous years and in *G. applanatum* there are distinct, visible layers of tissue between each tube layer, absent in *G. australe*. The spores of *G. applanatum* measure 5.8–7/0 x 4.5–5.2 µm and thus smaller than those of *G. australe*. *G. applanatum* is often parasitised by the Yellow Flat-footed Fly, *Agathomyia wankowiczii* which produces small, cone-like galls on the pore surface.

Ganoderma lucidum (Curtis: Fr.) P. Karst.
[lucidum = shining]
Fruitbody annual, starting as a hard, rounded knob, expanding to form a semicircular bracket, 5–20 cm across, usually with a distinct lateral stem although this varies enormously in length and thickness. Bracket surface smooth, shiny, lacquered, usually concentrically grooved or wrinkled and with radial grooves, easily depressed by finger pressure, deep plum-red, purplish, bay-brown to almost black at times, margin paler, yellowish, narrow. **Pores** creamy white to pale buff, 4–5 per mm. **Tubes** 5–10 mm deep. **Flesh** tough, corky, light brown. **Odour** not distinct. **Taste** slightly acidic. **Spores** ellipsoid, 10–13 x 7–8.5 µm, with coarse warts, thick, inner dark wall and a transparent exospore. **Spore deposit** rust brown.

Widespread in Britain but never very common, mainly southern or south-western, on broadleaved trees, especially *Quercus*, *Carpinus*, more rarely on conifers. The rare *G. carnosum* is very similar in appearance but is found on the trunks of *Taxus* in Britain, more often on *Abies* in continental Europe.

Ganoderma resinaceum Boud. in Pat.
[resinaceum = pertaining to resin]
Fruitbody perennial, starting as a hard, rounded knob, expanding to form a semicircular bracket, 5–35 cm across, shining, lacquered with a resinous, then dull with a waxy coating which can be scratched off and will melt if exposed to flame; bright plum-red, purplish, chestnut to rust, blackish with age, margin paler, white to yellowish, sometimes with a short stem. **Pores** whitish, 3–4 per mm. **Tubes** to 3 cm deep. **Flesh** firm, corky, pale brown. **Odour** spicy. **Taste** acidic. **Spores** 9–11.5 x 4.5–7 µm with fine warts. **Spore deposit** rust brown.

On trunks of broadleaved trees, especially *Quercus* or *Fagus*, widespread in England, locally frequent.

G. pfeifferi looks very similar but has darker sepia-brown flesh and broader spores (9–12 x 6–9 µm).

Family *Fomitopsidaceae*

Genus *Fomitopsis*

Fomitopsis betulina (Bull.) B.K. Cui *et al.*
[betulina = growing on *Betula*]
Syn. *Piptoporus betulinus*
Fruitbody annual, starting as a rounded cushion before expanding into a kidney-shaped bracket, 15–25 cm across, 4–6 cm thick, often with a short lateral stem, upper surface whitish, pale fawn to grey-brown, dry, smooth. **Pores** white to cream, circular to angular, 3–5 per mm. **Tubes** up to 1 cm thick, easily separated from the flesh. **Flesh** tough, corky, whitish. **Odour** not distinct. **Taste** mild. **Spores** narrowly cylindric, curved, 5–6 x 1.5–1.7 µm. **Spore deposit** white.

Only found on *Betula*, widespread and common everywhere. The change from the genus *Piptoporus* to *Fomitopsis* will surprise many but molecular studies places the fungus squarely into that genus; this species and *F. pinicola* even share the same parasitic fungus.

Fomitopsis pinicola (Fr.) P. Karst.
[pinicola = growing in pine woods]
Fruitbody perennial, forming a deep, hoof-shaped bracket 10–20 (30) cm across, surface lumpy, tuberculate, with each annual growth ring forming concentric zones, bright orange-brown to purple-brown when fresh and young, darkening to greyish brown or blackish with age, margin rounded, whitish; surface smooth to crustose, glossy when fresh, matt with age. **Pores** white to yellowish when young, browner when old, 3–4 per mm. **Tubes** 2–3 mm deep. **Flesh** very tough, woody, ochre-brown. **Odour** acidic, strong. **Taste** bitter. **Spores** narrowly ellipsoid, 6–8.5 x 3–4.5 µm. **Spore deposit** white.

On dead wood of conifers, usually *Pinus* or *Picea*, commonest in Scotland but also known from several English counties.

Ganoderma, Fomitopsis

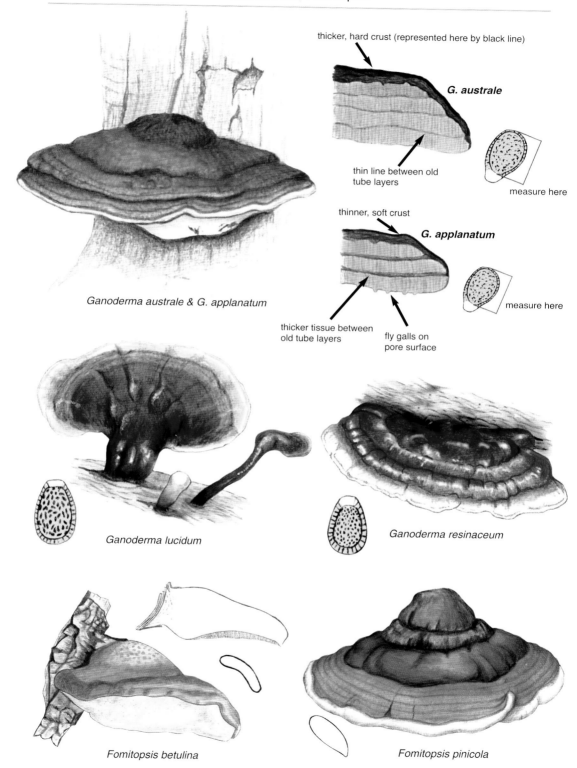

Ganoderma australe & G. applanatum

Ganoderma lucidum

Ganoderma resinaceum

Fomitopsis betulina

Fomitopsis pinicola

Genus *Daedalea*

Daedalea quercina L.: Fr.
[quercina = associated with *Quercus*]
Fruitbody perennial, 10–20 cm wide, ± semicircular, up to 8 cm thick, upper surface smooth to strongly rugulose with concentric zones and irregular nodules; reddish brown, ochraceous to greyish brown. **Pores** irregular, elongate-poroid, maze-like to almost lamellate, pore walls 1–3 mm thick. **Tubes** up to 4 cm deep, pale ochre. **Flesh** tough, corky, pale ochraceous-brown. **Odour** not distinct. **Taste** mild to peppery. **Spores** cylindric, 5.5–6 x 2.5–3.5 µm. **Spore deposit** white.

Mainly on *Quercus*, more rarely *Castanea*, often on hard, barely decayed wood. Widespread but not common, more frequently in older woodlands.

The fungus causes a progressive brown rot in the heartwood of the tree and this usually results in the death of the tree.

Genus *Postia*

Postia caesia (Schrad.) P. Karst.
[caesia = blue]
Syn. *Oligoporus caesius*
Fruitbody annual, often solitary, sessile then forming a rounded bracket, 5–6 cm across, 1–2 cm thick, upper surface tomentose-hairy, pale greyish to blue, bruising darker blue, colour often irregular, becoming more ochre with age. **Pores** angular, 3–6 per mm, whitish to pale grey or bluish, bruising blue. **Tubes** up to 6 mm deep, concolorous with pores. **Flesh** fibrous, soft, whitish. **Odour** pleasant, fungoid. **Taste** mild to bitter. **Spores** cylindric-ellipsoid to allantoid, 4.5–6 x 1.5–2 µm. **Spore deposit** whitish, amyloid.

Usually on conifers, especially *Pinus* but also *Larix* and *Picea*. Reported on hardwoods but British records are doubtful. Widely distributed but not common.

Often confused with the commoner *P. subcaesia* which differs in its paler, less blue-staining fruitbodies and narrower spores.

Postia subcaesia (A. David) Jülich
[subcaesia = less blue]
Syn. *Oligoporus subcaesius*
Fruitbody annual, semicircular, 2–6 cm across, 1–4 cm deep, upper surface hairy-tomentose, white to pale ochre with pale blue tones, especially at the margin, browner with age, often radially wrinkled. **Pores** angular, 5–6 per mm, whitish to greyish. **Tubes** 2–5 mm deep. **Flesh** fibrous, soft, white. **Odour** pleasant. **Taste** mild. **Spores** cylindric-allantoid, 4.5–5.5 x 1–1.2 µm. **Spore deposit** whitish, slightly amyloid.

Widespread and frequent on dead wood of a wide range of broadleaved trees, especially in the southern counties.

Might be confused with *P. caesia* which occurs on conifer wood and has broader spores as well as bluer colours, although the depth of colour can vary considerably in both species and some overlap in colour can occur.

Postia stiptica (Pers.) Jülich
[stiptica = astringent]
Syn. *Oligoporus stipticus*
Fruitbody annual, more or less semicircular, bracket-like; 3–12 cm across, up to 4 cm thick; upper surface tomentose or tuberculate, white then cream-ochre to greyish ochre when old. **Pores** rounded, 5–6 per mm, white. **Tubes** up to 1 cm deep, white. **Flesh** fibrous, soft, white. **Odour** pleasant, aromatic. **Taste** very bitter and astringent. **Spores** narrowly ellipsoid, 3.5–5 x 1.5–2 µm. **Spore deposit** whitish, inamyloid.

On dead conifer wood, more rarely hardwoods, widespread and common.

The flesh often holds large quantities of moisture and the fungus can be squeezed out like a sponge.

Postia tephroleuca (= *P. lacteus*), found on hardwoods, has similar white to greyish brown fruitbodies but differs in its lacerated or minutely dentate-angular pores, and in its mild-tasting flesh. It is common and widespread everywhere.

Genus *Ischnoderma*

Ischnoderma benzoinum (Wahl.: Fr.) P. Karst.
[benzoinum = from benzoin, a fragrant, resinous juice]
Fruitbody annual, forming irregular, often overlapping brackets, 5–20 cm across; upper surface minutely tomentose then later smoother and becoming radially wrinkled; dark red-brown to almost black with age, margin usually paler, more ochraceous. When young the fruitbody is soft and sappy becoming much firmer, more leathery when mature in the so-called 'fomitoid' stage. Often distinctly concentrically zonate with paler and darker zones. **Pores** rounded to distinctly angular, white to pale ochre, bruising brownish to pinkish brown. **Tubes** up to 5–8 mm deep, concolorous with the pores. **Flesh** pale ochre-brown becoming much darker blackish brown in the mature state. **Odour** mild, slightly aromatic. **Taste** mild. **Spores** cylindric-allantoid, smooth, 5.5–6 x 2–2.5 µm. **Spore deposit** white.

On fallen trunks and stumps of *Picea*, *Abies*, *Pinus* and other conifers, often in large groups of multiple brackets. Rather uncommon but widely distributed in southern counties of England.

The very similar *I. resinosum* is found on hardwoods such as *Fagus* and *Betula* and is almost identical morphologically. One of the principal differences apart from the host substrate is the change in colour of the flesh or context which in *I. benzoinum* becomes very dark brown with age while in *I. resinosum* it retains the pale brown of the younger state.

Rigidoporus, Hapalopilus, Daedaleopsis, Pycnoporus, Trametes

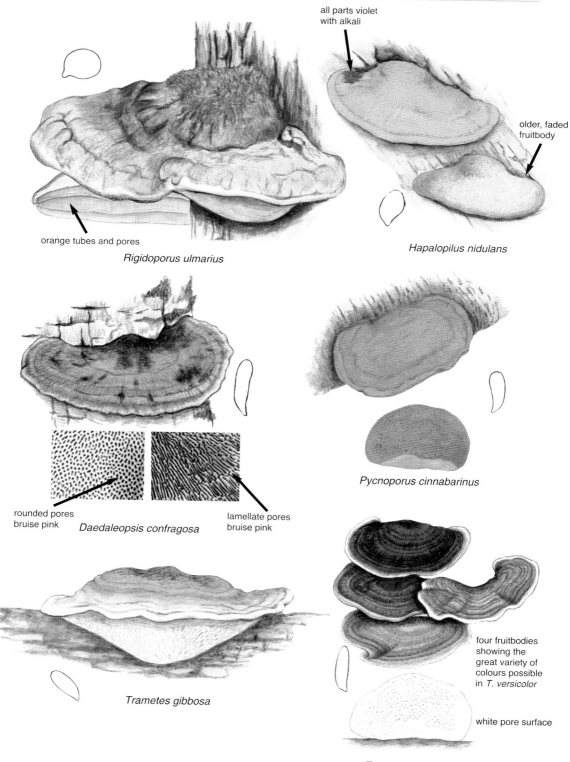

Genus *Lenzites*

Lenzites betulinus (L.) Fr.
[betulinus = associated with *Betula*]
Fruitbody annual, more or less semicircular. **Upper surface** tomentose-hairy, with concentric zones, greyish ochre, cream, greyish white or even greenish with algae. **Pores** form thin, gill-like plates, usually forked at the margin, up to 12 mm deep. **Flesh** thin, tough-fibrous, 1–2 mm thick, whitish. **Odour** ± nil. **Taste** mild. **Spores** ellipsoid-cylindric, often slightly curved, 5–6 x 2–3 µm. **Spore deposit** white.

On *Betula* wood but also on other broadleaved trees. Widespread and fairly common everywhere.

Genus *Datronia*

Datronia mollis (Sommerf.) Donk
[mollis = soft]
Fruitbody annual, starting resupinate, then upper margin growing over as a shallow bracket, projecting 2–3 cm. **Upper surface** ochre to reddish brown or blackish, tomentose, often zonate, undulate. **Pores** greyish white to pale brownish, rounded-angular or elongate when on vertical surface, 4–5 per mm. **Tubes** up to 7 mm thick, pale brownish. **Flesh** cream-ochre, tough then brittle when dry with dark line between the flesh and the tomentose cuticle. **Odour** pleasant. **Taste** mild. **Spores** cylindric-ellipsoid, smooth, 10–12 x 3–4 µm. **Spore deposit** whitish.

On fallen branches of broadleaved trees. Widespread, common everywhere.

Genus *Cerrena*

Cerrena unicolor (Bull.) Murrill
[unicolor = of one colour]
Fruitbody annual, forming elongate brackets projecting up to 5 cm. **Upper surface** tomentose, undulating, with concentric zones, pale grey, grey-brown, ochraceous, often with greenish algae at base. **Pores** labyrinthine, forming flattened 'teeth', 2–3 per mm, cream to greyish brown. **Tubes** to 1 cm deep, pale ochre-brown. **Flesh** tough, corky, pale brownish, separated from upper surface by a thin, dark line in section. **Odour** pleasant, fungoid. **Taste** unknown. **Spores** cylindric-ellipsoid, smooth, 5–7 x 2.5–4 µm. **Spore deposit** white. **Cystidia** absent.

On dead wood of deciduous trees. Widespread in England, occasional, much rarer elsewhere in Britain.

Family *Meruliaceae*

Genus *Bjerkandera*

Bjerkandera adusta (Willd.: Fr.) P. Karst.
[adusta = dusky, swarthy]
Fruitbody annual, fully resupinate to bracket-like with all stages between, often in extensive sheets of overlapping brackets, each bracket 10–15 cm across, projecting 5–8 cm, 2–3 cm thick. **Upper surface** finely velutinous, dull, ochre to pale brown, greyish brown to almost blackish brown, often concentrically zoned, margin usually whitish. **Pores** whitish then soon pale to dark grey, 4–6 per mm, darker when bruised. **Tubes** to 2 mm deep, grey-black. **Flesh** whitish, leathery, elastic. **Odour** pleasant, fungoid. **Taste** slightly sour. **Spores** ellipsoid, smooth, 4.5–5.5 x 2–3 µm. **Spore deposit** whitish.

On dead wood of a variety of broadleaved and coniferous trees. Very common everywhere in Britain.

B. fumosa lacks the contrast between pale flesh and dark tubes, is more ochre brown with odour of anise.

Genus *Phlebia*

Phlebia tremellosa (Schrad.) Nakasone & Burds.
[tremellosa = trembling or jelly-like]
Syn. *Merulius tremellosus*
Fruitbody annual, fully resupinate to bracket-like, often in large sheets, each bracket 5–15 cm across, projecting 2–4 cm. **Upper surface** hairy-tomentose, white to pinkish or ochraceous, margin undulating. **Hymenium** wrinkled and folded (merulioid), pinkish yellow to salmon. **Flesh**, soft, gelatinous, elastic, hard when dry. **Odour** nil. **Taste** mild. **Spores** cylindric-allantoid, smooth, 3.5–4 x 1–1.5 µm. **Spore deposit** white.

On decayed fallen wood of broadleaved trees, more rarely conifers. Widespread and common.

Genus *Abortiporus*

Abortiporus biennis (Bull.: Fr.) Sing.
[biennis = of biennial growth]
Fruitbody annual, forming a dense rosette of overlapping brackets, 8–15 cm across. **Upper surface** smooth to finely velutinous, radially wrinkled, whitish to reddish brown, bruising darker. **Pores** maze-like, 1–3 per mm, wrinkled, pinkish white, bruising reddish brown. **Tubes** 2–5 mm deep, whitish. **Stem** 4–7 x 2–3 cm, reddish ochre, buried in the soil. **Flesh** tough, softer on outside, white, reddening when cut. **Odour** rather unpleasant, sour. **Taste** mild. **Spores** elliptic, smooth, 4.5–6 x 3.5–4.5 µm. **Spore deposit** yellowish cream.

On soil on buried roots, in parks, woodland edges. Widespread and common in southern England.

Podoscypha multizonata differs in having no pores and even denser rosettes (see Fig. 8.4 on p. 61).

Family *Bondarzewiaceae*

Genus *Heterobasidion*

Heterobasidion annosum (Fr.) Bref.
[annosum = appearing for many years]
Fruitbody perennial, resupinate to fully bracket-like, 5–15 cm across, up to 3 cm thick. **Upper surface** slightly tomentose when young, glabrous with concentric ridges, dark red-brown to almost black, whiter on the marginal, growing edge. **Pores** rounded-angular or even slit-like, 3–4 per mm, white to pale ochre, bruising brownish, turning strongly brown with Melzer's Iodine. **Tubes** 3–6 mm deep, whitish, indistinctly stratified. **Flesh** pale ochre, tough and elastic. **Odour** strong, aromatic or fungoid. **Taste** unknown. **Spores** subglobose, finely warted, 4.5–6.2 x 4–5.2. **Spore deposit** white.

On living and dead conifers, especially *Pinus* spp., much rarer on hardwoods. It is a serious root-rot pathogen in conifer plantations.

Lenzites, Datronia, Cerrena, Bjerkandera, Phlebia, Heterobasidion, Abortiporus

Lenzites betulinus

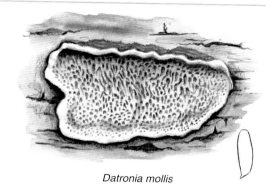

Datronia mollis

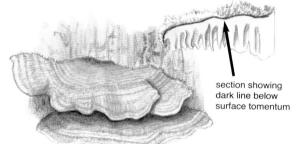

section showing dark line below surface tomentum

Cerrena unicolor

Bjerkandera adusta

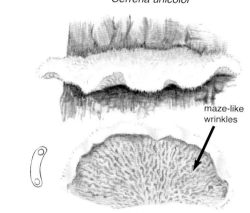

maze-like wrinkles

Phlebia tremellosa

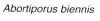

Abortiporus biennis

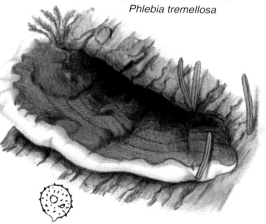

Heterobasidion annosum

Family *Hymenochaetaceae*

Genus *Pseudoinonotus*

Pseudoinonotus dryadeus (Pers.: Fr.) Murr.
[dryadeus = pertaining to oak]
Syn. Inonotus dryadeus
Fruitbody annual, forming a very thick, rounded bracket or cushion, 10–35 cm across and up to 15 cm deep, solitary or in imbricate layers, upper surface finely tomentose, ochre-buff to dark brown, margin very obtuse, rounded, often exuding droplets of amber liquid when fresh. **Pores** pale whitish buff, 4–6 per mm. **Tubes** up to 2 cm deep, buff. **Flesh** bright yellow-brown to reddish brown when old, soft and fibrous, zonate with darker streaks. **Odour** very unpleasant. **Taste** acidic. **Spores** subglobose-ovate, 7.5–8.5 x 5.5–6.5 µm. **Spore deposit** cream. **Hymenial setae** thick-walled, hooked, pointed, 22–40 x 9–16 µm.

Parasitic on *Quercus*, more rarely on *Castanea* and *Fagus*, even more rarely on *Abies*. Widespread but rather uncommon in Britain.

When fresh and moist it produces profuse amber droplets all over the upper surface of the fruitbody.

Genus *Inonotus*

Inonotus hispidus (Fr.) P. karst.
[hispidus = covered in hairs]
Fruitbody annual, forming semicircular bracket 10–30 cm across, up to 10 cm thick, upper surface densely hairy-tomentose, bright orange-brown when fresh to ochraceous, paler, more yellowish at the margin, becoming darker brown and almost black with age; margin rounded, obtuse. **Pores** rounded-angular, 2–3 per mm, whitish yellow, greyer with age, often exuding droplets when fresh. **Tubes** up to 4 cm deep, yellowish. **Flesh** thick, soft and spongy, yellow-ochre with darker zones, turning brown when cut. **Odour** pleasant or acidic. **Taste** mild to acidic. **Spores** ovate, smooth, thick-walled, 7–10 x 6–7.5 µm. **Spore deposit** brownish. **Hymenial setae** thick-walled, pointed, 20–24 x 6–8 µm.

Widespread and fairly common, often high up on the trunks of broadleaved trees, especially *Fraxinus*, *Malus*, *Ulmus*, more rarely other species.

Old fruitbodies fall to the ground and may be completely black. When wet and soft these dead fruitbodies are commonly the host for the tiny, white collybioid fungus, **Collybia cookei**.

Inonotus obliquus (Fr.) Pilát
[obliquus = slanted]
Chaga
Fruitbody annual, unusual in first forming a sterile, irregular growth or conk which erupts out of the bark of a standing tree; the rounded or amorphous conk is hard, blackish brown and prominently cracking but the interior is softer and rusty yellow-brown; this can be seen when it's removed from the tree and broken into chunks. When the tree dies it will then produce the fertile, poroid state which is fully resupinate and forms under the bark. **Pores** 6–8 per mm, at first whitish to golden brown then darker brown with age. **Tubes** up to 3 mm long, reddish brown. **Flesh** usually less than 1 mm thick, yellow-brown. **Odour** nil. **Taste** mild. **Spores** ellipsoid-ovoid, 9–10 x 5.5–6.5 µm. **Spore deposit** pale yellow. **Hymenial setae** ventricose, 14–30 x 4.5–7 µm.

On *Betula*, rare in England commoner in Scotland but also probably overlooked in the fertile stage unless peeling bark is pulled back. This fertile stage is readily consumed by insects and other invertebrates so is surprisingly difficult to find. The sterile, cinder-like conk is widely collected in both Europe and North America to make a traditional medicinal tea and is commonly referred to as Chaga. Powdered Chaga is available commercially from herbalists.

Inonotus nodulosus (Fr.) Pilát
[nodulosus = with small lumps or nodules]
Fruitbody annual, usually forming tiered small brackets fused together on a resupinate, often poroid crust; individual brackets 2–3 cm across, entire mass often tens of centimeters across. Upper surface of brackets tomentose, orange-ochre to orange-brown, darker when old. **Pores** cream to pale ochre, 3–4 per mm. **Tubes** 5–6 mm long, cream. **Flesh** cream to dark brown, tough. **Odour** pleasant, not distinct. **Taste** mild. **Spores** broadly ellipsoid, 4.5–5.5 x 3.5–4 µm. **Spore deposit** pale yellow. **Hymenial setae** brown, fusiform, 40–60 x 5–10 µm.

On dead wood, usually branches, of the common beech, *Fagus sylvaticus*. Widespread in Britain but not often reported. It appears to be fairly uncommon throughout its range in northern Europe down to Spain and across to Turkey.

I. nodulosus might be confused with **Mensularia radiata** which shares rather similar colours but that species normally has recurved setae and only rarely occurs on *Fagus*.

Inonotus cuticularis (Bull.: Fr.) P. Karst.
[cuticularis = pertaining to a cuticle or skin]
Fruitbody annual, forming tall tiers of crowded brackets, each bracket 5–11 cm across, up to 1.5 cm thick, upper surface tomentose-fibrillose or finally smooth when old, yellowish brown, cinnamon-brown, sometimes faintly zonate, margin rounded when young, acute when mature. **Pores** angular, 4–5 per mm, whitish to pale brown, 'glancing' or shimmering when tilted at certain angles. **Tubes** to 8 mm deep, whitish. **Flesh** fairly tough, fibrous, bright yellow-brown to reddish brown. **Odour** pleasant. **Taste** mild to sour. **Spores** broadly ellipsoid, 6–8 x 4.5–5.5 µm. **Spore deposit** yellowish brown. **Hymenial setae** simple, fusiform-ventricose. **Cuticular setae** from cap surface strikingly hooked and barbed.

On fallen and living trunks of *Fagus*, but also known from *Acer*, *Carpinus*, *Fraxinus*, *Quercus* and *Salix*. Widespread and occasional in England.

The rather similar *I. leporinus* is a boreal or montane species growing on *Picea abies* in Fennoscandia, northern and central Europe but has not yet been found in Britain. It might just occur in Scotland and should be looked out for on suitable hosts.

Pseudoinonotus, Inonotus

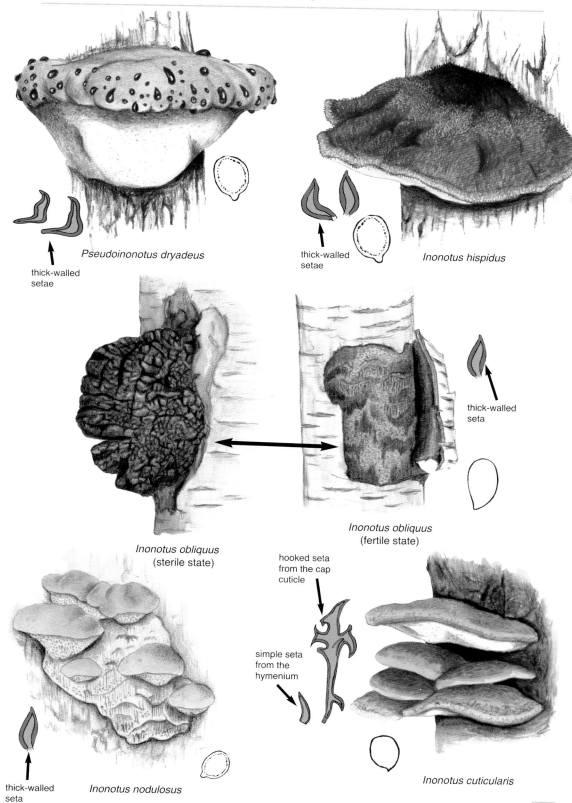

Genus *Mensularia*

Mensularia radiata (Sowerby) Lázaro Ibiza
[radiata = radiating outwards]
Syn. *Inonotus radiatus*
Fruitbody annual, semicircular, broadly attached to substrate, 3–10 cm across, 1–2 cm thick, upper surface finely tomentose to smooth or even lumpy, often concentrically zonate, yellow-brown to reddish brown, blacker with age. **Pores** angular, 2–5 per mm, whitish to pale yellowish brown with age. **Tubes** up to 1 cm deep, cream. **Flesh** tough, yellowish brown to reddish brown. **Odour** pleasant, fungoid. **Taste** mild, woody. **Spores** ovoid, 5–6.5 x 3–4.5 µm. **Spore deposit** yellowish. **Hymenial setae** thick-walled, pointed, 14–35 x 7–12 µm.

Usually on standing trunks of *Alnus*, but also on many other broadleaved trees, common and widespread.

Genus *Phellinus*

Phellinus igniarius (L.: Fr.) Quel.
[igniarius = pertaining to fire]
Fruitbody perennial, forming a deep, hoof-like bracket up to 20 cm across, 15 cm deep, upper surface with broad, concentric bulges, grey to blackish grey, crustose, often cracking with radial fissures looking as if burnt. **Pores** rounded, 5-6 per mm, pale cinnamon-brown to purplish brown. **Tubes** up to 5 mm deep, in several layers. **Flesh** tough, woody, dark brown, concentrically layered. **Odour** fungoid. **Taste** woody. **Spores** ellipsoid-subglobose, 5–7 x 4–6 µm, Melzer's Iodine negative. **Spore deposit** whitish. **Hymenial setae** thick-walled, pointed, 14–17 x 4–6 µm.

Widespread and occasional on a range of broadleaved trees, usually on the trunks of living *Salix*, also *Alnus*, *Betula* and *Malus*. **P. nigricans**, almost exclusively on *Betula*, is very similar but has narrower concentric zones, and a grey margin. It is known from Scotland but is probably very rare in Britain as a whole.

One of the largest of a great many species in this genus, many needing specialist literature for identification. *P. igniarius* may well be a species complex but distinguishing the different species has proven very difficult.

Phellinus hippophaeicola Jahn
[hippophaeicola = liking *Hippophae rhamnoides*]
Syn. *Fomitoporia hippophaeicola*
Fruitbody perennial, forming a broadly attached bracket, 3–6 cm across, 1.5–2 cm thick at the base, upper surface finely velvety to smooth, with concentric ripples, yellow-ochre to rust-brown, cinnamon-brown at the margin, centre often greyish brown and frequently with green algal growth. **Pores** rounded, 5–7 per mm, brownish to dark brown. **Tubes** 2–3 mm deep, concolorous with pores, stratified into annual layers. **Flesh** hard, cinnamon-brown, concentrically zoned. **Odour** nil. **Taste** unknown. **Spores** subglobose, 5.5–8 x 5–7 µm, dextrinoid. **Spore deposit** cream. **Hymenial setae** absent.

On dead branches and trunks of Sea Buckthorn, *Hippophae rhamnoides* on which it causes a white soft rot. Locally common along the English coast but also reported from Scotland so should be looked for wherever the host plant is found.

In mainland Europe it is also commonly found on *Eleagnus* species.

Phellinus punctatus (Fr.) Pilát
[punctatus = marked with dots]
Fruitbody perennial, fully resupinate, forming a long, cushion-like patch often tens of centimeters in length, up to 2 cm thick; surface finely pored, reddish brown to chesnut or cinnamon-brown. **Pores** rounded to elongate, 5–7 per mm. **Tubes** 1–3 mm long, cinnamon-brown, forming multiple, stratified layers in old fruitbodies. **Margin** paler whitish to ochre, without pores. **Flesh** tough, woody, golden brown to reddish. **Odour** nil. **Taste** not recorded. **Spores** subglobose-ellipsoid, 7–9 x 6–7.5 µm, sometimes thick-walled, strongly dextrinoid. **Spore deposit** cream. **Hymenial setae** absent.

On dead branches or trunks of *Crataegus*, *Fraxinus*, *Salix*, *Corylus* and sometimes other hardwoods. Recorded from a few English counties, rather rare.

Phellinus pomaceus (Pers.) Maire
[pomaceus = pertaining to fruit trees]
Syn. *P. tuberculatus*
Fruitbody perennial, sessile then forming a broad, rather flattened or hoof-shaped bracket up to 5 cm across, 5 cm deep, protruding just a cm or two; the upper surface is finely tomentose to smooth, irregularly grooved and ridged, often cracking, greyish brown to blackish with age, margin paler and rounded. **Pores** rounded, 7–9 per mm, yellowish to pinkish or reddish brown. **Tube layers** up to 1 cm deep, stratified, concolorous with the pores. **Flesh** yellowish brown to reddish brown, tough, zonate. **Odour** nil. **Taste** unknown. **Spores** ovoid to broadly ellipsoid, 4–5 x 3–4.5 µm, thick-walled, Melzer's Iodine negative. **Spore deposit** cream. **Hymenial setae** thick-walled, fusiform-ventricose.

On living and dead wood of *Prunus* spp., occasional and widespread in England, but not often reported in the south-eastern counties, or from Scotland or Wales.

Phellinus torulosus (Pers.) Bourdot & Galzin
[torulosus = cylindric with bulges]
Fruitbody perennial, forming a large bracket up to 45 cm across, 28 cm deep, margin obtuse, rounded, upper surface glabrous to finely tomentose, ochre-brown, reddish brown to blackened, often green with algae; margin tomentose, rust-brown. **Pores** rounded, 5–7 per mm, yellowish brown. **Tubes** up to 6 mm deep, yellowish, in distinct layers building up over several years. **Flesh** tough, yellowish brown, with one or more thin black lines visible on cut vertical surfaces, black with KOH. **Odour** musty. **Taste** unknown. **Spores** ellipsoid-ovoid, thick-walled, 4–6 x 3–4 µm. **Spore deposit** white. **Hymenial setae** infrequent, ventricose, thick-walled, yellowish brown, 20–50 x 6–11 µm.

Usually at the base of trunks of *Quercus*, *Castanea*, *Prunus* and *Crataegus*. Mainly in southern England, uncommon.

Mensularia, Phellinus

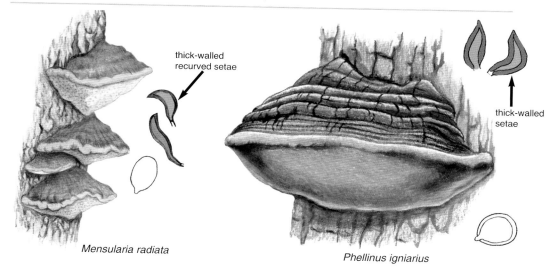

Mensularia radiata — thick-walled recurved setae

Phellinus igniarius — thick-walled setae

Phellinus hippophaeicola — Hymenial setae absent

Phellinus punctatus — Hymenial setae absent

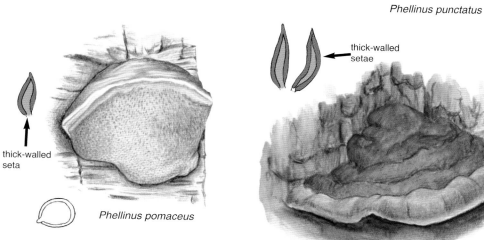

Phellinus pomaceus — thick-walled seta

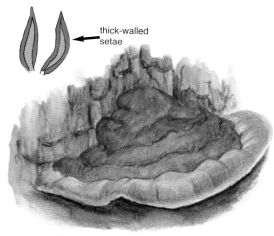

Phellinus torulosus — thick-walled setae

Phellinus tremulae (Bondartzev) Bondartzev & P.N.Borisov
[tremulae = associated with *Populus tremula*]
Fruitbody perennial, forming a blunt, rounded bracket up to 20 cm across, 15 cm thick, triangular in section, with the top and pore surfaces at 45 degrees from the horizontal axis. **Upper surface** pale grey-brown at margin, blackish elsewhere, crust-like and with numerous cracks. **Pores** circular, 5–7 per mm, purplish brown to reddish brown. **Tubes** up to 2 mm deep, with distinctly stratified layers. **Flesh** dark reddish brown, woody. **Odour** distinct, resinous, like wintergreen oil. **Taste** unknown. **Spores** subglobose, smooth, thick-walled, 4.8–5.6 x 3.5–5 µm. **Spore deposit** whitish. **Hymenial setae** infrequent to numerous, ventricose, thick-walled, brown in KOH, 12–30 x 6–7.5 µm.

Parasitic on trunks of *Populus tremula*. Recently recorded from Scotland where it appears to be frequent.

Genus *Porodaedalea*

Porodaedalea pini (Brot.) Murrill
[pini = pertaining to *Pinus*]
Syn. *Phellinus pini*
Fruitbody perennial, resupinate then forming a rounded bracket 8–15 cm across, up to 8 cm deep at the base, upper surface smooth to minutely tomentose, concentrically ridged, deep brown to blackish, brighter reddish brown at the margin. **Pores** yellowish brown, circular to slightly labyrinthine, 1–3 per mm. **Tubes** pale brown, stratified, each tube layer up to 6 mm deep. **Flesh** tough, corky, reddish brown. **Odour** pleasant, fungoid. **Taste** mild. **Spores** ovoid-ellipsoid, 6–7 x 4.5–6 µm, thick-walled. **Spore deposit** whitish. **Hymenial setae** thick-walled, ventricose-fusoid, 40–60 x 11–15 µm.

On living trunks of pines, especially *Pinus sylvestris*, uncommon, mostly recorded in Scotland but also with records from heathlands in southern England.

Genus *Phellinopsis*

Phellinopsis conchata (Pers.) Y.C. Dai
[conchata = resembling a conch shell]
Fruitbody perennial, semi-resupinate, forming imbricate brackets 5–15 cm across, projecting 1–5 cm, upper surface of brackets smooth, concentrically ridged and zoned, light brown, reddish brown to black when old, often with moss or algae coating, with paler, tomentose margin. **Pores** circular, 5–6 per mm, yellowish brown. **Tubes** rust-brown, up to 2 mm deep, in stratified layers. **Flesh** tough, rust-brown. **Odour** nil. **Taste** unknown. **Spores** ovoid, 5–6.5 x 4–4.5 µm. **Spore deposit** yellowish white. **Hymenial setae** abundant, thick-walled, ventricose-fusiform, often with the tips blunt as if broken off.

On dead or dying wood of broadleaved trees, usually on old trunks of *Salix* spp. but also *Betula*, *Populus* etc. Widespread in England but not often reported.

Genus *Phylloporia*

Phylloporia ribis (Schum.: Fr.) P. karst.
[ribis = associated with *Ribes*]
Fruitbody perennial, forming overlapping brackets, often circling the base of the shrub; upper surface smooth to lumpy with concentric ridges, rust-brown, yellow-brown, blackish with age, often with green algal growth. **Pores** rounded, 6–7 per mm, pale cinnamon-brown. **Tubes** up to 3 mm deep, pale brown, in layers in older fruitbodies. **Flesh** tough, corky, rust-brown, usually divided by a thin black line with the lower, thinner part harder and darker. **Odour** pleasant. **Taste** unknown. **Spores** subglobose, 3–4 x 2.5–3 µm. **Spore deposit** pale ochre. **Hymenial setae** absent.

On the base of living *Ribes* species (Red Currant) and *Euonymus europaea* (Spindle Tree). Widespread in southern England and quite frequent.

Genus *Coltricia*

Coltricia perennis (L.) Murrill
[perennis = perennial]
Fruitbody annual, forming a more or less round, funnel-shaped cap 3–5 cm across, 0.5–1 cm thick, margin usually undulating, upper surface cinnamon-brown, ochraceous, grey-brown, concentrically zoned. The hyphae of the cap surface are branched, antler-like. **Pores** greyish beige, rounded, 2–4 per mm, decurrent. **Stem** 3–4 x 0.3–0.6 cm, tomentose, rust-brown. **Flesh** tough, corky, ochraceous. **Odour** nil. **Taste** mild. **Spores** ellipsoid-ovoid, 6–9 (-10) x 3.5–5 (-5.5) µm, smooth. **Spore deposit** pale brown.

On the ground in conifer woods and sandy heaths, often on burnt sites; more rarely in broadleaved woods. Widespread and occasional in Britain. *C. cinnamomea* is rarer, with slightly shorter, broader spores (6.5–8 x 5–6 µm) and a darker brown cap with a radial silky sheen.

Coltricia confluens P.-J. Keizer
[confluens = merging into one]
Fruitbody annual, forming a more or less round, funnel-shaped cap with a lobed and incised margin, several caps often fused or growing into each other; upper surface velvety-tomentose, light to dark reddish brown, often with radial wrinkles and some zonation. **Pores** 2–4 per mm, slightly decurrent, reddish brown. **Tubes** to 2 mm deep. **Stem** 2–4 x 1–1.5 cm, reddish brown. **Flesh** tough, fibrous, reddish brown. **Odour** nil. **Taste** mild to sour. **Spores** ellipsoid to slightly cylindric, 6–9 (–10) x 3.5–5 (–5.5) µm. **Spore deposit** pale brown.

On the ground on rich soils under broadleaved trees. Very rare in Britain, known only from Surrey.

Phellinus, Phellinopsis, Phylloporia, Coltricia

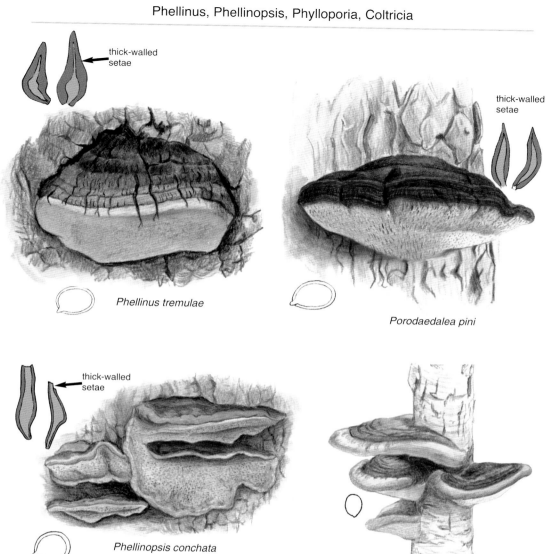

Phellinus tremulae

Porodaedalea pini

Phellinopsis conchata

Phylloporia ribis

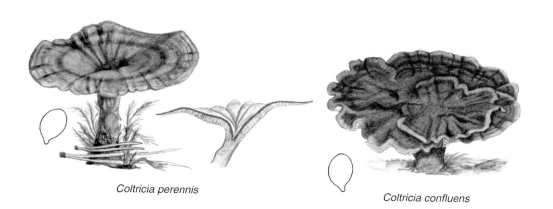

Coltricia perennis

Coltricia confluens

Genus *Hymenochaete*

Hymenochaete rubiginosa (Dicks.: Fr.) Lév.
[rubiginosa = rust-red]
Fruitbody perennial, partly sessile to bracket-like, often fused together in rows, upper surface undulating, with concentric ridges and wrinkles, minutely tomentose to smooth, dark red-brown to blackish, with reddish or ochre marginal zones. **Hymenium** ± smooth, bright reddish ochre, rust-red when fresh, darker brown to grey-brown when old. **Flesh** thin but tough, brown. **Odour** not distinct. **Taste** unknown. **Spores** ellipsoid, 4.5–6 x 2.5–3 µm. Spore deposit whitish. **Hymenial setae** dark brown, fusiform with sharp, narrow points, 40–60 x 5–7 µm.

On dead, usually barkless logs or stumps of *Quercus*, less often *Castanea* and other broadleaved trees. Widespread and rather common throughout Britain.

H. corrugata is equally common but very different in appearance and habitat, fruiting on branches of a number of broadleaved trees and shrubs, often gluing adjoining branches together. Its fruitbodies are fully resupinate forming grey-lilac to brownish lilac patches.

Genus *Trichaptum*

Trichaptum abietinum (Fr.) Ryvarden
[abietinum = inhabiting conifer (*Abies*) woods]
Syn. *Hirschioporus abietinus*
Fruitbody annual, from resupinate to partially or fully bracket-like, each bracket solitary to fused together into longer rows, projecting about 2.5 cm, surface tomentose to glabrous when old, undulating and zonate, greyish white to brownish with lavender margin and frequently green from algae and moss. **Pores** angular to slit-like to slightly jagged, almost toothed, 4–6 per mm, violet, brownish violet to yellow-brown when old. **Tubes** whitish, up to 2 mm deep. **Flesh** tough, elastic, usually less than 1 cm thick, duplex, upper layer whitish, floccose, soft, the lower layer white, firm, tough-fibrous. **Odour** nil. **Spores** cylindric-curved, smooth, 6–7.5 x 2–2.5 µm. **Spore deposit** white. **Cystidia** abundant, usually with encrusted apex, 18–30 x 3–6 µm.

On dead wood of conifers, especially *Picea*, *Abies*, *Pinus*, *Larix*, etc, often on stacked logs, more rarely on hardwoods. Widespread and very common.

Genus *Onnia*

Onnia tomentosa ** (Fr.) P. karst.
[tomentosa = covered with short, velvety hairs]
Fruitbody annual, terrestrial, with a cap and stem, cap round to irregularly lobed, sometimes fusing together with other caps. **Upper surface** tomentose, with indistinct zones, cinnamon-brown, ochre-brown, paler at the margin. **Pores** decurrent, rounded-angular, 2–4 per mm, pore mouths fringed, greyish to grey-brown. **Stem** cylindric-tapered, 3–4 x 1.5–2 cm, tomentose, rust-brown. **Flesh** yellow, corky, harder when old, with duplex structure in the cap. **Odour** curry-like. **Taste** mild. **Spores** ellipsoid, smooth, yellowish, 4–6 x 3–3.5 µm. **Spore deposit** pale yellow-brown. **Cystidia** none. **Setae** fusiform, sometimes bent at the base, thick-walled, dark brown, 40–75 x 12–15 µm.

In needle litter under conifers, especially *Pinus* and *Picea*; parasitic on the tree's roots. Usually in mountainous conifer woods. Not authentically British (one doubtful old record from the Channel Islands) but might occur in Scotland.

Family *Gloeophyllaceae*

Genus *Gloeophyllum*

Gloeophyllum sepiarium (Wulfen) P. Karst.
[sepiarium = growing in hedgerows]
Fruitbody annual, but fruitbodies often persisting for years, forming more or less semi-circular brackets, upper surface hairy-tomentose, concentrically ridged and zoned, dark brown to reddish brown, more orange-brown or yellowish at margin, almost black at base. **Pores** lamellate and slightly labyrinthine, ochre to grey-brown; 1–2 per mm, 0.5–1 cm deep. **Flesh** tough, fibrous, dark brown. **Odour** nil. **Taste** often slightly bitter. **Spores** cylindric-allantoid, 8.5–11.5 x 3.5–4.5 µm. **Spore deposit** white. **Cystidia** absent.

On dead wood of conifers. Common in Scotland, less frequent further south.

Gloeophyllum trabeum (Pers.) Murrill
[trabeum = pertaining to stumps]
Fruitbody annual, partly resupinate then forming semicircular to elongated brackets, sometimes several fusing together, 3–8 cm across, projecting 1–4 cm from the substrate. **Upper surface** finely tomentose when young then glabrous, lumpy and uneven, with concentric zones, cinnamon-brown to reddish brown, blackish at base. **Pores** elongate, often gill-like, more angular at the base, thick, pale ochre-brown. **Tubes** up to 4 mm deep. **Flesh** tough, rather elastic, 1–5 mm thick, cinnamon-brown. **Odour** nil. **Taste** mild. **Spores** ellipsoid, smooth, 7–8.5 x 3–4.5 µm. **Spore deposit** white. **Cystidia** absent.

On bare, often worked timber, structural timber in buildings, more rarely on old logs and stumps of conifers such as *Pinus* and *Taxus*. Widespread but uncommon.

Gloeophyllum odoratum (Wulfen) Imazeki
[odoratum = fragrant]
Fruitbody perennial, forming large, irregular brackets, semicircular to elongate or even almost circular, up to 20 cm across. **Upper surface** undulating, knobbly, tomentose or like leather, yellow-brown, orange-brown to dark brown or blackish when old. **Margin** obtuse, very broad, bright yellow-orange, rust. **Pores** rounded-angular to elongated, 1–2 per mm, yellowish and finally grey-brown. **Tubes** 1.5 cm deep, yellowish, in distinct layers in perennial growths. **Flesh** tough, corky, cinnamon-brown, black with KOH. **Odour** strong of anise/fennel. **Taste** mild or bitter. **Spores** cylindric-ellipsoid, smooth, 7.5–9.5 x 3–4 µm. **Spore deposit** whitish.

On decayed wood of conifers, especially *Picea* and *Abies*, also *Pinus* and *Larix*. Very rare in Britain with only one definite collection. The striking odour of anise and the cylindric spores should allow easy identification.

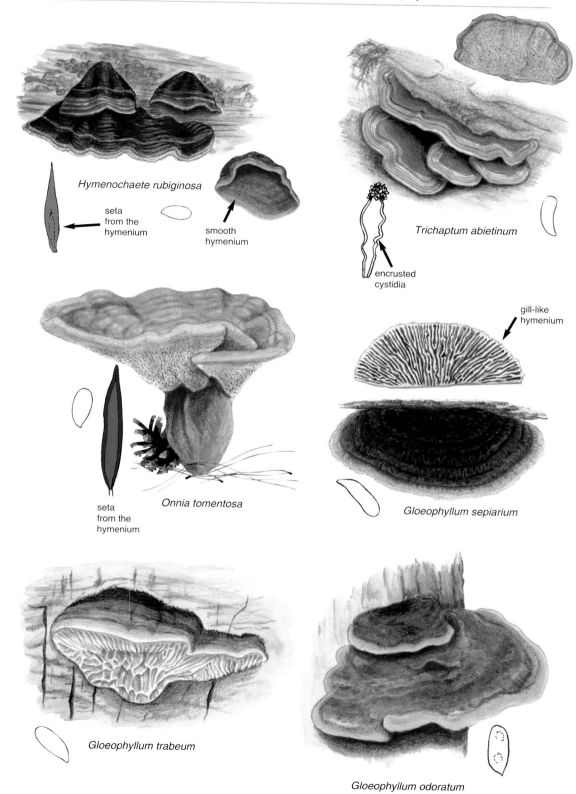

> **Family *Albatrellaceae***
> Fruitbody with a poroid hymenium, closely resembling *Polyporus* species but being entirely terrestrial and actually belongs in the order *Russulales*. Spores ellipsoid, smooth, slightly amyloid.

Genus *Albatrellus*

Albatrellus ovinus (Schaeff.) Murrill
[ovinus = pertaining to sheep]
Cap 5–18 cm across, circular to irregular in outline, dry, smooth, usually becoming strongly cracked with age; whitish to pale buff when young, darker, more greyish ochre or tan with age. **Tubes** very shallow, to 4 mm deep. **Pores** very small (2–5 per mm), circular, decurrent, white to cream-yellow with age. **Stem** cylindric-clavate, 3–10 x 1–4 cm, central to slightly off-centre, whitish to tan, smooth to finely velvety. **Flesh** firm, white to yellowish. **Odour** pleasant but not distinctive. **Taste** mild. **Spores** subglobose or broadly ellipsoid, 4–5 x 2.5–3.5 µm, smooth, inamyloid. **KOH** on flesh = golden yellow.

In mossy *Picea* woods, especially in hilly areas on sandy soils. Widespread in Europe it is uncertain if this species is present in Britain. At the moment it is excluded as very old records are without material to back them up.

A number of related species occur in Europe, none of them British as yet, including the macroscopically very similar *A. subrubescens*, with *Pinus*, differing in its amyloid spores and dark violet-tinted spots at cap centre, and *A. confluens* which has many caps and stems fused together.

Albatrellus pes-caprae (Schaeff.) Murrill
[pes-caprae = goats foot]
Cap 5–12 cm across, rounded-convex, dry and with finely scaly surface, rich reddish brown, sepia to dark brown with age. **Pores** angular-rounded, 6–10 per mm, somewhat decurrent, pale cream bruising yellowish. **Tubes** up to 5 mm deep, cream. **Stem** set to one side of the cap, cylindric-clavate, 3–6 x 1.5–3 cm, finely squamulose, yellow-ochre to brownish orange, firm. **Flesh** soft, rather fragile, whitish. **Odour** pleasant. **Taste** mild, nut-like. **Spores** ellipsoid, smooth, 8.5–10 x 4.5–6.5 µm, inamyloid. **Spore deposit** white.

In montane to subalpine conifer woods or *Fagus* woods. Widely distributed in Europe but not yet recorded in Britain.

> **Family *Bankeraceae***
> Most members of this family have their hymenium formed on spines or teeth and are dealt with in that section of this book, but *Boletopsis* is an exception, having instead tubes and pores like polypores so is placed here for convenience.

Genus *Boletopsis*

Boletopsis perplexa Watling & J. Milne
[perplexa = puzzling]
Cap 4–10 cm, rounded with incurved, undulating margin, smooth to innately fibrillose, dark brown to grey-black or even olivaceous. **Pores** whitish, decurrent, 1–3 per mm. **Stem** 3–7 x 1.5–3 cm, concolorous with the cap. **Flesh** firm, white turning pink when cut then greyish. **Odour** weak. **Taste** somewhat bitter. **Spores** ovate, with large warts, 4–6 x 3.5–4 µm. **Spore deposit** white.

Rare, in open pine woods on sandy soils in a few localities in Scotland.

> **Family *Mycenaceae***

Genus *Favolaschia*

Favolaschia calocera R. Heim
[calocera = a beautiful horn]
This subtropical/tropical species, originally described from Madagascar, is a recent arrival on our shores and is spreading across parts of Europe having already colonised many parts of New Zealand, Australasia, etc. It is quite unmistakable, looking like small (1–2 cm), bright orange ping-pong bats with large pores. Its spores are ellipsoid, 12–13 x 7–8 µm. It fruits on fallen wood and is turning up in numbers in Devon and Cornwall. Despite its pores it is related to the gilled genus *Mycena* in the *Agaricales*.

> **Family *Sparassidaceae***
> Members of this family are now placed in the *Polyporales* along with many of the polypores or bracket fungi, sharing their often parasitic and saprotrophic life style. In Britain and Europe we have just the one genus *Sparassis*.

Genus *Sparassis*

Sparassis crispa (Wulfen) Fr.
[crispa = crisped or irregularly kinked]
Cauliflower Fungus
Fruitbody annual, forming large, irregular, cauliflower-like masses up to 40 cm across, consisting of densely packed, convoluted, leaf-like lobes, often fusing together, arising from a basal rooting trunk. **Hymenium** spread over the surface of the lobes, pale cream at first, darkening to ochre-yellow, brownish when old. **Flesh** in the lobes tough-elastic to soft when old, whitish. **Odour** fungoid to cheesy, pleasant. **Taste** mild. **Spores** ovoid, smooth, 4.5–7 x 3.5–4.9 µm. **Spore deposit** cream. **Clamp connections** present in flesh.

Emerging from soil at the base of conifers, especially *Pinus* but also *Picea*, *Cedrus*, *Larix*, etc, usually close to the root buttresses. Widespread and quite common.

Sparassis brevipes Krombh.
[brevipes = short stem]
Syn. *S. spathulata*, *S. laminosa*, *S. simplex*?
Fruitbody annual, forming large, irregular, rounded masses up to 40 cm across, with broad, wavy or flattened lobes, not much crinkled or fused. **Hymenium** pale cream to ochre or pinkish ochre, arising from a basal rooting trunk. **Flesh** firm, crisp, whitish. **Odour** pleasant, fungoid or cheesy. **Taste** mild. **Spores** ovoid, smooth, 5–7 x 3.8–5.5 µm. **Spore deposit** cream. **Clamp connections** absent in flesh.

At the base of broadleaved trees, especially *Quercus*, usually around the root buttresses. Widespread but rarely reported, mainly in the southern counties.

This is a much confused species and is best distinguished by the absence of clamp connections in the hyphae of the flesh and the less undulating lobes.

Albatrellus, Boletopsis, Favolaschia, Sparassis

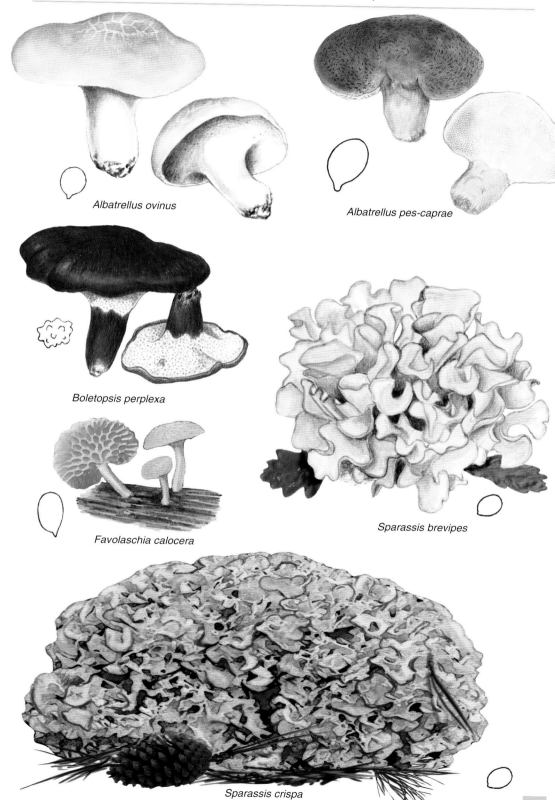

9. Resupinate Fungi

Resupinate is a general term describing the shape of fungi whose fruitbodies are in the form of thin sheets or crusts which spread over their chosen substrate.

In the early days of fungal classification resupinate species were grouped entirely on the basis of the form of their hymenial surface. The vast majority with a smooth or merely wrinkled hymenium were placed in *Corticium*, those with a poroid hymenium in *Poria*, and those with a hydnoid (toothed) hymenium, however fine or poorly developed, in *Hydnum*. All these genera have since been much further divided. In this book only a selection from the large number of possible species have been illustrated, including those which are either very common or those which are uncommon but with distinctive characters

These species are frequently referred to as corticioid fungi, being a group of fungi in the *Basidiomycota* typically having effused fruitbodies that are formed on the undersides of dead tree trunks or branches. They are often colloquially called crust fungi. Originally such fungi were referred to the single genus *Corticium* and subsequently to the family *Corticiaceae*, but it is now known that corticioid species are not necessarily closely related. The fact that they look similar is an example of convergent evolution – unrelated organisms evolving similar structures to cope with similar life strategies and problems.

Most corticioids are wood-rotting species, while a few are found on leaf litter and underneath fallen leaves. The majority form white rots with the few that form brown rots being found almost entirely on coniferous wood. Some very specialised species are found on dead herbaceous stems, grasses or rushes, especially in marshy areas. They are a very numerous group with around 1700 species found worldwide and over 300 recorded from Britain.

The hymenium of corticioid fungi is formed on the outer surface exposed to the air. This fertile layer may consist of a simple sheet of spore-producing basidia spread over the surface, while in other species the surface may be wrinkled, poroid or even spiny or 'toothed'.

Many species prefer fallen logs, branches and twigs on which to form their fruitbodies, while others may be found on standing trunks or stumps. Still others spend their lives high up in the branches of trees and are often only seen when such branches are blown down in a storm.

Microscopic characters

Apart from some striking species which are easily recognised in the field, the majority will require microscopy to be sure of their identity. The size and shape of the spores is of importance, along with any surface ornamention of warts or spines. Many species have striking cystidia of various types. These different types are sometimes named differently depending on the literature consulted; the most common terms are used here.

Lamprocystidia
Sometimes referred to as metuloid cystidia, they are thick-walled, frequently fusiform-clavate and the upper portion of the cystidium often has a granular coating of crystals.

Leptocystidia
Thin-walled, hyaline, rather featureless cells of varying shape depending on the species concerned.

Gloeocystidia
May resemble leptocystidia in shape but are frequently sinuous in outline and usually filled with granular or oily, droplet-like contents.

In the introduction to the bracket fungi the structure of the hyphae which make up the tissues of the fruitbody were discussed and the three possible types described. In corticioids it is usually much easier, the great majority being of one kind (monomitic), having no need for the strengthening elements so common in bracket fungi.

Because the tissues may be spread very thinly over the substrate, corticioids are subject to drying. They may then present a very different appearance from the normally moist, fresh tissues, frequently changing colour, hardening and cracking readily. They are seldom identifiable with certainty when in this condition. Some species may revive if kept in a humid collecting box for a day or two. This may also be useful to encourage production of basidia and mature spores.

Do be aware that many other microscopic fungi may be present on a sample of wood and therefore spores may be present of other species which can completely mislead you when it comes to identification. Wherever possible try to get a deposit of spores from the corticioid by placing the specimen hymenophore downwards on a glass slide in a humid box. This will help ensure you are seeing the correct spores.

As with all microscopy of fungi only a very tiny piece of tissue needs to be excised from the corticioid to assess the hyphae, cystidia, etc. Staining may be helpful to show up features. For specialist literature dealing solely with resupinates and corticioids see Further Reading in the introductory chapters.

Fig. 9.1. A great many genera and species in the *Polyporales* form resupinate, crust-like fruitbodies, some smooth or wrinkled, others with a poroid surface as seen here in *Physisporinus sanguinolentus*, a species that bruises strikingly red, quickly fading to brown.

Fig. 9.2. Some corticioids don't have a distinct margin, others, as seen here, have margins that differ in colour or texture from the rest of the fruitbody and this can be an important identification character. The photo shows *Hymenochaete corrugata*.

Fig. 9.3. Some corticioids are strikingly coloured, making identification a little easier, although microscopy is usually still required to confirm the identification. This bright yellow species is *Phlebia subochracea*.

Fig. 9.4. *Leucogyrophana romellii* has a complex, wrinkled and folded hymenium referred to as 'merulioid'.

Poroid Species

Genus *Ceriporiopsis*

Ceriporiopsis pannocincta (Romell) Gilb. & Ryvarden
[pannocincta = with a ragged girdle]
Syn. *Gloeoporus pannocinctus*
Fruitbody fully resupinate, tightly bonded to the substrate, forming irregular patches several centimeters across and 1–3 mm thick, surface finely pored, margin whitish, sterile and rather ragged. **Pores** rounded, 5–8 per mm, pale greyish ochre with a greenish or yellowish tint, separated from the subsurface by a thin black line. **Flesh** slightly gelatinous under the tube layer, greenish yellow. **Odour** strong, rather disagreeable of old cheese. **Spores** allantoid, smooth, (3)3.5–4.5 x 0.7–0.9 µm. **Spore deposit** whitish. **Cystidia** absent.

On fallen logs of *Fagus* but also known from *Quercus*, *Acer*, *Betula*, etc. First reported from Britain in 1991 but since then has spread rapidly and is now fairly common in southern England and known in Scotland.

Genus *Gloeoporus*

Gloeoporus taxicola (Pers.) Gilb. & Ryvarden
[taxicola = yew-dweller]
Fruitbody fully resupinate, forming irregular patches several centimeters across and up to 3 mm thick; firmly attached to the substrate. **Hymenium** is poroid to slightly merulioid, orange-ochre to reddish ochre to dark red-brown, brightly coloured; margin sterile, ragged, whitish. **Pores** rounded to irregular, 2–3 per mm. **Tubes** up to 1 mm deep, concolorous with pores, underlying layer white, cottony. **Odour** pleasant. **Spores** allantoid, smooth, 4–6 x 1.3–1.8 µm, containing two oil droplets. **Spore deposit** whitish. **Cystidia** absent.

Despite the specific epithet it is commonest on pines, especially *Pinus sylvestris* but also more rarely on *Picea* and even less commonly on wood of broadleaved trees. Widespread but not often reported.

Genus *Schizopora*

Schizopora paradoxa (Schrad.) Donk
[paradoxa = strange, referring to the tooth-like pores]
Fruitbody fully resupinate, forming rounded patches to long, coalesced sheets covering the substrate, often tens of centimeters across, up to 5 mm thick, white when young, becoming ochraceous-cream to ochre-yellow with age. **Pores** very variable, from rounded to lacerated and tooth-like or flattened, plate-like. **Tubes** 1–3 mm deep. **Flesh** rather soft when young, quite tough with age, hard when dry. **Odour** pleasant. **Spores** broadly ellipsoid, smooth, with some internal droplets, 4.5–6 x 3–4 µm. **Spore deposit** white. **Cystidia** absent. **Hyphae** of two sorts (dimitic), generative hyphae thin- to thick-walled, skeletal hyphae thick-walled and often with encrusted, rounded ends.

On dead wood of a variety of broadleaved trees, less commonly on conifer timber, also recorded from stems of *Clematis vitalba*, *Bambusa* spp. and even on dead fruitbodies of other polypores. Widespread and extremely common everywhere. Treated by some authors as a polypore, by others as a corticioid species.

S. radula is very similar with perhaps even more convoluted, plate-like pores and smaller spores, 4–5 x 3–4 µm. It is less common although perhaps overlooked.

Genus *Physisporinus*

Physisporinus sanguinolentus (Alb. & Schwein.) Pilát
[sanguinolentus = flowing with blood]
Fruitbody fully resupinate, forming irregular sheets, sometimes spread over a large area tens of centimeters across, surface often lumpy, white to cream. **Pores** rounded to narrow, slit-like, 3–5 per mm, bruising quickly red but soon brownish. **Tubes** 1–2 mm deep, concolorous with pores. **Flesh** soft, waxy to watery. **Odour** pleasant. **Spores** subglobose, smooth, 5.5–7 x 5–6 µm. **Spore deposit** white. **Cystidia** absent. **Hyphae** thin-walled with cystidia-like ends encrusted with crystals.

On dead and decayed wood of broadleaved trees, especially *Fagus* but on many other species also. Widespread and common.

Genus *Junghuhnia*

Junghuhnia nitida (Pers.) Ryvarden
[nitida = shining]
Fruitbody fully resupinate, forming irregular patches several centimeters across, up to 3 mm thick, tightly bonded to the substrate, pore surface pale salmon to ochraceous, margin ragged, whitish. **Pores** rounded-angular, 5–6 per mm. **Tubes** up to 2 mm deep, concolorous with pores. **Flesh** soft when fresh, hard and brittle when dry. **Odour** rather musty. **Spores** ellipsoid, smooth, 3.5–4.5 x 2–2.5 µm. **Spore deposit** cream. **Cystidia** (skeletocystidia) numerous, thick-walled, 40–100 x 5–10 µm, encrusted part 20–50 x 8–10 µm.

On decayed wood, mainly of *Fagus*, *Fraxinus* but also recorded from *Betula*, *Corylus*, *Populus*, etc. Widespread in England although not common. Rarely reported elsewhere.

J. lacera differs by its angular, almost tooth-like pores (very similar to ***Schizopora paradoxa***) and slightly larger spores (3.5–5 x 2.8–3.5 µm).

Genus *Skeletocutis*

Skeletocutis carneogrisea A. David
[carneogrisea = grey with a flesh-pink tint]
Fruitbody fully resupinate, margin distinctly bounded and lifting slightly when dry, forming irregular patches 5–10 cm across, 1–2 mm thick, greyish white, pinkish grey, paler at the margins. **Pores** rounded to angular, 4–6 per mm. **Tubes** to 1 mm deep, concolorous with pores. **Flesh** duplex with a gelatinous brownish layer immediately below the tubes and a whitish, tomentose layer lying directly on the wood. **Odour** nil. **Spores** strongly allantoid, smooth, 3–4 x 1–1.3 µm. **Spore deposit** whitish. **Cystidia** absent. **Hyphae** of two sorts (dimitic), with **generative hyphae** thin-walled, with clamp connections and often strongly encrusted toward the tube mouths, and with **skeletal hyphae** thick-walled, not encrusted.

On dead and decaying conifer wood, usually associated with dead fruitbodies of ***Trichaptum abietinum***. Occasional in southern England, possibly more widespread.

Ceriporiopsis, Gloeoporus, Schizopora, Physisporinus, Junghuhnia, Skeletocutis

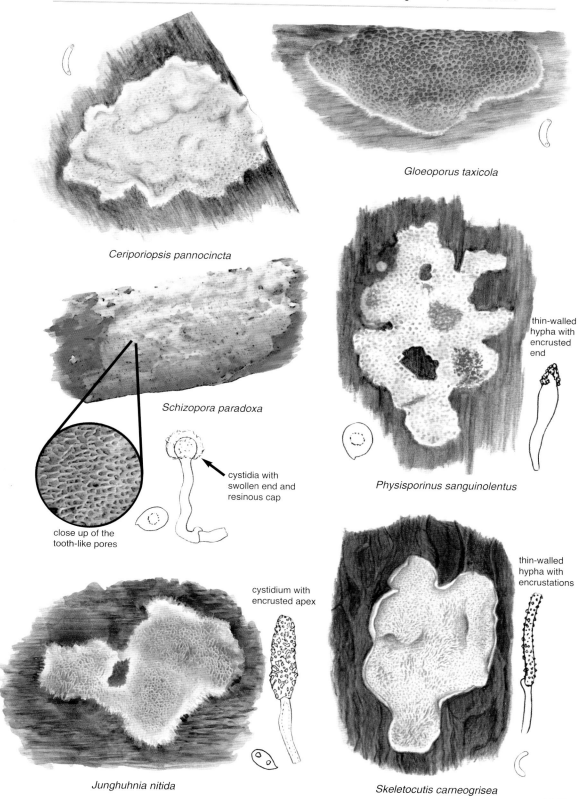

Genus *Fuscoporia*

Fuscoporia ferrea (Pers.) G. Cunn.
[ferrea = rust coloured]
Fruitbody perennial, fully resupinate, often forming very large, irregular patches tens of centimeters long, 1–8 mm thick. More or less uniformly cinnamon-brown irregularly tuberculose, margin fertile or narrowly sterile and paler. **Pores** 6–7 per mm. **Tubes** up to 0.8 mm long, concolorous with the pores. **Flesh** tough, ochre-brown. **Odour** nil to pleasant. **Taste** unknown. **Spores** cylindric, 5–7.5 x 2–2.5 µm, inamyloid. **Spore deposit** whitish. **Hymenial setae** thick-walled, 20–45 x 6–8 µm.

Weakly parasitic then saprotrophic, on fallen branches and trunks of many hardwood genera, rarely on conifers. Widespread and common but often confused with the very similar but less common ***F. ferruginosa***. That species has broader cylindric spores (5–7 x 3–3.5 µm), and has setal hyphae present in its margin up to 300 µm, as well as hymenial setae.

Genus *Antrodia*

Antrodia albida (Fr.) Donk
[albida = white]
Fruitbody usually resupinate or occasionally forming a narrow bracket, rounded to elongate or irregular. **Upper surface** initially finely hairy. **Hymenium** with large, rounded to elongate pores, 0.5–2 per mm, white to pale yellowish. **Tubes** up to 5 mm deep, white. **Flesh** tough, whitish. **Odour** pleasant. **Spores** cylindric, smooth, 8–12 x 3.5–4.5 µm. **Spore deposit** white.

On fallen or attached dead branches of e.g. *Salix*, *Fagus*, *Quercus*, *Fraxinus*, *Malus*, *Corylus*. Widespread and common in Britain.

Genus *Ceriporia*

Ceriporia mellita (Bourdot) Bondartsev & Singer
[mellita = pertaining to honey]
Syn? *C. herinkii*
Fruitbody fully resupinate, up to 20 cm across and 2.5 mm thick, surface pinkish cinnamon to purplish between the pores. **Pores** round to angular, 2–5 per mm, varying from pale sulphur-yellow to cinnamon or salmon pink, darkening with age to orange-brown. **Tubes** up to 2 mm deep, concolorous with the pores. **Flesh** soft, waxy or gelatinous, brittle when dry, pinkish cinnamon. **Odour** pleasant, spicy. **Spores** cylindric to slightly allantoid, smooth, 4.5–6 x 1.5–2 µm. **Spore deposit** whitish. **Cystidia** absent. **KOH** on pores = deep purple-red.

On the underside of fallen, rotten logs of *Fagus*, *Populus*, *Alnus*, etc. Very rare with only one British record from Essex. This species identification may well need revision as the entire genus is badly in need of molecular studies to sort out the species, many of which are very similar. It seems likely that more species remain to be described and that not all will stay in this genus.

Ceriporia purpurea (Fr.) Donk
[purpurea = purple in colour]
Fruitbody fully resupinate to slightly reflexed, in patches up to a few cm across and 2.5 mm thick. **Pores** rounded-angular, 3–4 per mm, pale to dark pinkish purple, purple, to brownish purple when old. **Tubes** to 1 mm thick, pale to dark brownish purple. **Flesh** soft when fresh, brittle when dry, purplish. **Odour** pleasant. **Spores** allantoid, smooth, 5.5–8 x 2–2.5 µm. **Spore deposit** whitish. **Cystidia** absent. **KOH** on pores = purple-violet.

On large, decayed logs of a variety of broadleaved trees, usually on calcareous soils.

Principally in southern England although also known from Scotland and Wales, rather rare.

Corticioid Resupinates

Genus *Coniophora*

Coniophora puteana (Schumach.) P. Karst.
[puteana = pertaining to wells]
Wet Rot
Fruitbody summer to autumn, forming small rounded patches which expand and merge into larger sheets of tissue reaching tens of centimeters across and up to 1–2 mm thick, firmly adhering to the substrate. **Surface** smooth to irregularly warted and wrinkled, cream at first then slowly yellow-ochre to reddish brown at centre, with fine radiating threads of white mycelium at the edges. **Flesh** soft, fibrous. **Odour** strongly fungoid. **Spores** ellipsoid, smooth, 10–13 x 7–8 µm. **Spore deposit** light brown. **Cystidia** absent.

On rotting wood of both broadleaved and coniferous trees. Widespread and common everywhere. The cause of Wet Rot in damp cellars and houses.

Genus *Leucogyrophana*

Leucogyrophana mollusca (Fr.) Pouzar
[mollusca = soft]
Fruitbody resupinate, loosely attached to the substrate, forming soft patches often many centimeters across and up to 2 mm thick. **Surface** white and smooth at first then soon wrinkled-merulioid, yellowish to orange or reddish; margin cobwebby, whitish. **Odour** pleasant. **Spores** ellipsoid, smooth, thick-walled, 6–7 x 4–4.5 µm, strongly dextrinoid. **Spore deposit** yellowish brown.

On very decayed stumps and logs of conifers, more rarely on broadleaved trees. Widespread but mainly in the southern English counties, occasional.

This genus is closely related to a familiar agaric, the False Chanterelle, *Hygrophoropsis aurantiaca* and is therefore actually a member of the *Boletales*.

Very similar is ***L. romellii*** (see photo p.85) which differs in its smaller spores (4.5–5.5 x 2–4.5 µm); it also grows on conifer wood and is a rare species in Britain.

Rather similar and often of very large size is the infamous Dry Rot fungus, ***Serpula lacrymans***, frequent in old houses in cellars and basements where it causes serious damage to timbers. It has much larger spores (11–13 x 5.5–8 µm).

Fuscoporia, Antrodia, Ceriporia, Coniophora, Leucogyrophana

Poroid species

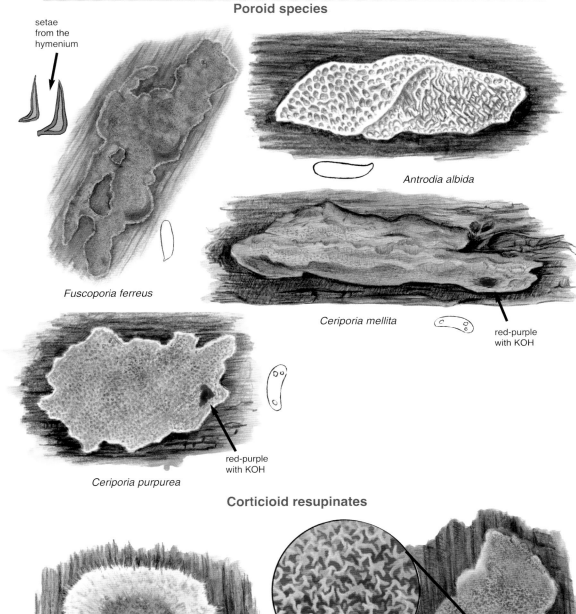

setae from the hymenium

Fuscoporia ferreus

Antrodia albida

Ceriporia mellita

red-purple with KOH

Ceriporia purpurea

red-purple with KOH

Corticioid resupinates

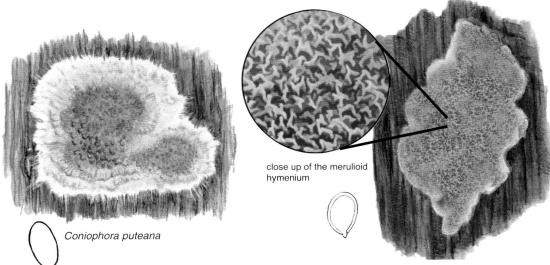

Coniophora puteana

close up of the merulioid hymenium

Leucogyrophana mollusca

Genus *Phlebia*

Phlebia radiata Fr.
[radiata = radiating]
Syn. *P. merismoides*
Fruitbody annual, fully resupinate, occasionally with the margin slightly reflexed, starting as a small spot then growing outward, often coalescing to form large sheets. **Surface** wrinkled, lumpy, usually radially furrowed, from bright pinkish orange when young to much paler, pinkish or salmon when old, or even greyish lilac in some stages or forms, marginal zone often paler. **Flesh** gelatinous and soft when fresh, much harder when dry. **Odour** nil. **Taste** mild. **Spores** cylindric-allantoid, smooth 4.5–5.5 x 1.5–2 µm, usually with two droplets. **Spore deposit** whitish.

On dead wood of broadleaved trees, more rarely on conifers. Widespread and common in Britain.

P. rufa, on fallen *Quercus*, differs in its more meruloid-wrinkled to almost poroid surface varying from cream to ochre or rust-brown, darker with age.

Vey different is the beautiful sulphur-yellow *P. subochracea* (see photo p.85) found on very rotten wood in damp places.

Genus *Chondrostereum*

Chondrostereum purpureum (Pers.) Pouzar
[purpureum = with purple colouration]
Fruitbody annual, resupinate to partially bracket-like, often in large clusters, upper surface hairy-tomentose, greyish white, vaguely zonate, paler at the margin. **Hymenium** smooth, undulating, smooth to wrinkled, pinkish violet to dark violet, browner with age. **Flesh** tough, rather elastic, greyish white, with a darker zone in cross section, separating the tomentose cuticle from the flesh. **Odour** nil. **Taste** mild. **Spores** cylindric-ellipsoid, 6.5–8 x 2.5–3.5 µm. **Spore deposit** white.

On dead or sick broadleaved trees, of commercial importance on fruit trees where it causes Silver Leaf disease. Widespread and common everywhere.

Genus *Stereum*

Stereum hirsutum (Willd.) Gray
[hirsutum = hairy]
Fruitbody annual, fully resupinate as rounded patches to bracket-like, 2–5 cm across. **Upper surface** hairy-tomentose, zonate, yellow-orange, ochre to greyish ochre. **Hymenium** smooth, yellow-ochre to yellow-brown, not turning red when injured. **Flesh**, tough, elastic, ochre. **Odour** nil. **Taste** mild. Spores ellipsoid-cylindric, 5.5–6.5 x 2–3 µm. **Spore deposit** white.

One of the commonest fungi on hardwoods everywhere, often in large sheets covering logs.

Stereum rameale (Pers. ex Fr.) Burt
[rameale = of twigs]
Syn. *S. ochraceoflavum* ss. European authors
Fruitbody annual, fully resupinate as rounded patches to semi bracket-like. **Upper surface** minutely downy with a fringed margin; pale ochraceous yellow to greyish yellow. **Hymenium** with faint concentric zones, pale ochre yellow to greyish yellow, does not bruise red when scratched. **Flesh** thin, leathery. **Odour** nil. **Spores** 7–9 x 2–3 µm. **Spore deposit** cream.

Usually growing on slender fallen twigs of hardwoods, especially *Quercus*. Widespread and probably common but often confused with resupinate forms of *S. hirsutum.*

Although commonly called *S. ochraceoflavum*, that species was described from N. America and almost certainly is something else. A new name is required for this common fungus however, as *S. rameale* is an invalid name having been used already for a different fungus.

Stereum gausapatum (Fr.) Fr.
[gausapatum = like a type of shaggy cloth]
Fruitbody resupinate with upper edge often reflexed to form a small bracket. **Surface** grey-brown and with a whitish margin. **Hymenium** turning bright red when scratched. **Spores** ellipsoid, 6.5–10 x 3.5–4.5 µm. Almost always on *Quercus* where it is quite frequent.

Two other red-staining species are:

S. rugosum (Pers.: Fr.) Fr.
Frequent on a range of broadleaved trees, more often fully resupinate, harder and sometimes perennial; ochraceous to pinkish buff. **Spores** 7–12 x 3.5–4.5 µm.

S. sanguinolentum (Alb. & Schw.: Fr.) Fr.
Very similar to *S. gausapatum* but confined to coniferous wood. **Spores** 6.5–7.5 x 2.5–3 µm.

Genus *Aleurodiscus*

Aleurodiscus wakefieldiae Boidin & Beller
[wakefieldii = after Elsie Wakefield, 1886–1972, mycologist]
Fruitbody forming rounded patches coalescing into larger groups 2–5 cm across, starting as small rounded shallow cups, then spreading out and flattening, 1–3 mm thick, bright pink with paler, narrow margin. **Flesh** pale brown, soft. **Odour** nil. **Spores** very large, broadly ellipsoid, 20–28 x 14–20 µm, amyloid, ornamented with numerous small warts and spines. **Spore deposit** white. **Acanthohyphidia** (cystidia-like cells in the hymenium) with finger-like outgrowths.

On attached or fallen branches of broadleaved trees, especially *Fagus*, *Alnus* and *Quercus*. Widely distributed, very conspicuous but uncommon.

Aleurodiscus amorphus (Fr.) Schroet.
[amorphus = of indefinite shape]
Fruitbody forming rounded discs, cushion-like to slightly cup-shaped, often fusing together into larger, more irregular shapes; breaking through bark and broadly attached at centre. **Upper surface** granular, pale cinnamon to orange-red fading to pinkish grey when old, margin fringed, hairy, whitish. **Lower surface** white, hairy. **Flesh** soft, waxy. **Odour** nil. **Spores** subglobose, very large, 25–30 x 22–25 µm, amyloid, with numerous blunt spines. **Spore deposit** whitish.

On dead or dying attached branches with the bark still attached, especially on *Picea*, more rarely *Abies*. Not often reported but widespread in Scotland and also recorded in some English counties. Frequently parasitised by the jelly fungus, ***Tremella simplex***.

Toothed species

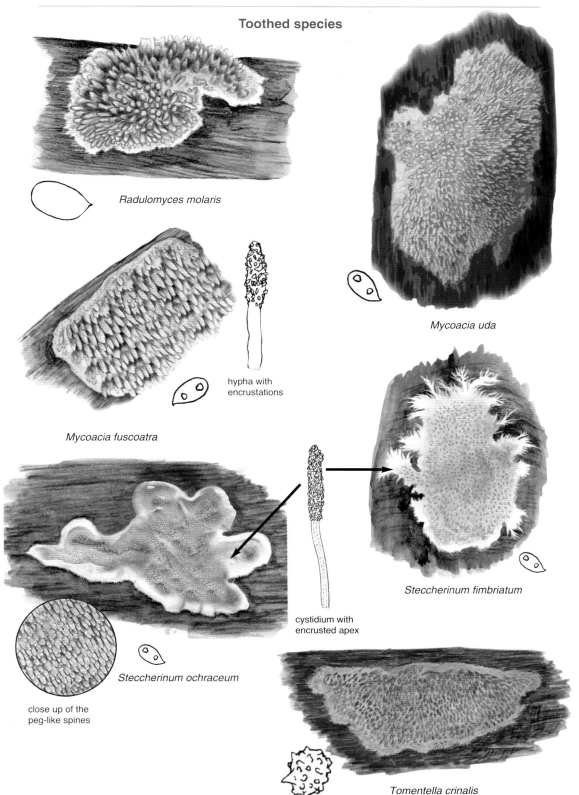

10. Jelly Fungi

Jelly fungi are another example of a group of fungi once thought to be closely related but now known to consist of several independently evolved lineages—this is known as a paraphyletic group. So, as with the polypores and the gasteromycetes they are placed together here as a useful convenience because they share similar morphological traits.

They are all members of heterobasidiomycete fungal orders from different classes of the subphylum *Agaricomycotina*: *Tremellales*, *Dacrymycetales*, *Auriculariales* and *Sebacinales*. These fungi are so named because their fruitbodies are, or appear to be, the consistency of jelly or rubber. When dried, jelly fungi become hard and shriveled, but when exposed to water many will return to their original form and may continue to produce spores. This appears to be an adaptation to often growing in exposed locations on branches, etc where they may be prone to drying out and having to wait for rainy periods to once more begin sporulating.

Jelly fungi come in a variety of shapes, having duplicated many of the forms familiar in other groups of fungi. Thus we have resupinate or corticioid forms, ones with spines and others with clubs, discs or lobed shapes. One (*Guepinia*) even has a chanterelle-like form.

Fruitbodies range in size from almost microscopic to many centimeters across. Several species are a common part of any foray through the woods. Others are rare and little-known and should be considered a prize when found and worthy of reporting to your local or national fungus group.

Many species appear to be saprotrophic on dead wood but species in the genus *Tremella* are unusual in being parasitic on a variety of other fungi, often being very specific about the choice of host. In *T. mesenterica* and *T. aurantiaca* for example, two species macroscopically very similar, they can easily be separated by host preference: *T. mesenterica* prefering *Peniophora* species while *T. aurantiaca* targets *Stereum hirsutum*.

Only a very few species are regarded as good edibles but anyone who has tried Chinese or other oriental cookery may have encountered species of tree ear (*Auricularia*) in their meal. *Tremella fuciformis*, a large, translucent, tropical species is not only edible but prized for use in soups and vegetable dishes. Most species are tasteless or perhaps bitter.

Microscopy

Jelly fungi share other unique characters apart from their bodily consistency, particularly when placed under a microscope. The most obvious is the structure of their basidia. These may be 2- to 4-spored but are uniquely divided internally with septa (internal walls) which may be either vertical or transverse depending upon the genus concerned. The sterigmata on which the spores are formed are often enormously long compared with the base of the basidium. Spores are frequently very large, smooth and commonly sausage-shaped (allantoid). In the *Dacrymycetales* the basidia are commonly referred to as tuning fork basidia for their resemblance to a musician's tuning fork (see figure below).

Basidium with vertical septa dividing into four partitions

Basidium with transverse septa dividing into four partitions

So-called 'tuning fork' basidium with two very large sterigmata on which the spores are produced

Fig. 10.1. Some jelly fungi, such as *Calocera viscosa* shown here, look very similar to the unrelated club and coral fungi but differ greatly in their texture and microscopy.

Fig. 10.2. *Pseudohydnum gelatinosum* has soft, jelly-like spines on which the spore producing basidia are located. This is a species of wet conifer woods.

Fig. 10.3. *Tremella* species range from extremely tiny to large masses such as this *T. foliacea*, some 15 cm across. It illustrates the gelatinous, floppy texture common to the group.

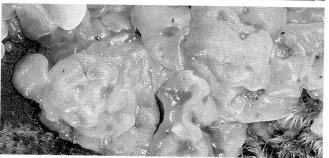

Fig. 10.4. *Exidia* species, such as this *E. thuretiana*, frequently form rather shapeless blobs on wood, often very small but some species reaching many centimeters across. Their colours range from translucent white through pink, brown and totally black.

Order: *Auriculariales*

Genus *Auricularia*

Auricularia mesenterica (Dicks.) Pers.
[mesenterica = resembling an intestine]
Tripe Fungus
Fruitbody bracket-shaped, often coalescing in long patches, projecting about 1–3 cm from the substrate. **Upper surface** densely tomentose-hispid, undulating, often concentrically zoned, pale grey, olivaceous, ochraceous. **Hymenium** on underside, veined-wrinkled, smooth, rubbery, purplish grey to purple-grey, sometimes white pruinose. **Flesh** soft, elastic, gelatinous, drying hard and shriveling. **Odour** pleasant. **Taste** mild. **Spores** cylindric-allantoid, smooth, 15–17.5 x 6–7 μm. **Spore deposit** white. **Basidia** with transverse septa, 2–4-spored.

On dead and decayed wood of broadleaved trees, stumps, etc. Widespread and common.

Auricularia auricula-judae (Bull.) Wettst.
[auricula-judae = Jew's ear, Judas's ear]
Tree Ears
Fruitbody ear-shaped, up to 9 cm across, attached laterally or sometimes with a short stalk, outer surface minutely pubescent, red-brown, olive-brown, blackish when dry. **Hymenium** on the inner surface of the 'ear' smooth, veined or wrinkled, slightly whitish pruinose from the spores. **Flesh** soft, gelatinous and rubbery, hard and crispy when fully dried but able to revive when wetted. **Odour** nil. **Taste** mild. **Spores** cylindric-allantoid, smooth, 17–19 x 6–7 μm. **Spore deposit** white. **Basidia** with transverse septa, 4-spored.

On living or dead branches of a wide variety of hardwoods but especially *Sambucus nigra* (Black Elder). Widespread and common everywhere. The common name refers to the legend that Judas hanged himself on an elder tree. This and related tropical species are popular edible fungi in oriental cooking.

Genus *Guepinia*

Guepinia helvelloides (DC.) Fr.
[helvelloides = resembling *Helvella*]
Syn. *Tremiscus helvelloides*
Fruitbody ear-shaped to trumpet-shaped, split down one side, up to 10 cm tall, 5 cm across, orange-pink to salmon, smooth on the inner surface, faintly pruinose. **Hymenium** on outer surface of upper trumpet faintly wrinkled-veined. **Flesh** rubbery, soft, gelatinous. **Odour** nil. **Taste** almost nil. **Spores** ellipsoid to rather irregular in outline, smooth, 9.5–11 x 5.5–6 μm. **Spore deposit** white. **Basidia** with vertical septa, with 2–4 spores.

On soil in grass or needle litter in woodlands. Widely scattered localities in England and Wales, rare overall but where found often in large numbers. Apparently spreading in recent years.

Genus *Exidia*

Exidia plana (F.H. Wigg.) Donk
[plana = flattened]
Witches' Butter
Syn. *Exidia glandulosa* of Fries and many authors
Fruitbody spread out to form a shapeless, wrinkled and puckered gelatinous patch, often many centimeters across and up to 2 cm thick, not attached to the substrate at the edges. **Surface** folded, convoluted, shiny, black and dotted with tiny glandular warts. **Flesh** gelatinous, very soft. **Odour** nil. **Taste** nil. **Spores** cylindric-allantoid, smooth, 12–14 x 4.5–5.5 μm. **Spore deposit** white. **Basidia** with vertical septa, 4-spored.

On dead wood, standing or fallen, most commonly *Fagus*, *Acer*, *Betula* and *Fraxinus*. Widespread and common.

Exidia glandulosa (Bull.) Fr.
[glandulosa = with small glands]
Syn. *Exidia spiculosa, E. truncata*
Fruitbody forming small, turbinate cushions, solitary or crowded together, each cushion 2–8 cm across, 1–3 cm high, with small basal stalk or joined directly to the substrate. **Hymenium** flattened or depressed, tuberculate or undulating, often with small furrows at the edges, with small glandular warts, black, shiny. **Flesh** soft, jelly-like, blackish, translucent. **Odour** nil. **Taste** nil. **Spores** cylindric, sausage-shaped, 9–12 x 5–7 μm. **Spore deposit** white. Also producing rod-like **conidia** 4–5 x 2 μm. **Basidia** with vertical septa, 4-spored

On dead wood of *Corylus*, *Quercus* and *Fagus*. Widespread and quite common.

This and the previous species have been confused for many years with the name *E. glandulosa* being applied to the more amorphous, brain-like *E. plana*. Both species have glandular dots on the hymenium.

Exidia nucleata (Schwein.) Burt.
[nucleata = possessing a nucleus]
Syn. *Myxarium nucleatum, Tremella nucleata*
Fruitbody forming small, rounded cushions 1–4 mm across, which can coalesce to form larger masses up to 4 cm across. Each cushion whitish translucent, sometimes tinted pinkish or violaceous, with a tiny crystal of calcium oxalate buried inside. **Flesh** soft, gelatinous. **Odour** nil. **Taste** unknown. **Spores** cylindric-allantoid, smooth, 10–18 x 4–6 μm. **Basidia** with vertical septa, 4-spored.

On fallen wood of *Fagus* and *Fraxinus*. Widespread and common.

There are several similar species but the presence of the calcium inclusion in the tissues is a good field character. This can sometimes be detected by pressing on the fungus but to be sure a section is recommended.

Exidia recisa (Ditmar) Fr.
[recisa = cut off, truncated]
Syn. *Tremella recisa*
Fruitbody forming small, turbinate cushions, solitary or crowded together, each cushion 2–3.5 cm across, 1–2 cm high, with small basal stalk or joined directly to the substrate. Each cushion amber to dark brown. **Flesh** gelatinous, fairly firm when fresh but shriveled and collapsing when dry, eventually forming just a thin brown crust. **Odour** nil. **Spores** cylindric-allantoid, smooth, 12–15 x 3–4 μm. **Basidia** with vertical septa, 4-spored.

On dead, attached twigs of *Salix* species. Widespread but not often reported.

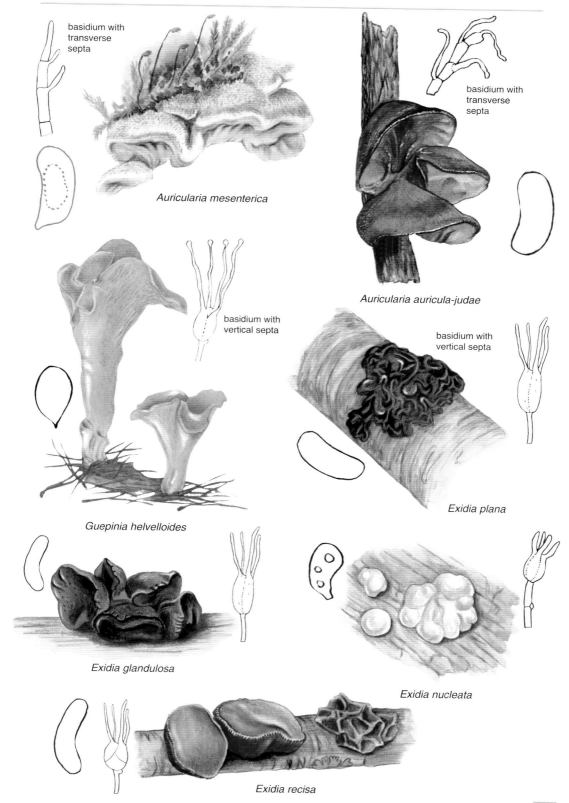

Genus *Pseudohydnum*

Pseudohydnum gelatinosum Pers.
[gelatinosum = jelly-like]
Fruitbody bracket to tongue-like, often with a short, eccentric stem. **Bracket** 2–6 cm across, 0.5–1 cm thick. **Upper surface** rough-tomentose, smoother at the margin, from pure white to grey, grey-brown. **Hymenium** on underside decurrent, formed of soft, gelatinous pegs or spines, 2–5 mm long, translucent whitish, greyish. **Flesh** soft, jelly-like, translucent. **Odour** and **taste** not distinct. **Spores** subglobose to globose, smooth, 5–6 x 4.5–5.5 µm. **Spore deposit** white. **Basidia** vertically septate, 4-spored.

On rotten coniferous wood, especially on stumps, often in small, imbricate groups. Widespread but not common.

Order: *Tremellales*

Genus *Tremella*

Tremella foliacea Pers.
[foliacea = leaf-like]
Fruitbody forming a mass of dense flattened lobes, each lobe twisted and convoluted to form a gelatinous mass up to 15 cm across. **Lobes** smooth, shiny to dull, deep reddish brown, ochraceous brown, soft and jelly-like. **Flesh** soft, gelatinous, rubbery, drying to a much smaller, hard mass. **Odour** and **taste** not distinctive. **Spores** ovoid, smooth 9–11 x 6–8 µm. **Basidia** subglobose, with vertical septa, 4-spored.

Parasitic on the mycelium of ***Stereum*** species on wood of broadleaved trees and more rarely on conifers (var. ***succinea***). Widespread and fairly common. With its large size, texture and complex structure it would be difficult to mistake this for any other species.

Tremella mesenterica Retz.
[mesenterica = like an intestine]
Syn. *Tremella lutescens*
Yellow Brain
Fruitbody forming an irregular, flabby, lobed or brain-like mass, 3–10 cm across, from bright golden yellow-orange to palest yellow or even white and translucent in some conditions. **Flesh** gelatinous and elastic in texture, smooth, shriveling to a small, hard, orange-brown, horny blob when dry, reviving when wet again. **Spores** ellipsoid, smooth, 10–16 x 7–8 µm. **Conidia** spherical to broadly ellipsoid, 3–5 x 2.5–3.5 µm. **Basidia** with vertical septa, 4-spored, typically sessile.

On dead branches of *Corylus* and *Ulex* among others, parasitic on the mycelium of ***Peniophora*** spp. (which may not be visible). This species of *Tremella* is unusual in asexually producing conidia before the spores mature.

It is easily confused with the rather less common (although perhaps just misidentified) ***T. aurantia*** which parasitises ***Stereum hirsutum*** and has smaller spores as well as basidia with a longer basal stem.

Tremella aurantia Schwein.
[aurantia = orange]
Misapplied name: *Tremella mesenterica*
Yellow Brain
Fruitbody 2-10 cm broad, consisting of clustered, convoluted folds with blunt margins; surface yellow, to yellowish orange, shiny when wet, otherwise dull. **Flesh** soft, gelatinous, drying to a stiff, hard crust, reviving after rain. **Odour** and **taste** not distinctive. **Spores** subglobose to ovoid, smooth, 6.0–9.5 x 6.0–7.5 µm. **Basidia** typically diagonally septate, typically stalked, 2–4-spored. **Conidia** 2–4 x 1.5–3 µm.

Parasitising fruitbodies of ***Stereum hirsutum***, on dead branches of broadleaved trees. Widespread and possibly frequent but not often recorded and frequently confused with *T. mesenterica*.

Tremella encephala Pers.
[encephala = brain-like]
Fruitbody forming a rounded cushion, 1–3 cm across, tightly bonded to the substrate. **Upper surface** undulating, brain-like, white to ochraceous or pinkish brown, translucent. **Flesh** gelatinous but in cross-section with a hard white core. **Odour** and **taste** not distinctive. **Spores** subglobose, smooth, 8–12 x 6–9 µm. **Basidia** subglobose with vertical septa, with two long sterigmata.

Parasitic on ***Stereum sanguinolentum*** on conifer logs or branches. Widespread and locally frequent.

According to the literature the hard white core is composed of compacted hyphae of the host fungus, *Stereum sanguinolentum*.

Much smaller and simpler in structure is *T. simplex*, which forms a tiny (1–5 mm) blob that may coalesce into large patches, see below.

Tremella simplex H.S. Jacks. & G.W. Martin
[simplex = of simple form]
Fruitbody forming small rounded to irregular gelatinous blobs, 1–5 mm across, sometimes fusing together into larger masses. **Surface** smooth to slightly bumpy, whitish to pale ochre or yellow-brown, translucent. **Flesh** soft, waxy or gelatinous, deliquescing and slimy with age. **Spores** subglobose, smooth, 5–7 µm. **Conidia** also present, subglobose, slightly smaller than the spores, 3–4 µm. **Basidia** divided into two, with one to two sterigmata.

Parasitic on the discs of the corticioid fungus ***Aleurodiscus amorphus***, on branches of *Abies* or *Picea*. Rather rare in Britain as the host fungus is rare but almost always parasitised when found.

Tremella globispora D.A. Reid
[globispora = round spore]
Fruitbody tiny gelatinous blobs, 2–5 mm across. **Surface** irregular, whitish and opalescent to slightly greenish, deliquescing in age. **Flesh** soft and gelatinous. **Spores** subglobose to ovoid, smooth, 6–7.5 x 5–7 µm. **Basidia** with vertical septa, 4-spored.

Parasitising dead fruiting bodies of pyrenomycetes, especially ***Diaporthe*** and ***Eutypella***, on small branches of various trees. Widespread but not often reported, although probably fairly common.

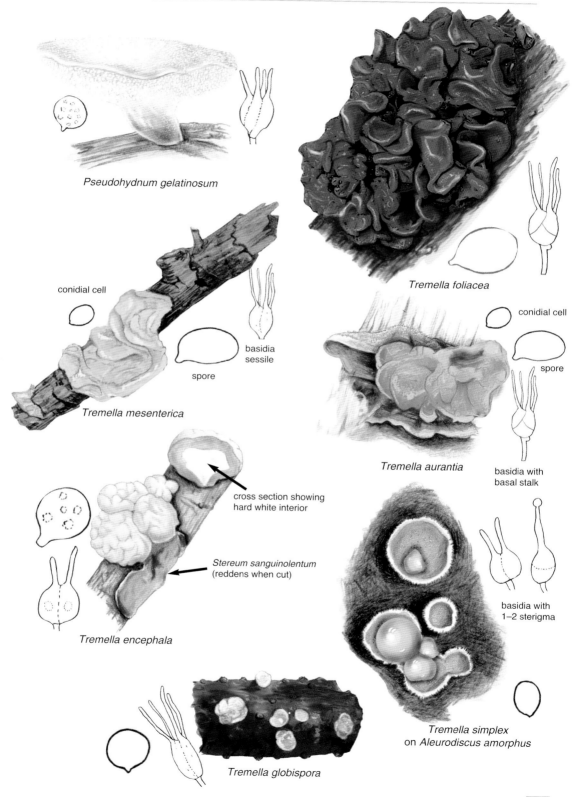

Order: *Dacrymycetales*

Genus *Calocera*

Calocera viscosa (Pers.) Fr.
[viscosa = sticky or viscid]
Fruitbody coral-like with erect branches which may end in unbranched to two- or three-branched tips, 2–8 cm tall, many branches fused together at the base, bright egg-yellow to orange, darkening to rust as it dries. **Surface** smooth, sticky-slippery when moist. **Flesh** gelatinous, rubbery-elastic. **Odour** and **taste** nil. **Spores** ellipsoid-allantoid, smooth, 8–10 x 3.5–4.5 μm, with one internal septum when mature. **Basidia** like a tuning fork.

On rotten conifer stumps and logs. Widespread and common everywhere.

Calocera cornea (Batsch) Fr.
[cornea = horn-like but not hard]
Fruitbody single pointed, unbranched clubs up to 1 cm high, 1–2 mm wide, often in rows or small groups, bright egg-yellow, yellow-orange, darker at the tips especially when drying out. **Surface** smooth to longitudinally furrowed. **Flesh** gelatinous but tough, cartilaginous, yellowish. **Odour** and **taste** nil. **Spores** ellipsoid-cylindric to slightly allantoid, smooth, with one septum, 7–10 x 2.5–4 μm. **Basidia** tuning fork-shaped.

On decayed wood, especially *Fagus* but a wide variety of other trees also, including conifers. Widespread and very common everywhere.

Calocera glossoides (Pers.) Fr.
[glossoides = shaped like a tongue]
Fruitbody single clubs up to 1 cm high, with stem distinct from the more swollen head, often in small to large groups. **Head** tongue-shaped to fusiform, often slightly flattened, usually distinctly longitudinally furrowed, bright yellow, yellow-orange to rust with age or on drying. **Stem** narrower than the head, brownish yellow. **Flesh** gelatinous, elastic, yellowish. **Odour** and **taste** nil. **Spores** ellipsoid to suballantoid, smooth, 12–14 x 3–4.5 μm, 1–3-septate, often budding off small rounded conidia. **Basidia** like a tuning-fork.

On decayed wood of various broadleaved trees. Records are mainly from southern England but it is frequently recorded in error, being confused with *C. cornea* and *C. pallidospathulata*.

Calocera pallidospathulata (Pers.) Fr.
[pallidospathulata = pale spade-like]
Fruitbody single clubs up to 1 cm high, with stem distinct from the more swollen head, often in small to large groups. **Head** tongue-shaped or very irregular club-like, often slightly forked, from translucent whitish to pale primrose-yellow. **Stem** narrower than the head, usually translucent, whitish. **Flesh** gelatinous, soft, elastic, whitish. **Odour** and **taste** nil. **Spores** allantoid, smooth, 10–15 x 3.5–4 μm, 1–3-septate, budding off small, rod-like conidia. **Basidia** like a long tuning fork.

Originally recorded on dead and decaying wood of conifers but more recently also on a variety of broadleaved trees. Widespread and now common in Britain, quite possibly an introduced alien species.

Calocera furcata (Pers.) Fr.
[furcata = forked]
Fruitbody usually single clubs or forking into two or more rarely three branches, up to 1.5 cm high, light yellow to yellow-orange or orange as it dries out, smooth, slightly sticky. **Flesh** gelatinous, tough, elastic, yellowish. **Odour** and **taste** nil. **Spores** ellipsoid-allantoid, smooth, 8–13 x 3–4 μm, 1–3-septate. **Basidia** like a tuning fork, with a diamond-shaped point between the sterigmata.

Confined to decayed coniferous wood. Widely distributed but very rarely reported with only a few British records.

Genus *Guepiniopsis*

Guepiniopsis buccina (Pers.) Fr.
[buccina = trumpet-like]
Fruitbody in the form of a stalked cup up to 1 cm high, 0.5–1 cm across. **Cup** may be flattened, somewhat lopsided or goblet-like, sometimes emerging directly from the substrate, more usually with a short stem. **Surface** smooth, yellow to yellowish orange. **Stem** often with longitudinal furrows. **Flesh** gelatinous but quite firm. **Odour** and **taste** nil. **Spores** cylindric-allantoid, smooth, 11–16 x 4–6 μm, only slowly developing septa as the spores mature, 1–3-septate. **Basidia** like a tuning fork.

On dead branches of broadleaved trees. Widely distributed in Britain but very rare as only a handful of records in the last 50 years. Its striking appearance, gelatinous texture and septate spores should serve to distinguish it from any ascomycete cup fungi that might be confused with it.

Genus *Dacrymyces*

Dacrymyces stillatus Nees
[stillatus = producing droplets]
Fruitbody rounded, cushion-like, each cushion up to 1 cm across, often coalesced into large masses. **Surface** smooth to wrinkled, from light yellow to dark orange, more rarely whitish, viscid when wet. **Flesh** soft, gelatinous, deliquescing into slime when old. Two forms often growing side by side, a lighter, yellow state forming basidiospores and a darker, more orange form which produces arthrospores. **Odour** and **taste** nil. **Spores** elliptic-cylindric, slightly curved, smooth, 14–17 x 5–6 μm, with 3 septa when mature. **Basidia** like a tuning fork. **Arthrospores** produced in chains, each arthrospore 9–12 x 3–4 μm. **Clamp connections** absent.

On dead wood of a wide variety of trees, both broadleaved and conifers, with or without bark, especially in wet weather. Widespread and common everywhere.

A number of similar species exist, several distinguished by the presence of clamp connections at hyphal septa such as ***D. variisporus*** with spores 15–19 (-27) x 4.5–7 μm and 3–7-septate.

Calocera, Guepiniopsis, Dacrymyces

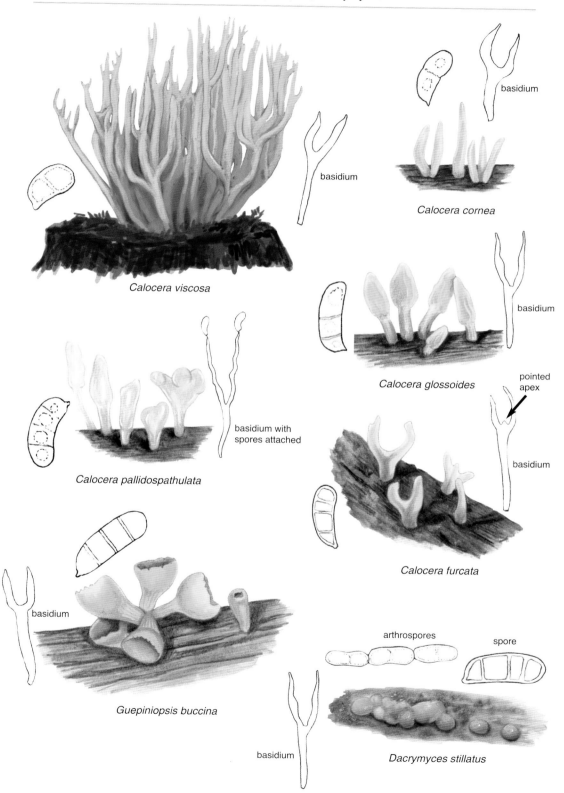

11. Boletes & Their Relatives

The order *Boletales* contains the familiar boletes, so popular with mycologists and mycophagists alike. The majority of these well known fungi are placed in the family *Boletaceae*. However, the order encompasses several different families apart from those with typical boletoid forms. These include, among others, the *Sclerodermataceae*, with puffball-like, tough fruitbodies and the *Paxillaceae*, *Hygrophoropsidaceae* and *Gomphidiaceae* which all have gills or lamellae under the cap, resembling members of the *Agaricales*. There are other genera where the spore-producing tissues are formed in truffle-like forms or even on simple, crust-like or resupinate sheets of tissue. This is because, as with many other fungi, the *Boletales* have evolved many different body types during their evolutionary history, but they are all united by common, morphological, chemical and molecular characters. The genera dealt with in this book are those with typical toadstool-like fruitbodies, with tubes or gills on the underside of the cap, plus the puffball-like forms.

In recent years molecular studies looking at the genetic makeup of the fungi have resulted in the formerly large genus *Boletus* being split into many smaller genera resulting in new and less familiar names. For example, species of the genus *Xerocomus* and *Xerocomellus* which were separated from the genus *Boletus* (and which appear in many field guides as such), based on morphological and molecular differences have in their turn also recently been shown to separate into distinct, smaller groups, equally worthy of generic rank. This process of separation of genera is ongoing and has accelerated in recent years. These new generic names are used in this book but in each case their older names are also given.

Typical boletes are characterized by having tubes instead of gills beneath the cap. The spore producing tissue—the hymenium—is formed on the inner surface of these tubes. The tubes have variously sized and shaped pores from which the spores will be ejected. These tubes are packed tightly together beneath the cap and appear sponge-like, being variously coloured from white-cream through to bright red. Some will bruise blue, greenish or other colours when handled and this can be a useful diagnostic character in species identification. The spore colour varies from ochre through brown all the way to black. Microscopically the spores of most boletes are quite distinctive; usually being relatively narrow and either spindle-shaped or ellipsoid, although as always with fungi there are exceptions.

Other genera in the *Boletales* such as *Paxillus*, *Gomphidius*, *Hygrophoropsis*, etc have gills (more properly referred to as lamellae but called gills throughout this book) but are not considered as being closely related to the true gilled fungi found in the order *Agaricales*. Their spores are formed on basidia which cover these gills in a hymenial layer.

Most genera within the *Boletales* form a mycorrhizal association with trees—both conifers and broadleaved—and more rarely shrubs such as *Helianthemum*, but some are parasitic, such as *Pseudoboletus parasiticus* on the common earthball, *Scleroderma citrinum*, while others are saprotrophic.

The species in the *Gomphidiaceae* all appear to have an association—perhaps parasitic?—with species of boletes in the genus *Suillus*.

The boletes contain some of the most highly prized and collected edible fungi such as the Cep or Porcini, *Boletus edulis*; in fact the majority of the boletes are edible to some degree. There are however, one or two exceptions, such as some of the dangerously toxic, red-pored species including *Rubroboletus satanas* and *R. legaliae*. Some species are not toxic *per se* but are inedible, such as the very bitter *Caloboletus radicans*, *C. calopus* and *Tylopilus felleus*, so be careful. As a general rule those boletes with red pores are best avoided, even though a few species are actually edible.

Other families such as the *Paxillaceae* contain species which are known to be dangerously toxic if consumed regularly over a period of time, although the exact mechanism and identity of the poison compounds involved is not entirely understood. As with all fungi you must always be certain of which species you have if you intend to eat them!

The boletes remain among the most beautiful and popular of larger fungi, not least because of their often large size, variety of textures and striking colours. They make an ideal group for beginners to study as they are relatively easy to identify compared with many other groups of fungi. Important field characters include colour changes in the cut flesh and bruised pores, presence or absence of a reticulation or squamules on the stem, pore colour and of course habitat. Microscopically the spore shape, size and any ornamentation is important to record, along with details of the cap cuticle and in some cases the size and shape of any hymenial cystidia. Many species are quite variable in colour, shape and size and if a selection of fruitbodies is available such variation should be carefully noted (see Figs 11.1–11.7 for important bolete characters).

Chemical reactions are perhaps less useful in the *Boletales* than in some other groups such as the *Russulales*, also included in this volume, but nevertheless a few are useful, in particular ammonia (NH_4OH) and potassium hydroxide (KOH), usually applied to the cap surface or the flesh. Such important reactions are noted in the text wherever they occur.

The gilled forms are easily confused with members of the order *Agaricales*, so much so that you might be tempted to look for them there. But they do have some significant differences. Their spores tend to be rather long, slender and fusoid (Fig. 11.7), while their gills are not firmly attached to the underlying flesh and are usually easily rubbed off using a thumbnail. The *Sclerodermatales* have strikingly ornamented spores combined with tough, leathery, puffball-like fruitbodies.

There are approximately 100 species of *Boletales* in Britain (not including those genera with underground fruitbodies or resupinate forms), plus many more species in mainland Europe, especially in the Mediterranean countries.

A number of non-British species are included for completeness and are indicated by two asterisks (**).

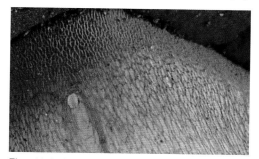

Fig. 11.1. Prominent reticulum on the stem of *Boletus legaliae*.

Fig. 11.2. Punctate stem of *Boletus luridiformis*.

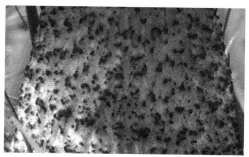

Fig. 11.3. Floccose-squamulose stem of *Leccinum aurantiacum*.

Fig. 11.4. Smooth stem of *Xerocomellus porosporus*.

Fig. 11.5. Hyphae of the cap cuticle of a *Xerocomellus* species showing encrusted surfaces. Photo © Alan Hills.

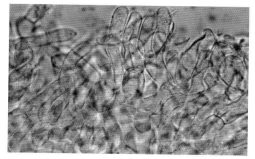

Fig. 11.6. Smooth hyphae of the cap cuticle of *Xerocomus subtomentosus*. Photograph © Alan Hills.

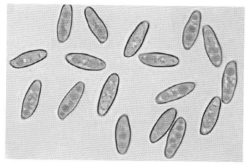

Fig. 11.7. Typical ellipsoid-subfusiform bolete spores of *Boletus impolitus*. Photograph © Malcolm Storey.

> **Family *Boletaceae***
>
> The spores are produced within tubes which form a spongy layer of pores on the underside of the cap. Pores commonly range from white through yellow to orange or blood red (but other colours occur also) and frequently change colour when touched. The stem may be almost smooth to distinctly minutely punctate, while many species have a raised network or mesh called a reticulum. The stem surface may be dry or sticky and some species have a partial veil in the form of a ring. Microscopically the spore shape and size is important. Spore deposits range from olive-brown, to pink, purplish brown or black.
>
> The majority of species are mycorrhizal but some are saprotrophs or even parasites.

Genus *Boletus*

Boletus edulis Bull.:Fr.
[edulis = edible]
Cap 5-25 cm, colour very variable, from dark brown to reddish brown, chestnut, to much paler, more fawn, buff or even yellowish.
Viscid when wet. **Tubes** and **pores** white at first then soon yellowish and finally greenish. **Stem** robust, clavate to cylindric, pale pinkish buff to almost white, with a more or less prominent white network, especially in the upper half. **Flesh** thick, white, usually flushed vinaceous immediately below the cap cuticle and more or less unchanging when cut. **Spores** olive-brown, subfusiform, 16–19 x 5–6 μm.

Widespread and common with a very wide range of host trees, both conifers and broadleaved; some of the varieties/species described below occur with a specific tree host. It is also occasionally found growing with *Helianthemum nummularium* in open calcareous grasslands. One of the best known and highly esteemed edible fungi in the world, it has a number of common names including Cep, Penny Bun, Steinpilz and Porcini.

Apart from (a) *B. edulis* in the strict sense a number of varieties exist, often regarded as good species, of which the most distinctive include: (b) var. ***fuscoruber*** with reddish cap and white network, (c) ***B. betulicola*** with greyish cap and growing with *Betula*, (d) ***B. persoonii*** with entirely white fruitbody (also called *B. edulis* var. *albus*) and (e) ***B. venturii*** with yellow cap (also called *B. edulis* forma *citrinus*). Recent molecular work indicates that these and other forms are hardly, if at all different from *B. edulis* although some appear to have consistent microscopic differences (e.g. *B. betulicola* has swollen end cells of the cap hyphae). Further molecular work is needed to ascertain if any of these forms can be maintained at specific rank.

A recently described species, ***B. pinetorum*** M. Korhonen (see Korhonen *et al.* 2009) resembles *B. edulis* but differs in its greyish brown cap with distinctively wrinkled margin and flask-shaped caulocystidia (cylindrical in *B. edulis*) and is found in dry, sandy *Pinus* heathland in Scandinavia, Finland and Estonia. It might well occur in Scotland and should be looked out for.

Boletus aereus Bull.:Fr.
[aereus = copper-coloured]
Cap 5–20 cm, dark, blackish brown when young, rich reddish brown to bay or umber when mature, sometimes mottled or faded cream in parts, surface dry and frequently minutely cracked and tessellated, often wrinkled. **Tubes** and **pores** white then soon cream to yellowish with age. **Stem** robust, cylindric-clavate, pale brown with a darker fine brown network at the apex. **Flesh** cream, to pale vinaceous in the cap, unchanging when cut. **Spores** olive-brown, subfusiform, 13.5–15.5 x 4.0–5.5 μm (smallest in this section).

Distinctly southern in Britain (although there is a Scottish record!), in warm, broadleaved woods, especially *Quercus* and *Fagus*, uncommon to locally frequent, e.g. common in parts of the New Forest, Hampshire.

Widely considered as one of the best edible species, even superior to *B. edulis*, the brown reticulum and dark brown, matt, cracking cap surface are characteristic. Ammonia and KOH turn the flesh pale rust-red, contrasting with the similar but paler *B. reticulatus* which has negative reactions to both chemicals.

Boletus reticulatus Schaeff.
[reticulatus = with a network, referring to the cap]
Syn. *B. aestivalis*
Cap 5-25 cm, slightly viscid when wet, soon dry and often matt, fawn-brown, ochre-brown to slightly greyish or reddish brown, usually minutely wrinkled and cracked-tessellated. **Tubes** and **pores** cream then yellowish when mature, finally greenish when old. **Stem** fleshy, cylindric-clavate, pale brown, ochre-brown, with a concolorous mesh or network over most of its length. **Flesh** whitish, unchanging when cut. **Spores** olive-brown, subfusiform, 13–16 x 4.5–5.5 μm, the smallest of this group of closely related species. **Cap cuticle hyphae** pale brownish or brownish granulose, easily breaking apart when prepared.

Widespread but uncommon in Britain, commonest in the south, although large collections have been made in Scotland also; with broadleaved trees, especially *Quercus*, *Fagus* and *Castanea*.

A good edible species widely collected for food. Often confused with *B. edulis* but that species has a smoother, more viscid cap and usually has a vinaceous flush immediately below the cap cuticle. There is some evidence that there may be a second species in this complex, differing in its darker brown cap when young and paler reticulum. It is also associated with *Quercus* as well as *Tilia*, *Corylus* and possibly *Betula*. The cap cuticle consists of a mixture of hyaline and dark brown hyphae and its spores are larger, 15.5–18 x 4.5–6 μm and more pointed at the apex.

Boletus pinophilus Pilát & Dermek
[pinophilus = pine loving]
Syn. *B. pinicola*
Cap 5–20 cm, rounded, smooth or more usually wrinkled, vinaceous-brown, reddish brown to liver-brown, viscid when wet. **Tubes** and **pores** whitish then yellowish to slightly rust-brown when old, pale bluish green then rust brown when bruised. **Stem** cylindric-clavate, concolorous with but paler than the cap and with a prominent paler network. **Flesh** white becoming pale vinaceous below the cap cuticle and stem cortex, unchanging when cut. **Spores** olive-brown, fusiform, 17–20 x 4.5–6 μm, more pointed at the ends than in *B. edulis*. Ammonia on stem = orange.

Exclusively with *Pinus sylvestris*, common in Scotland but also found occasionally in established pine plantations in southern England. A good edible species, it is usually easily distinguished from *B. edulis* by the vinaceous or liver tones of its cap and stem. It also differs from other species in the group by its swollen, inflated and encrusted end-cells of the cap cuticle hyphae.

Boletus

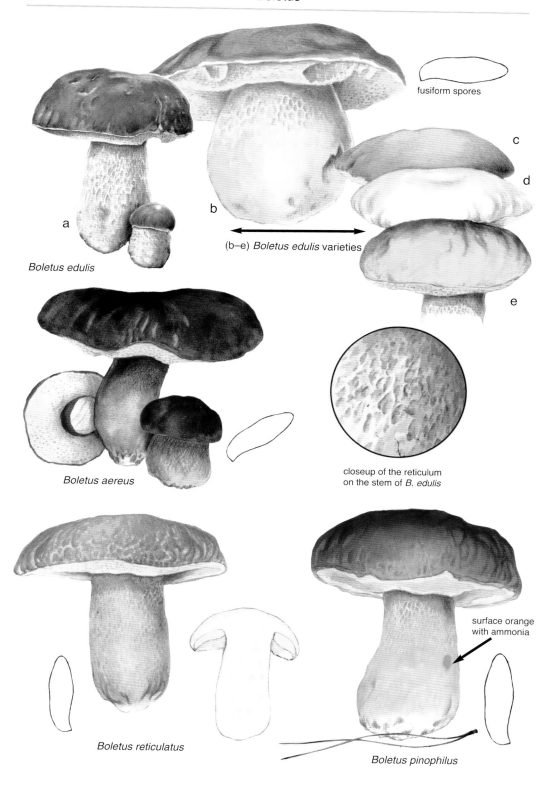

Genus *Caloboletus*

Caloboletus calopus (Pers.) Vizzini
[calopus = beautiful foot]
Syn. *Boletus calopus*
Cap 5–15 cm, broadly rounded, margin rather inrolled, whitish then soon greyish buff, pale ochraceous-brown, sometimes slightly olivaceous, surface dry, felty, matt then smooth, often cracking minutely in dry weather. **Tubes** and **pores** bright yellow, bruising blue-green to olive-grey when injured. **Stem** cylindric-clavate, varying from squat to rather tall, yellow above shading to carmine red below, with a fine white, raised reticulum above, becoming redder below. **Flesh** whitish yellow, flushing pale blue when cut. **Odour** faint, fruity or sour. **Taste** extremely bitter. **Spores** olive-brown, subfusiform, 12–16 x 4.5–5.5 μm.

Widespread and frequent, commoner in the northern counties and Scotland; in mixed woods.

The bright red colours in the stem, combined with the positive Melzer's iodine reaction easily separates *C. calopus* from the rather similar *C. radicans* which lacks red tones and has a negative iodine reaction. *C. calopus* is the type of this newly erected genus.

Caloboletus radicans (Pers.) Vizzini
[radicans = rooting]
Syn. *Boletus albidus, B. pachypus?*
Cap 5-20 cm, rounded then flattening, ivory white when young, greyish ochre, beige to clay-buff with age, minutely velvety when young, smoother and often cracking when old. **Tubes** and **pores** bright yellow, olivaceous when old, bruising blue. **Stem** robust, clavate, pale yellow, spotted ochraceous at base, sometimes with a reddish zone at the base; surface with a fine whitish or concolorous reticulum. **Flesh** white to pale lemon-yellow, flushing greyish blue to pale blue when cut. **Odour** pleasant. **Taste** extremely bitter and unpleasant. **Spores** olive-brown, subfusiform, 12–16 x 4.5–6 μm.

Widespread and common especially in southern England, with *Quercus* and *Fagus*, often in large numbers.

Caloboletus kluzakii ** (Šutara & Špinar) Vizzini
[kluzakii = after the mycologist Zdeněk Kluzák]
Syn. *Boletus fallax* Kluzák (illegitimate, later homonym of *B. fallax* Corner)
Cap 5–20 cm, rounded then flat-convex, dry, velvety, then smooth, whitish with pinkish tint, later almost entirely pink, conspicuously red to reddish vinaceous where bruised or rubbed. **Tubes** and **pores** lemon-yellow, later with olivaceous tint, blueing when bruised. **Stem** cylindrical or clavate, often rooting, lemon-yellow to yellow, sometimes discolouring to yellowish with brownish spots, with a fine concolorous reticulum, at least at the extreme apex. **Flesh** lemon-yellow or yellowish, sometimes reddish or brownish in the stem base, flushing blue throughout when cut. **Odour** not distinctive. **Taste** very bitter. **Spores** olive-brown, subfusiform, 10–16 × 4.5–6.5 μm.

Extremely rare, in broadleaf or mixed forests, probably mycorrhizal with *Quercus* but also found near *Pinus*. So far known only from the Czech Republic.

Most closely resembling *C. radicans* but differing in the pinkish red layer beneath the outer cap cutis which is exposed by scratching and tends to colour the entire cap pinkish with age.

Genus *Neoboletus*

Neoboletus xanthopus ** (Klofac & A. Urb.) Klofac & A. Urb.
[xanthopus = yellow stem]
Syn? *Boletus erythropus var. discolor*
Cap 5–12 cm broad, mottled deep brown, light brown, ochraceous or with shades of yellow. **Stem** yellowish, rarely with orange or reddish tones at different places, nearly smooth to finely, inconspicuously floccose, floccules mostly concolorous with stem surface, later darkening; base of stem purplish. All parts weakly to strongly blue when bruised. **Pores** red or orange, yellower at the margin. Flesh rapidly turning blue when cut. **Spores** olive-brown, fusiform, (10−) 13.5−15 (−16) × (3.8−) 4−5 (−5.5) μm. **Melzer's iodine** negative.

Recently described this species may be the same as *Boletus erythropus* var. *discolor*. Its presence in Britain is uncertain owing to confusion with varieties of *N. praestigiator* and possible yet to be defined species.

Neoboletus praestigiator (R. Schulz) Svetasheva, Gelardi, Simonini & Vizzini
[praestigiator = conjurer or magician]
Syn. *Boletus erythropus ss. auct., B. luridiformis ss. auct.*
Cap 5–15 cm, very variable in colour, from almost blackish brown to the more usual reddish brown through to forms with entirely yellow caps (var. *discolor*); at first velvety-plush then finally smooth when old; bruising blackish blue. **Tubes** yellow. **Pores** from yellow-orange to deep blood-red, bruising blue-black. **Stem** cylindric-clavate, with orange-red to deep red minute floccules overlaying a yellow ground, bruising blue. **Odour** not distinct. **Taste** pleasant. **Flesh** yellow, flushing almost instantly deep blue throughout when cut, more rarely with a red flush in the stem base, remaining yellow in *Boletus erythropus* var. *immutatus* (not yet combined under *N. praestigiator*). **Spores** olive-brown, subfusiform, 12–15 x 4–6 μm. **Melzer's iodine** negative.

Widespread and common everywhere in mixed woodlands, parks and hedgerows, usually on acid soils. This appears to be the earliest valid name for this mushroom; *B. erythropus* of Fries being what we currently call *S. queletii* and *B. luridiformis* being a different, poorly defined species. The numerous distinctively coloured forms, often regarded as species, may just be varieties and are treated here as such but see the recently described *N. xanthopus* above.

Neoboletus junquilleus (Quél.) Gelardi *et al.*
[junquilleus = jonquil yellow]
Cap 5–12 cm, rounded, velvety when young, smoother with age, bright primrose to lemon-yellow, instantly bruising blue then black when touched, often with rust-brown discolourations. **Tubes** and **pores** bright yellow, blueing instantly when bruised. **Stem** cylindric-clavate, lemon yellow, surface with minute, concolorous floccules, all parts bruising blue-black when touched, often with rust-brown stains and spots. **Flesh** pale yellow blueing throughout when cut. **Odour** slightly acidic or rubber-like (of *Scleroderma*). **Taste** pleasant or acidic. **Spores** olive-brown, subfusiform, 12.5–15 x 4.5–6 μm. **Melzer's iodine** negative on flesh, blue-green on pores.

Very rare in Britain, but apparently widespread, usually with *Quercus*, *Fagus* or *Tilia*.

This is almost certainly a purely yellow form of *N. xanthopus* in which case the correct name might be *N. xanthopus* forma *junquilleus* although I don't believe that combination has yet been officially made.

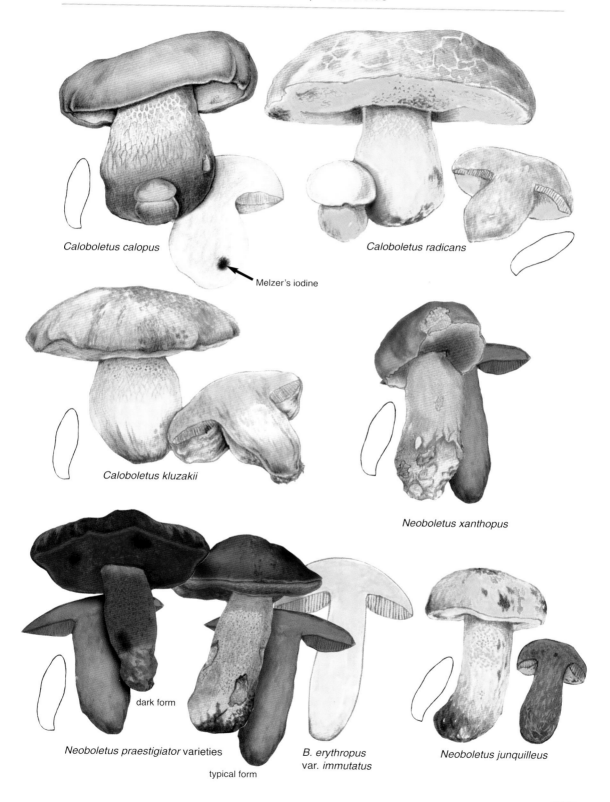

Genus *Suillellus*

Suillellus dupainii* (Boud.) Blanco-Dios
[dupainii = after Victor Dupain, French pharmacist]
Cap 5–12 cm, rounded to flattened, smooth and viscid, shiny when fresh, blood-red to scarlet, yellower at the margin, bruising blue-black. **Tubes** yellowish. **Pores** deep blood-red, scarlet, yellow-orange at the extreme margin of the cap, bruising blue-black. **Stem** cylindric-clavate, yellow overlaid with minute scarlet floccules, often forming a slight reticulum above. **Flesh** yellow, bruising blue. **Odour** fruity, aromatic. **Taste** pleasant. **Spores** olive-brown, subfusiform, 10–16 x 4–6 μm.

Rare, not authentically British, grows with *Quercus* and *Castanea* on calcareous soils in southern and central Europe.

The blood-red, viscid cap makes this species almost unmistakable. It has been reported - but without any voucher material - from southern England.

Suillellus queletii (Schulzer) Vizzini *et al.*
[queletii = after Lucien Quélet, mycologist]
Syn. *Boletus queletii, B. erythropus* ss. Fries
Cap 5–15 cm, rounded to flattened, velvety-plush when young, more polished with age, very variable in colour from ochre-yellow to orange-brown or even deep brick-red (var. *lateritius*) or garnet red (var. *rubicundus*), bruising blue-black. **Tubes** yellowish. **Pores** apricot-orange, bruising blue. **Stem** clavate with a pointed, radicant base, yellow with fine reddish orange floccules over the lower half, shading to deep beetroot-red to blackish red at the base. **Flesh** pale yellow, beetroot-red in the stem base, flushing quickly blue when cut. **Odour** pleasant, taste slightly acidic. **Spores** olive-brown, subfusiform, 11–14 x 5–6.5 (7) μm. **Melzer's iodine** positive, blue-black.

Uncommon but widely distributed in southern England, associated with broadleaved trees, especially *Quercus*, *Tilia* and *Fagus* on calcareous soils.

The beetroot-red flesh in the rather pointed stem base, combined with the orange pore mouths and broader spores than other members of this group make *S. queletii* fairly easy to recognise, despite the variable cap colours.

Suillellus luridus (Schaeff.) Murrill
[luridus = lurid in colour]
Syn. *Boletus luridus*
Cap 5–16 cm, colour very variable from peach to yellow ochre, ochre-brown, reddish brown with olive tones, or even entirely yellow (forma ***primulicolor***) or garnet-red (var. ***rubriceps***). **Tubes** yellow. **Pores** orange to deep red, bruising dark blue; if the pores are pulled away from the cap a layer of red flesh is exposed. **Stem** has a fine, red or orange raised network over most of the surface, bruising blue. **Flesh** blue in cap and upper half of stem, beetroot-red in the stem base. **Melzer's iodine** on the flesh = dark blue. **Spores** olive-brown, subfusiform, 12–15 x 5–6.5 μm.

Widespread in open, mixed broadleaf woods and parks, often on calcareous soils.

Remarkably variable in colour and even in the discolouration of the flesh. In the var. ***erythroteron*** (not yet British, very rare) the cut flesh is reddish, staining purple throughout and the network on the stem is a brighter red. It is possible that this wide variation indicates a species complex but a detailed molecular study is required to ascertain the exact relationship between these varieties. Compare with the very similar *S. mendax* below.

Suillellus mendax (Simonini & Vizzini) Vizzini *et al.*
[mendax = deceiving]
Syn. *Boletus mendax*
Cap 5–15 cm, pale brick-red, pinkish red, velvety then polished with age, cracking when dry. **Tubes** yellowish. **Pores** orange-red bruising deep blue; if the pores are pulled away from the cap a layer of yellow flesh is exposed. **Stem** cylindric-clavate, pale yellow with pink overtones, with a red reticulum confined to the extreme apex, the remainder with minute red floccules overall. **Flesh** yellowish, flushing blue in cap and upper stem, beetroot-red in the stem base. **Odour** pleasant. **Taste** slightly acidic. **Spores** olive-brown, subfusiform, 12–14 x 5.5–6.5 μm. **Melzer's iodine** positive on the flesh.

Rare in Britain with *Quercus* and *Fagus*, although probably overlooked because of its similarity to *S. luridus*. Distinguished from that species by the partially punctate ras well as reticulate stem and narrower spores.

Suillellus poikilochromus* (Pöder et al.) Blanco-Dios
[poikilochromus = of varying colours]
Syn. *Boletus poikilochromus*
Cap 5–15 cm, convex with rather inrolled margin, then flattening, smooth and viscid when wet, polished or felty when dry, very variable in colour, from dark chestnut brown to ochre-brown, ochre-yellow, sometimes with olivaceous tones or mottled with vinaceous-brown, bruising deep blue-black. **Tubes** and **pores** yellow, pores often developing with rust-brown patches, bruising intensely blue. **Stem** cylindric-clavate, pale yellow, more rust-brown below, with a reddish brown reticulum overall, bruising intensely blue-black. **Flesh** pale yellow flushing deep blue throughout. **Odour** strong and persistent of fermenting fruit. **Taste** mild. **Spores** olive-brown, subfusiform, 11–13.5 x 4.5.–5.5 μm.

Not yet British, very uncommon; growing with *Quercus* and *Pinus*; known in the Mediterranean area.

Genus *Rubroboletus*

Rubroboletus satanas (Lenz) Kuan Zhao & Zhu L. Yang
[satanas = pertaining to the Devil]
Syn. *Boletus satanas*
Cap 6–25 (-30) cm, subglobose then expanding to almost flat, dry and slightly felty, somewhat viscid when wet, white to greyish white, flushed olivaceous-grey when aging, more rarely with a slight pinkish tone at the margin, frequently cracking with age and exposure. **Tubes** yellow. **Pores** from orange-red to deep blood-red, often yellower at the margin, bruising blue. **Stem** robust, fleshy, often very swollen-clavate, pale yellow above flushed redder below to carmine-red overall; with a fine concolorous network overall or especially at the apex. **Flesh** is whitish yellow, flushed pale blue when cut, especially in the cap and upper stem. **Odour** mild at first but soon becoming repulsive of rotting garlic with age. **Taste** mild but widely regarded as quite toxic. **Spores** olive-brown, subfusiform, 11–15 x 5–7 μm.

Uncommon to rare in Britain, found in southern counties with *Fagus*, *Quercus*, *Carpinus*, etc and apparently also with *Helianthemum*, always on calcareous soils.

A large and infamous species because of its striking colours and toxicity. Although not usually fatal it can cause severe gastric upsets if eaten. A rare form is known (forma ***crataegi***) which entirely lacks the red pigments, so that its stem and pores are yellow, at which time it can look very like *Caloboletus radicans*. From other species in this group the repulsive odour of *R. satanas* that develops as it ages is a good distinguishing character.

Suillellus, Rubroboletus

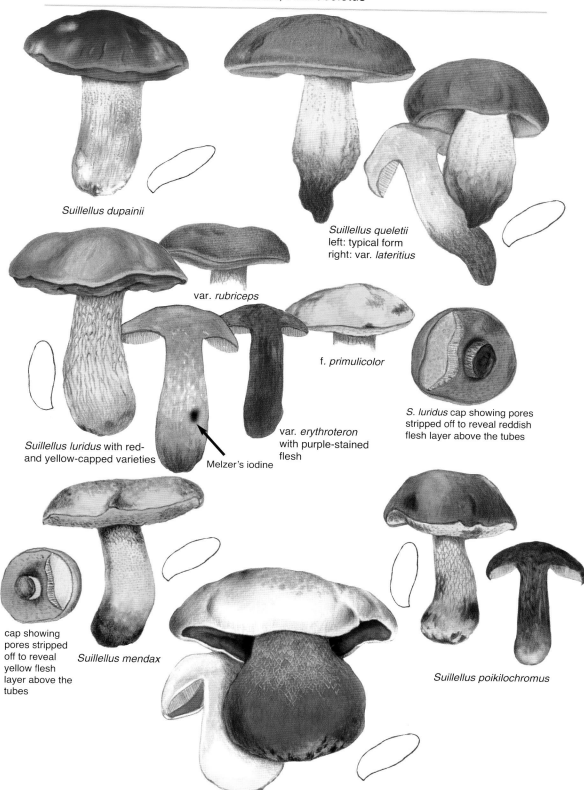

Suillellus dupainii

Suillellus queletii
left: typical form
right: var. *lateritius*

var. *rubriceps*

f. *primulicolor*

S. luridus cap showing pores stripped off to reveal reddish flesh layer above the tubes

Suillellus luridus with red- and yellow-capped varieties

var. *erythroteron* with purple-stained flesh

Melzer's iodine

cap showing pores stripped off to reveal yellow flesh layer above the tubes

Suillellus mendax

Suillellus poikilochromus

Rubroboletus satanas

Rubroboletus legaliae Pilát ex Mikšík
[legaliae = after Mme M. Le Gal, French mycologist]
Syn. *Boletus satanoides*
Cap 8–20 cm, smooth to finely felty, very variable in colour, ivory white to ochre then soon flushed pinkish purple, finally pinkish brown, bruising blue or blue-black. **Tubes** yellowish. **Pores** from yellow-orange to deep red bruising blue. **Stem** robust, cylindric-clavate, pale, whitish or whitish yellow, with fine red network overall, bruising blue. **Flesh** whitish then soon pale blue throughout when cut. **Odour** pleasant, said to be of chicory. **Taste** mild. **Spores** olive-brown, subfusiform, 11–13 x 4.5–5.5 μm. **Melzer's Iodine** on stem flesh amyloid.

Mainly in southern England in open broadleaf woods and parks, or along road edges, where it can occur in some numbers, usually with *Quercus*, on neutral to acid soils. Recently recorded from Wales.

Rubroboletus satanas differs in its persistently white or greyish white cap without pinkish purple tones, often very swollen, rounded stem and flesh developing a repulsive odour of rotten garlic.

Rubroboletus pulchrotinctus ** (Alessio) Zhao & Yang
[pulchrotinctus = beautiful colours]
Syn. *Boletus pulchrotinctus*
Cap 8–20 cm, dry, felty then finally smooth, pinkish white to lilac-pink, darker at the margin, greyer with age. **Tubes** and **pores** usually yellow, occasionally the pores flush orange, blueing when bruised. **Stem** clavate-swollen, pale yellow with a concolorous network, flushing pinkish red with age. **Flesh** pale yellow, pinkish below the cap cuticle, blueing faintly in the cap. **Odour** pleasant. **Taste** mild. **Spores** olive-brown, subfusiform, 12–15 × 4.5–6 μm.

Not yet British, an uncommon species found with mixed broadleaved trees on calcareous soils in the Mediterranean.

Rubroboletus lupinus ** (Fr.) Costanzo *et al.*
[lupinus = from the Lupin flower, genus *Lupinus*]
Syn. *Boletus lupinus*
Cap 5–15 cm, rounded to broadly expanded, dry and felty, pale whitish buff to slightly olivaceous-buff, flushing increasingly pink to carmine as the cap matures. **Tubes** yellow. **Pores** yellow when very young then soon bright carmine red to blood red, bruising deep blue. **Stem** cylindric-clavate, bright yellow to orange-yellow, with a fine network only at the extreme apex of the stem, minutely floccose over the remainder of the stem, bruising blue. **Flesh** yellowish bruising pale blue throughout. **Odour** distinct, rubber-like, acrid, of *Scleroderma* (Common Earthball). **Taste** rather acidic. **Spores** olive-brown, subfusiform, 11–14 x 4.5–6 μm.

This species has not yet been found in Britain, preferring the rather hotter and drier regions of the Mediterranean and up through Central Europe, growing in association with *Quercus* on calcareous soils. It might possibly occur in some of our coastal counties or off-shore islands.

As with other species in this group the cap colour can change enormously from young to old, specimens being found which are entirely olivaceous buff with just a flush of pink to those which are entirely pinkish red.

Rubroboletus rhodoxanthus (Krombh.) Zhao & Yang
[rhodoxanthus = rose and yellow]
Syn. *Boletus rhodoxanthus*
Cap 6–20 cm, viscid when wet, finely felty-tomentose when dry, broadly rounded, white when young flushing greyish ochre, often with pink showing through at the margin or where bruised. **Tubes** yellow. **Pores** blood-red, carmine, blueing where bruised. **Stem** cylindric-clavate, with a yellow ground overlaid with a deep red network, darker red below, yellower at the apex, bruising blue. **Flesh** bright, golden yellow flushed pale blue in cap and upper stem. **Odour** pleasant, slightly fruity. **Taste** mild. **Spores** olive-brown, subfusiform, 11–18 x 5–7 μm.

Rare, as yet only known in Northern Ireland, commoner on the Continent, with *Quercus* and *Fagus* on calcareous soils. It is surprising that this species is not more common in Britain, at least in the southern counties, as it is recorded in most other European countries. Like other species in the group the white cap changes colour with age but the golden flesh, blueing only in the cap and upper stem is a good, distinctive character. Recently placed in the new genus ***Rubroboletus***.

The very similar ***R. rubrosanguineus*** is a conifer associate, not known in Britain, which differs in its cap becoming entirely red with age and flesh blueing throughout.

Genus *Imperator*

Imperator rhodopurpureus (Smotl.) Assyov *et al.*
[rhodopurpureus = rose and purple]
Syn. *Boletus rhodopurpureus*
Cap 5–15(-20) cm, broadly rounded, slightly viscid when wet, dry and felty with age, often rather wrinkled or puckered, very variable in colour, faded rose-red, flushing vinaceous but often with yellower areas or even olivaceous with age, bruising deep blue-black. **Tubes** bright yellow. **Pores** from golden yellow or more usually orange to blood-red, instantly blue-black when touched. **Stem** robust, clavate, bright yellow ground overlaid with a red reticulum, often vinaceous red at the base. **Flesh** firm, yellow, flushing instantly blue-black when cut, with a deep beetroot-red patch at the extreme base of the stem. **Odour** pleasant or slightly acidic. **Taste** mild. **Spores** olive-brown, subfusiform, 11–15 x 5–6 μm.

Rare overall in Britain but locally common in some warmer southern counties of England such as Hampshire. Usually with *Quercus* on acid to neutral soils.

Numerous colour forms exist and some have been given formal names: forma ***polypurpureus*** Smotl. is deep pinkish crimson when young, var. ***gallicus*** (Romagn.) Redeuilh is golden-yellow mottled with reddish and forma ***xanthopurpureus*** Smotl. is lemon-yellow flushing vinaceous.

Imperator luteocupreus ** (Bertéa & Estades) Assyov *et al.*
[luteocupreus = yellow and copper]
Syn. *Boletus luteocupreus*
Cap 5–15 cm, rounded and slowly expanded, smooth and slightly viscid when wet, dry and felty with age, often wrinkled-puckered, bright chrome yellow flushed with orange or copper-red as it ages, strongly blue-black when bruised. **Tubes** yellow. **Pores** bright scarlet bruising blue-black. **Stem** clavate, robust, bright golden yellow overlaid by a dense scarlet reticulum, bruising blue-black. **Flesh** yellow flushing instantly blue-black when cut, slowly fading to reddish. **Odour** rubber-like or like *Scleroderma*. **Taste** acidic. **Spores** olive-brown, subfusiform, 12–16(-18) x 5–6 μm.

Not yet British but probably here, confused with yellow forms of *B. rhodopurpureus*, but has an inamyloid reaction in the stem flesh to Melzer's iodine solution.

Rubroboletus, Imperator

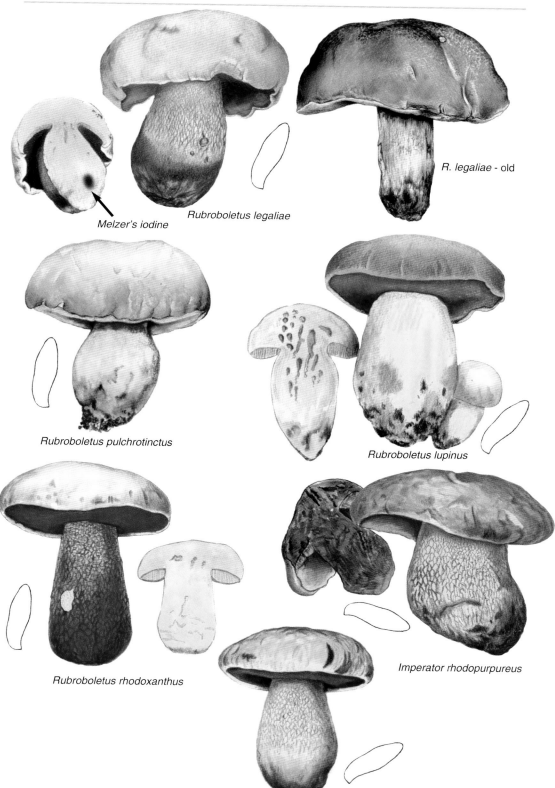

Melzer's iodine

Rubroboletus legaliae

R. legaliae - old

Rubroboletus pulchrotinctus

Rubroboletus lupinus

Rubroboletus rhodoxanthus

Imperator rhodopurpureus

Imperator luteocupreus

Imperator torosus ****** (Fr.) Assyov *et al.*
[torosus = cylindrical with bulges]
Cap 6–20 cm, rounded-globular then expanded, slightly viscid when wet then dry and felted to polished; very variable in colour from yellow-ochre, yellowish grey to olivaceous, frequently with rust-brown areas, sometimes with a violaceous-red flush, finally dull olivaceous, bruising blue-black at the slightest touch. **Tubes** and **pores** golden yellow, becoming reddish or rust with maturity, bruising intensely blue. **Stem** robust, clavate, yellow, more reddish at the base, with a fine more or less concolorous network overall, bruising blue-black when touched. **Flesh** very firm, dense, golden yellow flushing deep blue when cut. **Odour** characteristic, medicinal or fermenting. **Taste** mild. **Spores** olive-brown, subfusiform, 12.5–16 x 5–7 μm.

Uncommon to rare, with mixed broadleaved trees, on calcareous soils. Without any authentic records in the UK. Recently made the type of this new genus.

Boletus xanthocyaneus (Ramain) Romagn.
[xanthocyaneus = yellow and blue]
As above, but entirely yellow and with fruity smell. This species, although recorded in Britain, seems likely to be a colour form of *I. rhodopurpureus* or perhaps *I. torosus*.

Genus *Butyriboletus*

Butyriboletus appendiculatus (Schaeff.) D. Arora & J.L. Frank
[appendiculatus = with overhanging margin]
Cap 6–16 (-20) cm, convex to expanded, viscid when wet, the dry and slightly fibrillose-felty, cinnamon-brown, ochre-brown, tawny to chestnut; cap margin slightly overhanging. **Tubes** and **pores** bright yellow, pores bruising blue. **Stem** cylindric-clavate, usually with a pointed base, pale to bright yellow overall with a reddish brown flush at the base, with a fine yellow network especially on the upper half. **Flesh** yellowish white, flushing pale blue in upper half, slightly reddish brown in base of stem. **Odour** slightly pungent of *Scleroderma*. **Taste** mild. **KOH** on cuticle and flesh = reddish brown. **Spores** olive-brown, subfusiform, 11–14 x 4.5–5.5 μm.

Widespread and fairly common in summer and early autumn, especially in the southern counties, associated with *Quercus* and *Fagus*, on acid soils.

The blueing of the flesh may be quite faint but is a useful character to distinguish it from the similar *B. subappendiculatus* combined with its different spore quotient (see below).

Butyriboletus subappendiculatus (Dermek *et al.*) D. Arora & J.L. Frank
[subappendiculatus = less than appendiculatus, referring to the bolete of that name]
Cap 6–12 cm, rounded to slightly flattened, viscid when wet, felty-fibrillose when dry, ochre-brown, tawny, yellow-brown. **Tubes** and **pores** bright yellow, pores bruising blue. **Stem** cylindric-clavate, usually with a rounded base, pale to bright yellow, browner below, with a concolorous network overall. **Flesh** yellowish white, hardly if at all flushing blue above the tubes. **Odour** earthy to slightly fruity. **Taste** mild. **KOH** on cuticle and flesh = bright red. **Spores** olive-brown, subfusiform, 12–17 x 4–5 μm.

Rare, recorded from Scotland with conifers, especially *Abies*.

Formerly much confused with *B. appendiculatus* from which it differs in slightly paler cap, tendency for the stem to be rounded at the base rather than pointed, the longer, narrower spores, ± unchanging flesh and stronger KOH reaction.

Butyriboletus fechtneri (Velen.) D. Arora & J.L. Frank
[fechtneri = after František Fechtner, 1883–1967, mycologist]
Cap 5–15 cm, convex to expanded, smooth to slightly felty, silvery-grey, ochre-grey, pale beige, sometimes with a pinkish tint. **Tubes** and **pores** bright yellow, pores bruising blue. **Stem** cylindric-clavate, bright yellow suffused with a reddish pink zone above the base, with a concolorous fine network overall, bruising blue when bruised. **Flesh** whitish yellow flushing pale blue in the cap and upper stem, pale reddish brown in base. **Odour** characteristic, slightly fermented or like acrylic paint. **Taste** mild. **Spores** olive-brown, subfusiform, 11–14(-17) x (4.5-)5–6(-6.5) μm.

Rare, mainly in southern England but recorded from Wales and reported from Scotland, associated with *Quercus*, *Fagus* and *Castanea* on calcareous soils.

B. appendiculatus has a darker, more reddish brown cap and lacks the reddish zone around the stem centre, *B. fuscoroseus* has a pinkish red tone to the cap and stem base.

Butyriboletus fuscoroseus (Smotl.) Vizzini & Gelardi
[fuscoroseus = brown and rose]
Syn. *B. pseudoregius*, *B. speciosus* ss. Singer non orig.
Cap 5–15 cm, rose-pink, pinkish brown, pinkish beige, darker where rubbed, felty-fibrillose. **Tubes** and **pores** bright yellow, bruising blue. **Stem** cylindric-clavate, bright yellow with a pinkish red zone above the stem base, with a concolorous network on the upper half. **Flesh** whitish yellow, flushed blue in the cap and upper stem, pinkish brown in the base. **Odour** in young fruit bodies not distinctive, later medicinal, of tempera paints, or resembling smoked meat, and of chicory when dried. **Taste** mild. **Spores** olive-brown, subfusiform, 11–15 x 4.5–5.5 μm.

Rare, known from a very few sites in southern England, associated with *Quercus* (also recorded with *Fagus* and *Castanea* on the Continent) on warm sites.

Formerly called *B. speciosus*, a species described from North America but when it was realised that it was a different species the name *B. pseudoregius* was adopted. However, the earlier name of *B. fuscoroseus* should take precedence. Much confused with *B. regius* which has almost unchanging flesh, brighter cap colours and narrower spores.

Butyriboletus regius ****** (Krombh.) D. Arora & J.L. Frank
[regius = royal]
Cap 5–15 cm, convex to expanded, dry and felty to fibrillose-scaly, beautiful rose-red, crimson, sometimes rather mottled. **Tubes** and **pores** bright yellow, unchanging when bruised. **Stem** cylindric-clavate, bright yellow, often with some pinkish red marks at the base, with a fine concolorous network on the upper half. **Flesh** pale to bright yellow, unchanging when cut to very slightly blue just above the tubes. **Odour** in young fruit bodies not distinctive, later medicinal, of tempera paints, or resembling smoked meat, and of chicory when dried. **Taste** mild. **Spores** olive-brown, subfusiform, 11–14 x 3.5–4.5 μm.

Very rare, reported in error from a single location in southern England; probably not yet British. It prefers warm sites with *Fagus*, *Quercus* and *Castanea*.

Some records of this species are *B. fuscoroseus* with pinker than usual caps or where the flesh has not stained as blue as normal.

Imperator, Butyriboletus

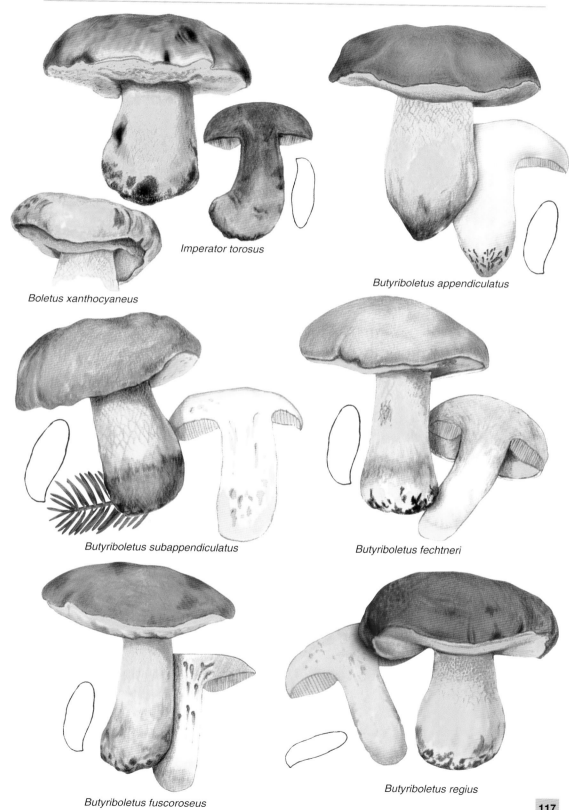

Genus *Lanmaoa*

Lanmaoa fragrans (Vittad.) Vizzini *et al.*
[fragrans = fragrant]
Cap 60–15 cm, rounded, margin often lobed or irregular, bay brown, reddish brown, chestnut or maroon, velvety-plush when young, more polished with age. **Tubes** and **pores** bright yellow, lightly olivaceous with age, often stained rust brown on the pore mouths, bruising blue. **Stem** robust, fleshy, cylindric, usually pointed at base and forming clumps with adjoining fruitbodies, yellow above, reddish brown below, irregularly streaked or spotted. **Flesh** pale yellowish white, flushing sky blue when cut. **Odour** fragrant of mixed fruit or chicory. **Taste** slightly bitter or unpleasant. **Spores** olive-brown, subfusiform, 10–15 x 4–5 μm.

A Mediterranean species, often growing in large clumps; very rare in Britain with just a handful of records, associated with *Quercus*. Formerly *Boletus fragrans*, it was recently placed in the new genus **Lanmaoa**.

Genus *Baorangia*

Baorangia emilei ** (Barbier) Vizzini *et al.*
[emilei = after the Aemilii, Roman patricians]
Syn. *Boletus spretus*, *B. aemilii*
Cap 5–15 cm, rounded then slowly expanding, with a slightly inrolled margin, deep crimson red, carmine, streaked or spotted blackish when touched, surface felty to finely scaly then more polished. **Tubes** and **pores** yellow, decurrent and rather short, bruising blue-green. **Stem** fleshy, often tapered or pointed, rather short, flecked with deep crimson on a yellowish ground, paler yellow at the apex. **Flesh** whitish yellow flushing blue-green then after several hours dark red. **Odour** earthy. **Taste** rather acidic. **Spores** olive-brown, subfusiform, 9–13 x 3.5–4.5 μm.

Growing with *Quercus ilex*, *Q. suber*, *Castanea* and *Arbutus* on acid soils in the Mediterranean, uncommon to rare, not yet British.

Genus *Imleria*

Imleria badia (Fr.) Vizzini
[badia = bay-brown]
Syn. *Boletus badius*
Cap 5–12 cm, rounded to plane, rich bay-brown to chestnut or maroon, often more yellow-brown, orange-brown in collections from broadleaved woods, surface plush-velvety when young then polished and often viscid when wet. **Tubes** and **pores** pale yellow, bruising greyish blue. **Stem** cylindric or slightly clavate to pointed at the base, pale cream-yellow flushed reddish brown to yellowish brown, often irregularly streaked with darker brown, smooth. **Flesh** whitish, slightly reddish brown immediately below the cuticle, flush pale blue when cut, especially in the cap. **Odour** pleasant, fruity. **Taste** mild. **Spores** olive-brown, subfusiform, 12.5–16 x 4–5 μm.

Widespread and common everywhere in both coniferous and broadleaved woods, especially *Fagus*. The form in coniferous woods tends to be darker and more strongly blueing than the broadleaf form which is often much more orange-brown.

The closely related *I. heteroderma* differs in its more velvety, cinnamon-brown cap and stem and more especially in its cuticular hyphae which have plaques staining in Congo Red dye (congophilous). It has not yet been found here but might occur in the southern counties.

Imleria heteroderma ** (J. Blum) T. Rödig
[heteroderma = mixed cuticle]
Cap 5–10 cm, broadly convex to plane with slightly inrolled margin, ochre-brown, cinnamon-brown, fulvous, velvety-plush when fresh, polished with age. **Tubes** and **pores** lemon yellow flushed greyish with age, not bruising blue. **Stem** fleshy, cylindric-clavate, concolorous but paler than the cap, smooth to minutely fibrillose-punctate, usually pointed at the base. **Flesh** whitish to pale yellow, not blueing when cut. **Odour** earthy. **Taste** pleasant. **Spores** olive-brown, subfusiform, 12-15 x 4–4.5(5) μm.

Under *Cedrus* in the Mediterranean, not yet British but might occur in the southern counties.

Described in 1969 this species has remained almost forgotten until recently redescribed. It is close to *Boletus badius* but differs in both colour and structure of the cap cuticle hyphae. In *I. heteroderma* the end cells are covered in small encrustations or plaques which stain in Congo Red dye.

Genus *Hemileccinum*

Hemileccinum depilatum (Redeuilh) Šutara
[depilatum = plucked or hairless]
Syn. *Boletus depilatus*, *Leccinum depilatum*
Cap 5–15 cm, rounded then flattening, umber brown to nut brown fading to beige brown, darker at centre, surface distinctly lumpy and dented as if hammered or planished. **Tubes** and **pores** yellow to greenish yellow with age, not bruising blue. **Stem** clavate-pointed, often curved at the base, pale yellow frequently flecked with rust at the base, surface with fine woolly flecks overall, often deeply buried in the soil. **Flesh** pale whitish yellow, more or less unchanging when cut. **Odour** distinct of iodoform especially in the cut flesh of the stem base. **Taste** mild. **Spores** olive-brown, subfusiform, 12–15 x 5–6 μm.

Associated with *Carpinus* and *Ostrya carpinifolia* on calcareous soils, apparently very rare with only one record in southern England.

Microscopically the species is well separated from *H. impolitum* by the structure of the cap cuticle. In *H. depilatum* the cuticle consists of a palisade of swollen, inflated cells, while in *H. impolitum* the cells form a simple cutis of slender, only slightly clavate cells.

Hemileccinum impolitum (Fr.) Šutara
[impolitum = unpolished, rough]
Syn. *Boletus impolitus*, *Leccinum impolitum*
Cap 5-15 (20) cm, broadly rounded, reddish brown, sienna, ochre-brown, sometimes with a grey-olive flush, with an overall greyish flush when young, paler ochre with age, surface velvety-felty when young, more fibrillose with age. **Tubes** and **pores** bright yellow, unchanging when bruised. **Stem** cylindric-clavate, pale to brighter yellow with some rust brown stains below, surface with fine, minute concolorous floccules. **Flesh** pale whitish yellow, slightly reddish below the cap cuticle, generally unchanging when cut. **Odour** distinct of iodoform in the cut flesh of the stem base, occasionally difficult to detect. **Taste** mild. **Spores** olive-brown, subfusiform, 9–16 x 4.5–6 μm.

Widespread and fairly common in southern England, with broadleaved trees, mainly *Quercus* on sandy or calcareous soils.

Despite the odd odour of iodoform it is usually regarded as an excellent edible species. If collections are made growing with *Carpinus* you may have the much rarer *H. depilatum*. That species will have a palisade of broad, inflated end cells unlike the more typical cutis of slender cells in *H. impolitum*.

Lanmaoa, Baorangia, Imleria, Hemileccinum

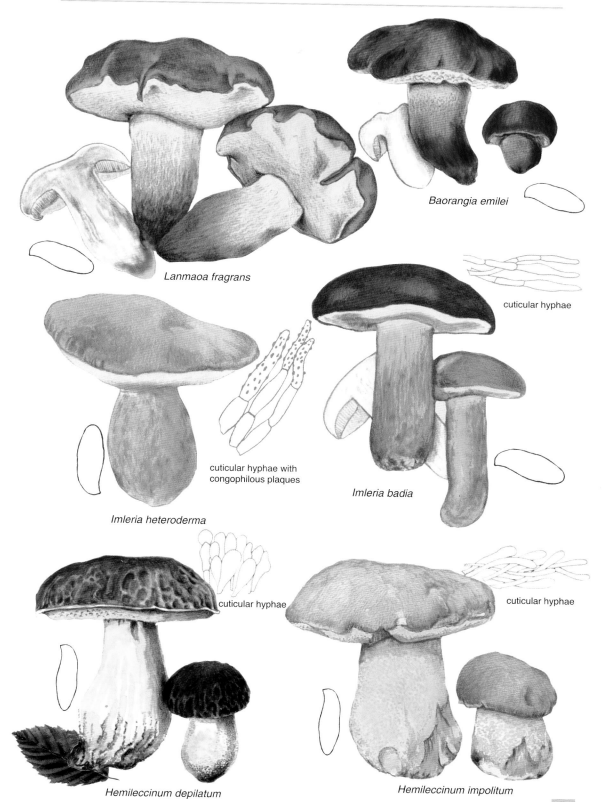

Genus *Cyanoboletus*

Cyanoboletus pulverulentus (Opat.) Gelardi *et al.*
[pulverulentus = covered with powder]
Syn. *Boletus pulverulentus*
Cap 4–8(10) cm, slightly sticky when wet, normally dry and felty to polished, reddish brown, ochre-brown, copper to vinaceous, bruising deep blue-black. **Tubes** and **pores** yellow, instantly blue-black to touch. **Stem** cylindric-tapered, yellow above, reddish below, sometimes entirely yellow, smooth to finely pruinose, bruising deep blue-black. **Flesh** pale yellow, almost instantly azure-blue then blue-black throughout when cut. **Odour** pleasant, not distinct. **Taste** mild. **Spores** olive-brown, subfusiform, 10–15 x 4–6 μm. **Cap cuticle** with heavily encrusted cells.

Uncommon to locally frequent, widespread across Britain, usually with broadleaved trees, especially *Quercus* and *Fagus* but also occasionally with conifers. No other British bolete bruises so intensely deep blue.

Genus *Aureoboletus*

Aureoboletus gentilis (Quél.) Pouzar
[gentilis = of the same genus or race]
Syn. *A. cramesinus*
Cap 3–5 cm, viscid when wet, glossy and smooth when dry, strawberry-pink to pinkish brown, pinkish ochre, with slightly darker radial streaks or fibrils. **Tubes** and **pores** brilliant golden yellow, unchanging when bruised. **Stem** slender, cylindric with a tapering base, bright yellow above, pinkish brown below, smooth and sticky. **Flesh** white, pink below the cap cuticle, unchanging when cut or slightly yellow above the tubes. **Odour** pleasant. **Taste** mild to slightly acidic. **Spores** ochraceous-buff, subfusiform, 11–15 x 4.5–5.5 μm. **Tube mouths** lined with large, clavate cystidia, filled with bright golden sap.

Uncommon but widespread, especially in southern England, with *Fagus*, *Quercus* and *Castanea*.

Some forms of this beautiful little species can be quite reddish, others more ochre with just a hint of pink, but the sticky cap, yellow unchanging pores and yellow-filled cystidia are distinctive characters.

Aureoboletus moravicus (Vaček) Klofac
[moravicus = after Moravia, Czech Republic]
Syn. *Boletus leonis, B. moravicus, B. tumidus, Xerocomus moravicus*
Cap 3–6 cm, convex-expanded, smooth and slightly sticky when wet, soon dry and felty-floccose, especially at the centre, tawny, ochre-yellow, ochre-brown, cinnamon brown. **Tubes** and **pores** yellow to yellow-ochre, unchanging when bruised. **Stem** tapering below, often slightly swollen at centre, smooth to slightly longitudinally ribbed, concolorous with or slightly paler than the cap. **Flesh** whitish to pale cream-beige, not blueing when cut. **Odour** pleasant, slightly fruity. **Taste** mild to slightly acidic. **Spores** ochraceous olive, ellipsoid, 10–13 x 5–6.5 μm.

Rare, mainly from southern England but there is a record from Scotland also, associated with *Quercus*.

This species has been the subject of much confusion, naming and re-naming, and moving from genus to genus. The majority of authors now accept that *Boletus leonis* Reid, a name commonly used for this species, is a synonym. For a long time this species has been placed in the genus *Xerocomus* but molecular studies do not support that placement. Following the recent monographic treatment of the genus *Aureoboletus* by Šutara it is placed in that genus.

Genus *Buchwaldoboletus*

Buchwaldoboletus lignicola (Kallenb.) Pilát
[lignicola = wood loving]
Syn. *Pulveroboletus lignicola*
Cap 3–6 cm, convex-expanded, margin inrolled, dry, woolly-floccose, breaking up into woolly patches or scales, bright tawny, ochre-brown, sienna. **Tubes** and **pores** yellow, decurrent, bruising faintly bluish. **Stem** cylindric-tapered, concolorous with or paler than the cap, smooth to fibrillose or minutely floccose. **Flesh** pale yellowish cream, unchanging when cut except immediately above the tubes where it turns faintly blue. **Odour** not distinct or faintly resinous. **Taste** mild to acidic. **Spores** olive-brown, subfusiform, 6.5–10 x 3–4.5 μm.

Rare but widespread in Britain, growing on or adjacent to conifer stumps or trees, including *Pinus*, *Larix*, *Picea*, etc. Often found growing next to the polypore *Phaeolus schweinitzii* and it has been supposed that the bolete might be a mycoparasite of the polypore. However that does not seem to be the case according to recent research.

Buchwaldoboletus sphaerocephalus (Barla) Watling & T.H. Li
[sphaerocephalus = rounded head]
Syn. *Buchwaldoboletus hemichrysus*
Very similar to the above but the **Cap** is a brighter, lemon-yellow and with its flesh bruising stronger blue and reddish. **Spores** olive-brown, subfusiform, 6–8.5 x 3–3.5 μm.

Very rare in Britain, growing on conifer sawdust, very rotten wood or stumps, often ± caespitose.

Genus *Pseudoboletus*

Pseudoboletus parasiticus (Bull.) Šutara
[parasiticus = parasitic on another organism]
Syn. *Xerocomus parasiticus*
Cap 2–6 cm, convex-expanded, felty-tomentose when dry, ochraceous-yellow, ochre-brown to slightly olivaceous. **Tubes** and **pores** yellow, decurrent, rather large and angular-irregular, pore mouths often flushing rust-red, not blueing when bruised. **Stem** tapered below, concolorous with the cap, surface smooth to irregularly cracked-scaly. **Flesh** whitish buff, unchanging when cut. **Odour** pleasant. **Taste** mild. **Spores** olive-brown, elongate-ellipsoid, 11–21 x 3.5–5 μm.

Uncommon to locally frequent, in some years quite common, widespread everywhere, always growing attached to *Scleroderma citrinum*, causing the fruitbody of the earthball to become hollow and partially collapsed. Often numerous bolete fruitbodies will grow from one ball, always starting at the junction with the soil. Interestingly the earthball still seems able to produce spores, although not in the usual quantity.

Related species of bolete occur on different earthballs in other parts of the world, principally in Asia.

Cyanoboletus, Aureoboletus, Buchwaldoboletus, Pseudoboletus

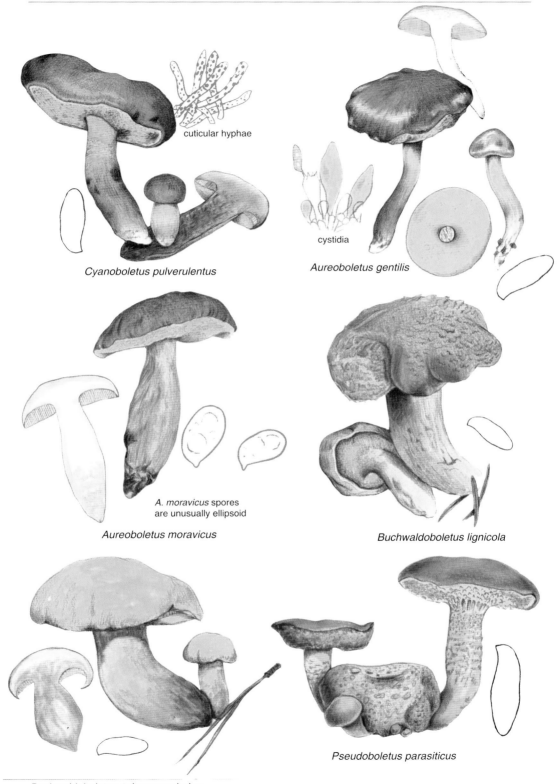

Genus *Chalciporus*

Chalciporus rubinus (W.G. Smith) Singer
[rubinus = ruby-red]
Syn. *Rubinoboletus rubinus*
Cap 2.5–8 cm, convex-expanded, slightly viscid when wet, felty-tomentose when dry, tawny brown, ochraceous-brown, often flushed with red especially at the margin. **Tubes** decurrent, ochre-buff, flushed carmine-red towards the pore mouths. **Pores** bright carmine-red, unchanging when bruised. **Stem** cylindric to tapered at base, bright carmine-red above with minute red punctae, more ochre below and often bright lemon-yellow at extreme base. **Flesh** white in cap centre, flushed pinkish towards the cap margin and down into the stem, bright lemon-yellow in the stem base. **Odour** pleasant. **Taste** more or less mild. **Spores** pale reddish brown, broadly ellipsoid, 6–8 x 4–5.5 µm.

Widespread in Britain from Yorkshire southwards, mostly rare although locally frequent in some counties of southern England, associated with *Quercus*.

Unique amongst British boletes by its combination of bright carmine tubes and broadly ellipsoid spores. It is sometimes placed in a genus of its own, *Rubinoboletus*, but molecular studies suggest that it belongs with other species of the genus *Chalciporus*. It shares the bright yellow flesh in the stem of those species and brightly coloured pores.

Chalciporus piperatus (Bull.:Fr.) Bataille
[piperatus = peppery]
Syn. *Boletus piperatus*
Cap 3-8 cm, tawny, ochre-brown, rich reddish brown, rust-red, smooth to slightly tomentose, dry or viscid when wet. **Tubes** and **pores** rich cinnamon to copper-red, unchanging when bruised. **Stem** ochre-brown to cinnamon-brown with bright chrome yellow base, smooth. **Flesh** chrome-yellow in base of stem, much paler above. **Odour** pleasant. **Taste** extremely acrid-peppery. **Spores** cinnamon-brown, subfusiform, 8–13 × 3.5–5 µm.

Common everywhere, it grows especially under *Betula* but occasionally with other broadleaved trees and sometimes conifers. It is possible that it may have a close association with *Amanita muscaria*, rarely fruiting more than a few inches away.

Chalciporus amarellus** (Quélet) Bataille
[amarellus = somewhat bitter]
Syn. *C. pierrhuguesii* (Boud.) Bon?
Cap 3–8 cm, rounded then expanding to convex or almost flat, dull ivory, cream, pale ochre, yellowish, rarely spotted pinkish or reddish. **Tubes** carmine or pinkish red. **Pores** carmine or pinkish red, unchanging when bruised. **Stem** cylindrical or ventricose, usually tapering towards the base, yellowish, tinted carmine immediately below the tubes. **Flesh** cream or yellowish, yellow in the stem base, unchanging when exposed to air. **Odour** not distinctive. **Taste** slightly acrid or bitter. **Spores** cinnamon-brown, subfusiform, 10–13 × 4–5 µm.

Rare, in coniferous forests, mycorrhizal with *Picea*, *Pinus* and probably *Abies*. Widespread in Europe, not British but might occur here. There is some question as to the correct name for this fungus, *C. pierrhuguesii*, described from the Mediterranean, being a possible alternative. For the moment I retain the better-known *C. amarellus*.

Genus *Xerocomus*

Xerocomus subtomentosus (l.: Fr.) Quél.
[subtomentosus = a little less than velvety]
Syn. *Boletus subtomentosus*
Cap 5–12 cm, convex to flattened, colour very variable, ochraceous, tawny, sometimes with olivaceous tint, pale to dark brown or brick-red, sometimes bright yellow or even red, dry, tomentose, later smooth, sometimes cracking and yellowish or whitish flesh is then seen in the cracks. **Tubes** and **pores** pale to bright yellow, blueing when bruised. **Stem** cylindrical, tapered to almost clavate, often widening at the apex, pale yellow or yellow, sometimes becoming brownish below, sometimes striate, often covered with scattered fine brownish or reddish floccules, sometimes forming a dotted network, unchanging when bruised; basal mycelium whitish. **Flesh** whitish or yellowish in the cap and in the stem, pinkish or pinkish brown in the lower parts of the stem, blueing faintly in the cap or not blueing at all when exposed to air. **Odour** somewhat acidic. **Taste** slightly sharp, acidic. **Spores** olive-brown, subfusiform, 10–15 × 4–6 µm, smooth.

Widespread and in some years fairly common, associated with a wide range of broadleaved and coniferous trees.

This is the type species of the genus *Xerocomus*. It is variable in colour and often difficult to separate from *X. ferrugineus*. Good field characters include the lemon-tinted flesh, tendency of the stem to widen at the apex and the frequent faint blue flushing of the context. Like *X. ferrugineus* it can have a variable amount of punctate network on the stem apex, from almost completely absent to very prominent. As with all xerocomoid boletes the tubes when torn (not cut!) split lengthwise to leave numerous half tubes.

Xerocomus ferrugineus (Schaeff.) Bon
[ferrugineus = rust-red]
Syn. *Boletus citrinovirens, B. lanatus*
Cap 4–10 cm, usually plush, velvety when young, smoother and polished with age, variable, from bright yellowish to reddish brown to vivid green. **Tubes** and **pores** lemon-chrome fading to olivaceous-yellow, pores angular, large, sometimes blueing when bruised. **Stem** cylindric-swollen, usually not enlarged at the apex, dull yellow with variable amounts of coarse brown floccules, often forming a network of ridges of varying length. **Flesh** white to pinkish cream, slightly straw-yellow over the tubes, unchanging. **Spores** olive-brown, subfusiform, 10–16.5 × 4–6 µm, smooth.

Widespread and common everywhere, usually with conifers but with broadleaf trees also on occasion. Good field characters are the whitish (not yellowish) context and the stem usually narrowed or at least parallel at the apex (not widened).

Ammonia fumes or liquid on the cap surface often produces an intense green 'flash' and these forms were formerly called *Boletus lanatus*. A form with very green cap and a robust, boletoid stature was described by Roy Watling as *B. citrinovirens* - now usually regarded as just a colour form of *X. ferrugineus*.

As with *X. subtomentosus* the amount of reticulation on the stem - if any - is extremely variable, sometimes being confined to the stem apex, sometimes over the entire length. *X. ferrugineus* is rather commoner in Britain than *X. subtomentosus* for the most part but the two have long been confused. Compare also with the recently described *X. silwoodensis* and *X. chrysonemus*.

Chalciporus, Xerocomus

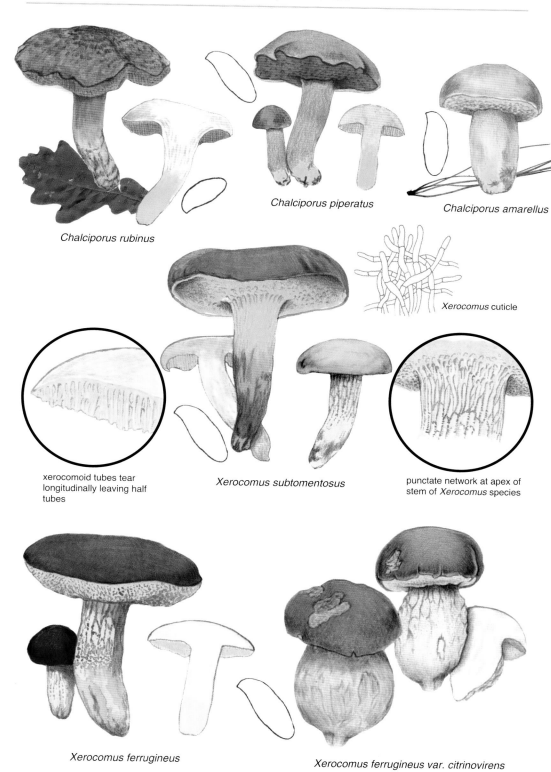

Chalciporus rubinus

Chalciporus piperatus

Chalciporus amarellus

Xerocomus cuticle

xerocomoid tubes tear longitudinally leaving half tubes

Xerocomus subtomentosus

punctate network at apex of stem of *Xerocomus* species

Xerocomus ferrugineus

Xerocomus ferrugineus var. *citrinovirens*

Xerocomus silwoodensis A.E. Hills, U. Eberh. & A.F.S. Taylor
[silwoodensis = after Silwood Park, Berkshire]
Syn. *Boletus hieroglyphicus* Rostk.?
Cap 4–11 cm, rounded to flattened, tomentose to smooth, tawny, reddish brown to chestnut. **Tubes** and **pores** bright yellow to straw yellow, sometimes spotted with rust. **Stem** rather stout, often tapered and deeply rooting at the base, yellow at the apex, buff to red-brown lower down, with coarse red-brown network of ridges in the upper half. **Flesh** whitish to pale yellow, flushing reddish to greyish rose in the lower half of the stem. **Spores** olive-brown, subfusiform, $9.5–14.6 \times 4.0–5.8$ µm, smooth.

Rare, although possibly overlooked, this species was recently described and seems to be strictly associated with *Populus*.

Xerocomus chrysonemus A.E. Hills & A.F.S. Taylor
[chrysonemus = golden mycelium]
Cap 4–7 cm, convex then flattened, tomentose to polished with age, greyish golden-yellow, mustard-yellow to olivaceous-yellow, some darker reddish tones with age. **Tubes** and **pores** bright yellow to amber, finally dull straw-yellow, unchanging when bruised. **Stem** slender to robust, tapered below, bright yellow when young, finally dull straw with red-brown floccules sometimes forming an incomplete rough reticulum; with bright golden yellow mycelium at the base. **Flesh** off-white to pale lemon, brighter more golden yellow in base of stem, not turning blue when cut. **Spores** olive-brown, ellipsoid, $10–12 \times 4.5–6$ µm, smooth.

Uncommon to rare, an associate of *Quercus* in southern England. The species probably formed part of the general muddle around *X. subtomentosus et al.* in the past.

The bright yellow mycelium and yellow flesh at the base of the stem combined with the unchanging flesh and broad spores in relation to their length (spore quotient ≤ 2.3) are good distinguishing characters. *X. ferrugineus* also has yellowish mycelium but differs in its unchanging to pinkish flesh and a larger spore quotient (≥ 2.5).

Genus *Hortiboletus*

Hortiboletus rubellus (Krombh.) Simonini *et al.*
[rubellus = reddish]
Syn. *Boletus versicolor*
Cap 4–8 cm, remaining convex until quite mature, tomentose or plush then finally smooth, a deep blood red to carmine red it can fade to olivaceous-brown with extreme age. The cuticle rarely cracks to any great extent although fine marginal tessellation may be present. **Tubes** and **pores** bright yellow then slowly greenish yellow with age. **Stem** cylindric to clavate, often rather robust, yellow above, concolorous with the pileus, paler, more yellow at the base with cream-coloured mycelium. **Flesh** pale yellow to ochre throughout, purple-red below the pileus cuticle, slightly blueing above the tubes and down into the upper stem. The flesh in the stem base can have tiny orange-red spots very like those in *X. engelii* to which it is closely related. **Odour** pleasant. **Taste** mild. **Spores** olive-brown, subfusiform, $11–13 \times 4.5–5.7$ µm, smooth.

Widespread and common, in open grassland or grassy pathsides with *Quercus*. Appearing in late summer and early autumn, this used to be easy to identify but with the discovery of more red-capped species extra care is now needed before pronouncing on any collection.

Hortiboletus engelii (Hlaváček) Biketova & Wasser
[engelii = after Heinz Engel, (b. 1954), mycologist]
Syn. *B. communis*, *B. declivitatum*.
Cap 3–8 cm, soon flattened, tomentose to quite polished with age, olive-brown, buff-brown to ochre, flushing slightly pinkish or reddish with age, cracking with age to reveal pallid flesh beneath. **Tubes** and **pores** yellow then dull olive-yellow, angular with age, bruising dull blue. **Stem** mostly yellowish with some red punctae or striae in the middle, flushing reddish overall with age. **Flesh** distinctive since it stays mostly pale yellow, flushing blue only in the upper stem and above the tubes. At the base of the stem may be seen tiny, intensely orange-red dots which may sometimes coalesce to form a bright orange patch. **Odour** pleasant. **Taste** mild. **Spores** olive-brown, subfusiform, $10.3–14.3 \times 4–6.4$ µm, smooth.

The species is common in open urban sites, parks, etc, wherever *Quercus* grows, less so in woodlands, especially in southern England; usually one of the first species to appear in the summer.

Sometimes confused with *X. rubellus*, but that species starts bright red and fades to brown while *X. engelii* starts brown and only flushes reddish with age.

Hortiboletus bubalinus (Oolbekk. & Duin) L. Albert & Dima
[bubalinus = of cattle, i.e. dull brown or buff]
Cap 4-7 cm, rounded-flattened, velvety to smooth with age, pinkish brown, reddish brown to dull apricot with pinkish margin, often cracking. **Stem** and **pores** bright yellow when young, greenish yellow when old, bruising blue. **Stem** cylindric, yellowish at the apex, flushed red below. **Flesh** in the cap flushing distinctly pink below the cuticle with a narrow blue zone just above the tubes and the upper stem, yellowish in the remainder of the stem where there may be odd tiny reddish spots in the stem base. **Odour** pleasant. **Taste** mild. **Spores** olive-brown, subfusiform, $10–14.2 \times 4.3–5.2$ µm, smooth.

Recently recorded in Britain, mainly in southern England, this species appears to be quite widespread and even common in suitable habitats. It prefers open, parkland habitats and seems to have a preference for *Populus* and *Tilia* although collections have also been made under *Quercus* and *Pinus*.

Genus *Rheubarbariboletus*

Rheubarbariboletus armeniacus (Quél.) Vizzini *et al.*
[armeniacus = coloured like a peach]
Syn. *Xerocomus armeniacus*
Cap 5–11 cm, rounded then soon flattened, tomentose to smooth, varies from bright cherry red to pinkish orange and the surface can crack readily. **Tubes** and **pores** lemon yellow to greenish yellow with age, blueing when bruised, pores large, angular. **Stem** cylindric, bright red, yellow at the apex, with fine floccose striations. **Flesh** pale yellow above, usually a bright apricot yellow in the stem base, blueing faintly in the cap. **Odour** pleasant. **Taste** mild. When $FeSO_4$ is applied to this context there is an instant deep blue-green reaction, also on the cap surface (dull grey-green in other species). The cuticular hyphae of the cap show a variable amount of large, red-stained plaques adhering to the terminal cells when stained with Congo Red. **Spores** olive-brown, subfusiform, $11.6–13.9 \times 5.0–5.8$ µm.

The species seems to prefer *Quercus* or *Fagus* and is reported from the warmer, southern counties of England, though there are as yet very few confirmed records. The generic name refers to the rhubarb-yellow flesh in the stem.

Xerocomus, Hortiboletus, Rheubarbariboletus

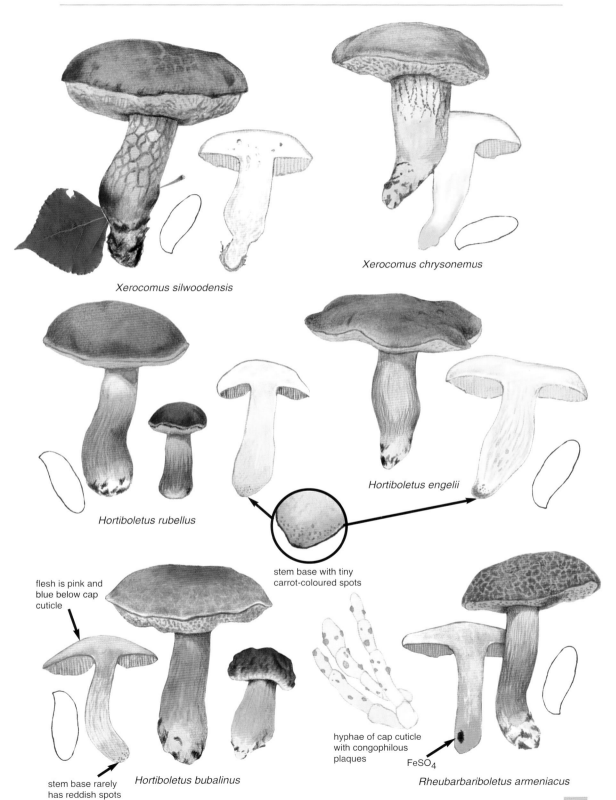

Rheubarbariboletus persicolor (Engel *et al.*) Vizzini *et al.*
[persicolor = peach coloured]
Syn. *Xerocomus persicolor*
Cap 6–12 cm, convex-expanded, felty-tomentose, often cracking slightly when older and dry, pale pink, rose-red, browner with age, blueing if bruised. **Tubes** and **pores** yellow, blueing when cut or bruised. **Stem** cylindric-clavate, surface smooth, pruinose, pale to brighter yellow above, flushed with orange-red to rust-brown below, bruising blue. **Flesh** yellowish in the cap, yellow in the stem, orange yellow to apricot in the stem base (this colour is persistent and remains after drying), blueing mostly in the cap when exposed to air. **Odour** not distinct. **Taste** mild. **Spores** olive-brown, subfusiform, 11–15 × 4.5–5.5 µm. Cap cuticle hyphae minutely ornamented.

Not found in Britain, growing with *Quercus*, especially *Q. ilex*, so far known only from Bulgaria, Croatia, Greece, Italy and Spain.

Genus *Xerocomellus*

Originally included in *Xerocomus*, recent molecular studies have shown quite clearly that they really form a discrete group of their own.

Xerocomellus chrysenteron (Bull.) Šutara
[chrysenteron = golden insides, literally 'intestines']
Syn. *Boletus chrysenteron*
Cap 3–8 cm, convex then soon flattened, tomentose when young, polished with age, almost black when very young, then sepia to olive-brown, finally paler, more ochre with age and dessication, frequently cracking and cracks often tinted pinkish red. **Tubes** and **pores** yellow when young, more olivaceous yellow with age, bruising slowly blue-green. **Stem** cylindric, lemon-yellow at the apex, flushed or streaked with scarlet over the remainder. **Flesh** cream to dirty white, distinctive in flushing deep red below the cortex of the stem and with hardly any blueing. **Odour** not distinct. **Taste** mild. **Spores** olive-brown, subfusiform, 11.8–16.6 × 4.8–6.8 µm.

In Britain it appears to be confined almost entirely to coniferous woods although very occasionally it can appear with *Fagus*. It does not seem to occur with *Quercus* at all, so collections from that host are likely to be something else.

From being once the commonest species in popular guide books we now know that this species is in fact quite restricted in habitat and that other species are much more frequently encountered. It is much confused with *X. cisalpinu*s, collections of which can readily be found with distinctly red cracks although this does vary considerably.

Xerocomellus cisalpinus Simonini, H. Ladurner & Peintner
[cisalpinus = below or south of the Alps]
Cap 4–10 cm, convex then soon flattened, very variable in colour, from almost black-brown to quite red or dull red-brown in some forms, dull ochre when very old, the surface cracks extensively, even when quite young to reveal pinkish red flesh beneath. **Tubes** and **pores** yellow when young, olivaceous yellow with age, bruising blue-green. **Stem** cylindric, when fresh very distinctive, almost bicoloured showing a bright golden yellow upper half and strongly red punctate-fibrous below; the surface of the stem usually bruises intensely blue in 5 minutes or so, which is a good field character. **Flesh** of the stem and cap flushes slowly deep blue throughout, often with deep purplish red stains at the base of the stem. **Odour** pleasant, not distinct. **Taste** mild. **Spores** olive-brown, subfusiform, 11.4–15 × 4.5–5.7 µm, very finely longitudinally striated but this needs a very high quality microscope, preferably with ×1000 oil immersion objective and ×15 eye pieces to stand a chance of seeing the striations.

The species has proved to be one of the commonest in southern England under *Quercus* from summer onwards.

Xerocomellus sarnarii Simonini, Vizzini & Eberhardt
[sarnarii = after Mauro Sarnari, mycologist]
Described from Italy in 2015 this species is very similar to *X. cisalpinus* but differs in the striking red flesh in the lower half of the stem, plus its spores are smooth and very slightly truncated. The cap surface is sepia brown and usually cracked without any red tones (rather like *X. porosporus* to which it is closely related). It is associated with *Quercus* and although not known in Britain might just occur here in the southern counties.

Xerocomellus ripariellus (Redeuilh) Šutara
[ripariellus = on banks of small streams]
Cap 3–8 cm, convex-expanded, slightly tomentose when young then soon smooth, bright vermilion to deep red, fading to pale carmine, finally reddish buff, the cuticle soon cracking extensively to expose pale flesh beneath. **Tubes** and **pores** bright yellow fading to dull yellow-ochre, bruising blue. **Stem** cylindric, bright yellow when young, especially at the apex, flushed or streaked with bright carmine red punctae, bruising deep blue on handling. **Flesh** yellow throughout, blueing very strongly and quickly in the lower stem. **Odour** pleasant. **Taste** mild. **Spores** olive-brown, subfusiform, 11–13 × 3.5–4.5 µm, finely longitudinally striate.

A beautiful and striking species, it is found in wet, riverine soils or around pond edges under *Alnus*, *Salix*, etc but it can also occur in parklands or woodland wherever the soil tends to stay wet and has also been recorded with *Quercus*. The striations on the spore surface are slightly easier to see in this than the other striate-spored species.

Xerocomellus pruinatus (Fr.) Šutara
[pruinatus = with a white bloom or pruina]
Syn. *Boletus pruinatus*
Cap 4–12 cm, rounded to flattened, tomentose to smooth, often with a rather lumpy or irregular surface, deep red-black, vinaceous-chestnut or plum with a thin, bright red marginal zone, with a fine hoary white bloom when very fresh, fading eventually to a dull ochre-brown. **Tubes** and **pores** bright golden yellow then finally olive-yellow, bruising blue. **Stem** cylindric-clavate, rather robust, bright golden yellow when young, often with fine red punctae scattered over the lower half, flushing red overall when mature, with prominent yellow mycelium at the base, often binding the leaf litter together. **Flesh** bright golden yellow throughout in young specimens, blueing in the stem base and apex. **Odour** pleasant to slightly metallic. **Taste** slightly acidic. **Spores** olive-brown, subfusiform, 10.8–14.8 × 4.3–5.7 µm, with fine longitudinal striae. Fruiting especially late in the year (although it can occur in early autumn also) it is a common and beautiful species widespread in Britain in *Fagus* woods, more rarely with conifers.

The bright yellow flesh with the blueing mainly confined to the stem base, combined with finely striate spores and the presence of thick-walled, amyloid hyphae in the stem flesh are together diagnostic.

Rheubarbariboletus, Xerocomellus

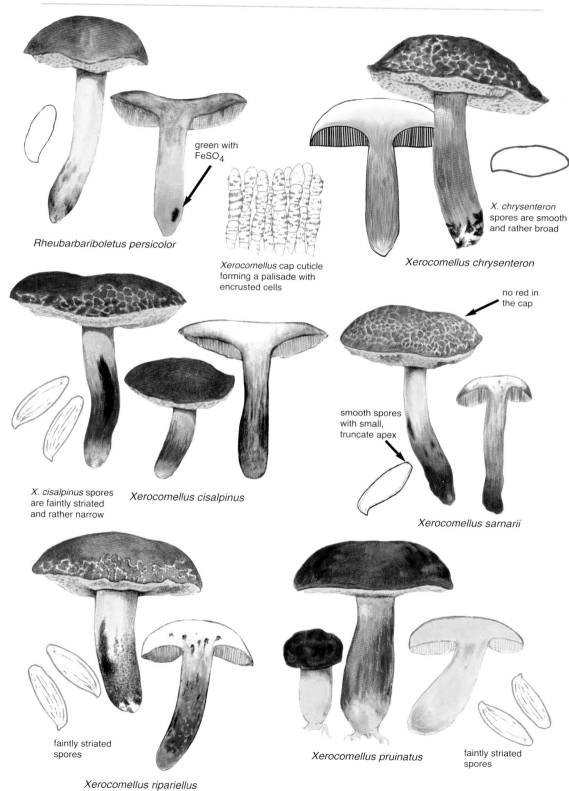

Xerocomellus porosporus (Imler ex G.Moreno & Bon) Šutara
[porosporus = spores with an apical pore]
Syn. *Boletus porosporus*
Cap 4–8 cm, soon flattened, subtomentose when young then smoother with age, dull sepia brown, olive-brown to cigar brown, soon cracking to expose pale cream flesh. **Tubes** and **pores** pale yellow, more olive-yellow with age. **Stem** cylindric, dull greyish sepia over most of its length, having a fine red zone just above the sepia portion with a brighter yellow zone at the apex. **Flesh** in the stem yellowish but soon staining dull brownish red with very rarely any blue. **Odour** pleasant. **Taste** mild. **Spores** olive-brown, subfusiform, 11.4–17.2 × 4.5–6.5 µm, with a high proportion truncated or flattened at the distal end, best observed in spores taken from a spore deposit, when a higher proportion of spores will exhibit the truncation.

It is fairly frequent growing under *Quercus* in particular. One of the easier species to identify, perhaps the best descriptive term is 'ugly'! The entire bolete tends to become a dirty sepia brown and the fruitbodies are very soft and rapidly affected by attacks of white mould.

Genus *Phylloporus*

Phylloporus pelletieri (Lév.) Quél.
[pelletieri = after Pelletier de Guernisac, French pharmacist]
Syn. *Phylloporus rhodoxanthus* of European authors
Cap 5–8 cm, velvety to smooth, rounded to flattened, reddish–brown, brown. **Gills** thick, waxy, bright yellow with numerous transverse connections or wrinkles. **Stem** cylindric to tapered, yellowish with fine reddish brown floccules, with yellow mycelium at base. **Flesh** yellow, reddish below the cap cuticle. **Odour** not distinctive. **Taste** agreeable. **Spores** olive-brown, subfusiform, 10–14.5 × 4–5 µm.

Rather rare overall, but widespread in Britain, it may be locally common in suitable habitats, growing with a wide variety of broadleaf trees.

Unique amongst European boletes for its yellow, gill-like hymenophore this species is unmistakable. DNA analysis shows that it is closely related to *Xerocomus subtomentosus* (even sharing the green ammonia reaction on the cap cuticle) hence some authors prefer to place it into that genus. However, as the genus as a whole around the world has not been investigated in depth it is perhaps safer to maintain it in its current status.

Genus *Tylopilus*

Tylopilus felleus (Bull.: Fr.) P. Karst.
[felleus = bitter tasting]
Cap 6–15 cm, yellow-tan to ochre or sienna, dry, smooth to slightly velvety. **Tubes** and **pores** cream when young but maturing to a strong pinkish flesh tone. **Stem** cylindric to broadly clavate, swollen, concolorous with pileus or a little paler, with a pronounced concolorous network covering the stem. **Flesh** white, unchanging when cut. **Odour** pleasant. **Taste** extremely bitter but it is not poisonous. **Spores** pinkish, subfusiform, 11–17.5 × 3–5 µm.

Common in both broadleaved and coniferous woods throughout Britain, often very early in the season.

Often mistaken when young for *Boletus edulis* but easily distinguished by the pore colour, taste and colour of spores.

Tylopilus felleus var. ***alutarius*** (Fr.) P. Karst.
[alutarius = leather coloured]
The variety **alutarius** differs in its duller, more greyish brown cap and much reduced network which is confined to the top of the stem. It also is less bitter in taste, sometimes completely mild. It appears to be confined to conifer woods in upland areas and might be worthy of investigation by molecular methods to ascertain its status compared to the type variety.

Genus *Porphyrellus*

Porphyrellus porphyrosporus (Fr.) E.-J. Gilbert
[porphyrosporus = with purplish spores]
Syn. *Porphyrellus pseudoscaber*
Cap 5–10 cm, dark grey-brown, sepia-brown to umber, dry, velvety. **Tubes** and **pores** dull cream-brown to vinaceous-buff, pore mouths dark sienna when old, bruising slowly blue-green. **Stem** cylindric-clavate, ± concolorous with pileus, dry and slightly velvety when young. **Flesh** cream to buff becoming vinaceous-grey to dark bluish green above tubes and in stem apex. **Odour** not distinctive or somewhat acid. **Taste** unpleasant, sour. **Spores** vinaceous-brown, subfusiform, 12–18 × 5.5–7.5 µm.

Mainly northern in Britain, scarce, under both conifers and broadleaved trees, usually on acid soils.

It would be difficult to confuse this species with any other in Britain, its very dark colours and vinaceous spore deposit are quite distinct. The flesh will stain white paper slowly green. The genus *Porphyrellus* was at one time included within the genus *Tylopilus* but molecular studies support its separation.

Genus *Strobilomyces*

Strobilomyces strobilaceus (Scop.: Fr.) Berk.
[strobilaceus = with overlapping scales]
Syn. *Strobilomyces floccopus*
Cap 5–15 cm, blackish grey, densely woolly-scaly, almost white between the scales. **Tubes** and **pores** whitish then grey to black, staining red then black when bruised. **Stem** cylindrical, shaggy-scaly up to the remains of the partial veil. **Flesh** white staining red then black. **Odour** not distinctive. **Taste** mild. **Spores** purplish black, subglobose, distinctly reticulate, 8.5–14.5 × 7–11 µm.

In mixed woods and hedgerows, often in deep shade, rather uncommon, it is widely distributed from Scotland to the southern counties of England. Unmistakable, the only species in the genus in Britain but with many more species in Asia, North America, etc, e.g. **S. confusus**, common in eastern North America is macroscopically identical but differs in its spores which have ridges but no complete reticulum. It has not been found in Europe but then mycologists rarely check the spores of *S. strobilaceus*!

Despite the blackening flesh *S. strobilaceus* is edible and highly regarded in some parts. The tough, fibrous nature of the fruitbodies means that old, dry fruitbodies may persist for some weeks after fruiting and hence should be avoided in this state.

Xerocomellus, Phylloporus, Tylopilus, Porphyrellus, Strobilomyces

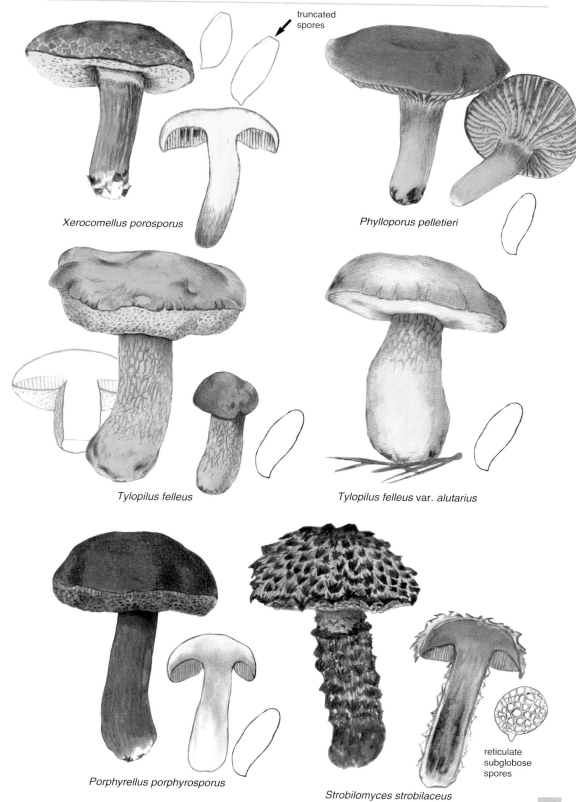

Genus *Leccinellum*

Leccinellum crocipodium (Letell.) Della Maggiora & Trassin.
[crocipodium = yellow leg]
Syn. *Leccinum nigrescens*
Cap 5–15 cm, rounded, slowly expanding, bright yellow to yellow-ochre, rapidly fading and blackening, frequently cracked and tessellated. **Tubes** pale yellow. **Pores** bright golden yellow, bruising black. **Stem** robust, cylindric or swollen, fusiform, pale to bright yellow with concolorous flocculess, bruising black. **Flesh** thick, firm, pale yellow or cream-coloured, staining reddish brown to violaceous-grey when bruised, blackening in damaged parts. **Odour** indistinct. **Taste** mild. **Spores** olive-ochre, subfusiform, 12–15 × 5.0–6.5.0 µm.

Found under *Quercus*, it is fairly common from summer onwards in the warmer, southern counties of England. A form with bright orange pores (**L. nigrescens var. luteoporus** J. Blum) has been found in Epping Forest and needs to be re-collected to ascertain its true status.

The cap cuticle of this species is composed of more or less erect chains of short, irregularly swollen cells overlying filamentous hyphae.

Leccinellum lepidum ** (P. Bouchet) Bresinsky & Manfr. Binder.
[lepidum = scaly]
Cap 5–15 cm, convex-expanded, viscid when wet, smooth, surface often rather irregular, dented, tawny yellow, yellow-brown, ochre to darker, reddish brown. **Tubes** and **pores** pale yellow, darker ochre when bruised. **Stem** often robust, clavate, pale yellow, flushed reddish below with age, surface with concolorous floccose scales. **Flesh** pale cream-yellow, flushing pinkish then violaceous-grey when cut. **Odour** pleasant. **Taste** mild. **Spores** snuff-brown, subfusiform, 13.5–21 × 5–6 µm.

Not British. Found with *Quercus*, especially *Q. ilex* in the Mediterranean, uncommon.

Resembling *L. crocipodium* but lacking the swollen, sub-spherical cells in the cap cuticle of that species.

Leccinellum pseudoscabrum (Kallenb.) Mikšík
[pseudoscabrum = false scabrum]
Syn. *Leccinum carpini*
Cap 3–7 (-10) cm broad, rounded to broadly expanded, often rather lumpy or pitted, pale to moderately dark grey-brown, sometimes with slight olivaceous tinge, dry, dull, initially velvety to rugulose or veined, usually cracked with age with concentric fissures, particularly near margin showing pale pinkish context. **Tubes** and **pores** pale, whitish buff, pores very small, 1-2 per mm, rounded, staining brown-grey when bruised. **Stem** cylindrical to subclavate, sometimes flexuous, whitish to pale grey ochre, entirely covered with brownish black flocculess, which become somewhat coarser towards the base. **Flesh** dirty white at first, on cutting first slowly staining pink to purple then grayish black with a purple tinge. **Odour** not distinct. **Taste** mild. **Spores** ochre-brown, subfusiform (12.5-)13–18.5 × 4.5–6.0 µm.

Found under *Carpinus* and *Corylus*, this is a very easily recognised species; the brown, smooth, but often cracked and 'lumpy' cap, combined with flesh staining violet-black are very distinctive. It is one of the earliest of boletes to appear in summer and widely distributed, though commonest in the south.

The cap cuticle of this species is composed of more or less erect chains of short, irregularly swollen cells overlying filamentous hyphae.

Genus *Leccinum*

Leccinum scabrum (Bull.: Fr.) Gray
[scabrum = rough, with small flecks]
Syn. *L. roseofractum*, *L. rigidipes*, *L. avellaneum*, *L. subcinnamomeum*
Cap 5–15 cm, convex, expanding to plano-convex, pale biscuit to yellowish brown to dark brown, reddish brown or olivaceous brown when old, minutely tomentose, breaking up in minute, adpressed squamules with age, somewhat viscid when moist. **Tubes** and **pores** greyish white, pores often with brownish spots, discolouring brownish when bruised. **Stem** cylindrical to clavate, whitish, often more brownish towards apex, entirely covered with blackish to greyish, sometimes pale brownish flocculess, fine above, becoming gradually coarser towards base, sometimes almost forming a network. **Flesh** whitish, usually not changing colour when bruised but sometimes discolouring strongly pinkish or reddish, finally often brownish or slightly greyish; *never blue*. **Odour** not distinctive. **Taste** mild. **Spores** dull brown, subfusiform, 14.5–19.0 × 5.0–6.5 µm. **Cap cuticle** consists of long, flexuose smooth hyphae, from clear to pale brown.

Widespread and common everywhere with *Betula*, usually in slightly drier areas on acid soils.

Although the species most often mentioned in books, it is frequently misidentified. It is best distinguished by the usually coarse dark squamules on the lower stem against pale background tissue, never having bluish discolourations in the flesh, very large, usually clavate caulocystidia and long, flexuose hyphae in the pileipellis. Flesh colour is now known to vary from unchanging to distinctly pinkish red.

Leccinum melaneum (Smotl.) Pilát & Dermek
[melaneum = blackish]
Almost identical to *L. scabrum* but with very dark, blackish brown cap and black stem scales set on a greyish or blackish background. It can only be separated with certainty by molecular means.

Leccinum variicolor Watling
[variicolor = of variable colours]
Syn. *L. oxydabile* by some authors
Cap 3.5–10 cm, convex to broadly rounded-conical when mature, dark brown, grey to grey-black with a radial pattern of lighter spots, sometimes almost whitish with dark spots or entirely dark brown; very finely tomentose all over, often subviscid with age. **Tubes** and **pores** creamy white, tubes often with yellowish brown spots, discolouring brownish when bruised. **Stem** cylindrical to clavate, whitish to greyish white, often with a distinct greenish blue discolouration in the lower half of the stem, covered with brown to black flocculess, fine at apex, coarser towards the base. **Flesh** white, when bruised often staining bright pinkish in pileus and upper half of the stem, usually discolouring greenish blue in the lower half of the stem, often taking several hours to do so; when dried the discoloration in the lower half of the stem often turns yellow which remains visible in the herbarium specimens. **Odour** pleasant. **Taste** mild. **Spores** dull brown, subfusiform, (10-) 13.5–17.5 (-20) × 5.0–6.5 µm. **Cap cuticle** is distinctive with numerous, short, broad, easily disarticulating cells (cylindrocysts). Forms with uniformly dark or pale caps can occur. The cystidia on the stem can be extraordinarily long and flexuose.

Common and widespread under *Betula*, often in rather wet areas. This can be the most striking and beautiful species with its dark mouse-grey to blackish, frequently mottled and streaked cap.

Leccinum, Suillus

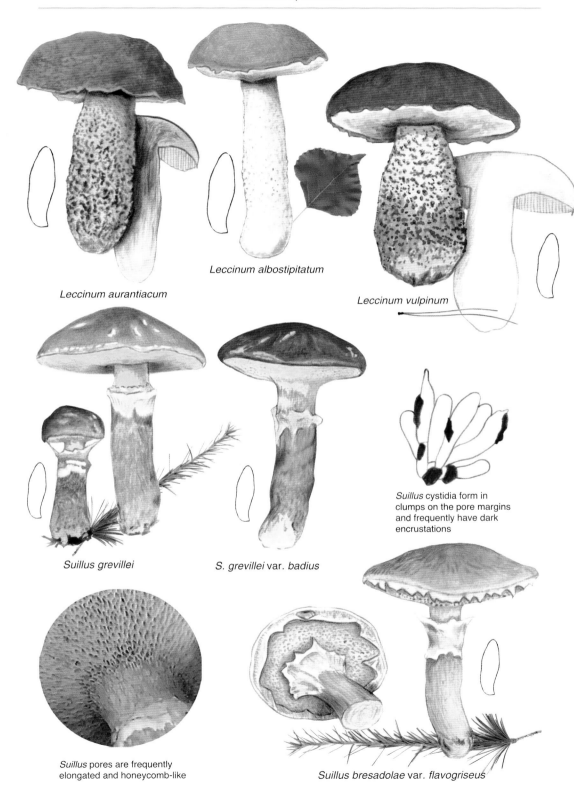

Leccinum aurantiacum

Leccinum albostipitatum

Leccinum vulpinum

Suillus grevillei

S. grevillei var. *badius*

Suillus cystidia form in clumps on the pore margins and frequently have dark encrustations

Suillus pores are frequently elongated and honeycomb-like

Suillus bresadolae var. *flavogriseus*

Suillus viscidus (L.: Fr.) Roussel
[viscidus = glutinous, sticky]
Syn. *S. aeruginascens*
Cap 6–10 cm, convex then expanding to broadly conical-convex, at first off-white, then pale yellowish ochraceous, ochraceous-grey, usually becoming olivaceous with age, viscid-glutinous. **Tubes** greyish white. **Pores** medium sized, concolorous with the tubes. **Stem** cylindrical or clavate, concolorous with the cap, with more or less distinct ring which soon collapses and darkens. **Flesh** whitish, often slightly green when bruised. **Odour** not distinctive. **Taste** slightly acid. **Spores** reddish brown, subfusiform, 8–14 × 4–5 µm.

Uncommon but widely distributed in Britain, with *Larix* on calcareous or sandy soils. The dull, pallid colours and sticky cap make it one of the easiest *Suillus* species to recognise.

Suillus tridentinus (Bres.) Singer
[tridentinus = three-toothed]
Cap 6–10 cm, rounded then expanding to convex or bluntly conical, yellowish orange, orange-yellow, orange to brownish orange, viscid, with numerous darker scales or fibrils near the cap margin. **Tubes** and **pores** orange yellow to orange, pores medium sized, orange yellow to orange with rusty or brownish tint, rather large and angular, decurrent. **Stem** cylindric-clavate, almost concolorous with the cap, with more or less distinct whitish ring, although the ring may be almost absent in older material. **Flesh** pale orange or yellowish orange in the stem, paler in the cap. **Odour** not distinctive. **Taste** slightly acid. **Spores** yellowish brown, subfusiform, 9.5–13.5 × 4–5 µm.

Widespread but rare in Britain, mainly in the southern counties of England, associated with *Larix* on calcareous soils. The bright colours, rather scaly cap at the margin and the large, honeycomb-like pores all help to distinguish this species.

Suillus cavipes (Opat.) Smith & Theirs
[cavipes = with hollow stem]
Syn. *Boletinus cavipes*
Cap 5–10 cm, fibrillose-scaly, rust-brown, pale to dark brown (forma *cavipes*) or yellow (forma *aureus*). **Tubes** short, decurrent, pale yellow to olivaceous yellow. **Pores** large, angular, pale yellow to olivaceous yellow, with smaller pores inside the larger pores, unchanging when bruised. **Stem** cylindrical, with partial cavities at least in the lower part, whitish or concolorous with the cap, with distinct fibrillose ring. **Flesh** whitish or yellowish, unchanging when exposed to air. **Odour** not distinctive. **Taste** not distinctive. **Spores** yellowish brown, subfusiform, 7–10.5 × 3.5–4.5 µm.

Rather rare but widespread, strictly associated with *Larix*. **Suillus lakei** var. **landkammeri**, extremely rare in Britain, is very similar but is associated with *Pseudotsuga*. It has a very hairy, fibrillose-scaly cap, usually with redder, orange-red tones.

Suillus asiaticus ** Singer
[asiaticus = from Asia]
Cap 5–8 cm, fibrillose-scaly, dry, bright vinaceous-crimson to russet-red, scales darker at the centre, radially arranged, margin ragged, overhanging. **Tubes** olivaceous-yellow. **Pores** large, angular, compound, yellowish to olivaceous, decurrent. **Stem** cylindric-clavate, concolorous with the cap, with a distinct ragged ring zone at the apex. **Flesh** rather soft, whitish to yellowish, unchanging. **Odour** pleasant. **Taste** slightly acidic. **Spores** yellowish brown, subfusiform, 11–12 × 4.5–5 µm. Not yet recorded in Britain, this is a rare species associated with *Larix* and found in a very few European countries, outside of its native Asia.

Suillus flavidus (Fr.: Fr.) J.S. Presl.
[flavidus = yellowish]
Cap 2–5 cm, at first rounded-conical then expanding, ochre-brown to fulvous when very young, then pale yellowish, pale tawny-ochre, sometimes spotted olivaceous or entirely with somewhat olive tint, viscid. **Tubes** pale yellow to olivaceous yellow. **Pores** large, angular, pale yellow to olivaceous yellow. **Stem** cylindrical, more or less dull white to pale yellowish, with more or less distinct viscid ring, basal mycelium pale whitish. **Flesh** pale yellow to whitish. **Odour** not distinctive. **Taste** not distinctive. **Spores** yellowish brown, subfusiform, 7.5–11 × 3–4 µm.

Uncommon in the south, mainly northern, especially in Scotland, associated with *Pinus* spp. in boggy areas.

Suillus collinitus (Fr.) O. Kuntze
[collinitus = besmeared]
Syn. *S. fluryi*
Cap 4–10 cm, rounded to slightly conical then expanding, brown to dark brown, reddish brown, with innate fibrils, viscid when wet. **Tubes** pale yellow to olivaceous yellow. **Pores** fine, rounded, pale yellow to olivaceous yellow. **Stem** cylindrical, often tapering below, more or less yellowish, without ring, covered with numerous small pale to brownish glandular dots; stem base and the basal mycelium pinkish. **Flesh** whitish to yellowish, somewhat pinkish in the stem base. **Odour** not distinctive. **Taste** not distinctive. **Spores** yellow-brown, subfusiform, 8–10 × 3.5–4 µm.

Locally common and widespread in Britain, associated with *Pinus* species.

Suillus granulatus (L.: Fr.) Roussel
[granulatus = with granules]
Cap 4–10 cm, rounded then expanding, ochraceous to brownish or orange brown, smooth, viscid when wet. **Tubes** pale yellow to olivaceous yellow. **Pores** fine, rounded, pale yellow to olivaceous yellow. **Stem** cylindrical, often tapering below, more or less yellowish, without ring, covered with numerous small whitish to brown glandular dots. **Flesh** whitish. **Odour** not distinctive. **Taste** not distinctive. **Spores** yellow-brown, subfusiform, 8–10 × 3–4 µm.

Common and widespread everywhere with *Pinus sylvestris*. Easily confused with *S. collinitus* but that species has a darker cap and pinkish mycelium at the stem base.

Suillus placidus (Bonord.) Singer
[placidus = gentle]
Cap 6–10 cm, rounded then expanding to somewhat flattened, whitish, ivory or yellowish, sometimes spotted ochraceous. **Tubes** pale yellow to olivaceous-yellow, decurrent. **Pores** fine, rounded, pale yellow to olivaceous-yellow. **Stem** cylindrical, more or less concolorous with the cap, without ring, covered with numerous brown glandular dots. **Flesh** whitish. **Odour** not distinctive. **Taste** not distinctive. **Spores** yellow-brown, subfusiform, 7.5–10 × 3–4 µm.

Rare, known from only one site in East Kent, associated with 5-needle pine, *Pinus strobus*. A striking species unlike any other British species. From other species it may be distinguished by its association with *P. strobus* and by the tendency for the glandular dots on the stem to coalesce into larger brown marks.

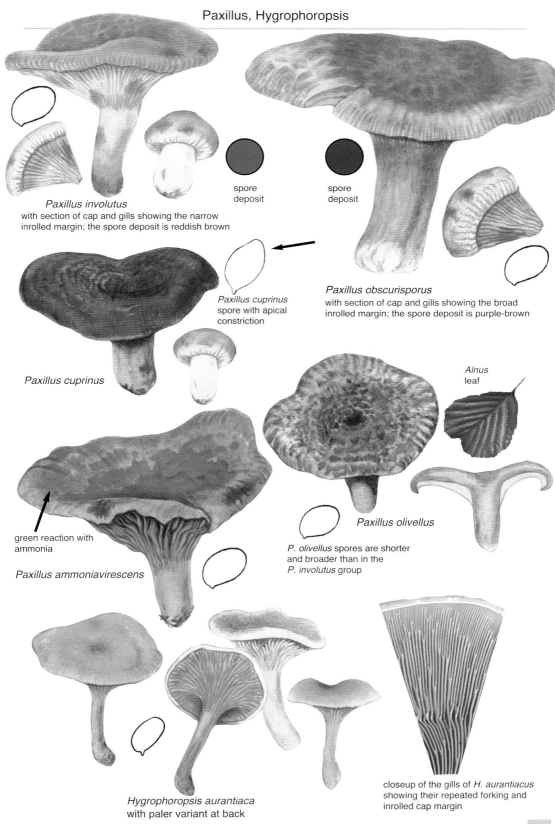

Paxillus, Hygrophoropsis

Paxillus involutus with section of cap and gills showing the narrow inrolled margin; the spore deposit is reddish brown

spore deposit

spore deposit

Paxillus cuprinus spore with apical constriction

Paxillus obscurisporus with section of cap and gills showing the broad inrolled margin; the spore deposit is purple-brown

Paxillus cuprinus

green reaction with ammonia

Paxillus ammoniavirescens

Alnus leaf

Paxillus olivellus

P. olivellus spores are shorter and broader than in the *P. involutus* group

Hygrophoropsis aurantiaca with paler variant at back

closeup of the gills of *H. aurantiacus* showing their repeated forking and inrolled cap margin

Hygrophoropsis rufa (D.A. Reid) Knudsen
[rufa = reddish brown or fox-red]
Cap 3–8 cm, flattened then soon depressed, funnel-shaped, with inrolled margin, from deep reddish brown, umber to apricot, always with a fine coating of darker hairs, plush to matt. **Gills** decurrent, crowded, pale yellowish apricot, cream-apricot to pale orange, repeatedly forking. **Stem** cylindric-clavate, 3–5 x 0.7–1.5 cm, more or less concolorous with the cap. **Flesh** thin, soft, whitish, often rather hollow, pithy in the stem. **Odour** quite strong, of ozone, hot photocopiers or *Hamamelis* blossom. **Taste** rather astringent. **Spores** white, ellipsoid, 6–7 x 3–4.5 µm, weakly dextrinoid. **Cuticle** with distinct tufts of vertical cells forming a palisade, some swollen-clavate, others slender and others with yellow-brown contents. **Basal mycelium** pale sulphur-yellow, coated with large (1–3 µm) crystals.

Probably widespread and frequent on and around rotting conifer stumps and also on woodchip mulch, but certainly under-recorded in Britain. The cap colour varies widely from very dark brown when young to pale apricot when expanded. Collections fruiting in the open generally appear lighter in tone. The best character to confirm the identity is the cuticle structure. Similar crystals on the mycelium are found in the genus *Paxillus*.

Hygrophoropsis macrospora (D.A. Reid) Kuyper
[macrospora = large-spored]
Cap 3–6 cm, soon flattened to depressed, funnel-shaped, surface felty-tomentose, pale ivory, cream-buff, slightly darker yellowish buff where bruised, margin inrolled. **Gills** pale cream, ivory, decurrent, repeatedly forked. **Stem** cylindric-clavate, pale yellowish ivory, darker below or where bruised. **Flesh** whitish to pale brownish when cut, slightly hollow in the stem. **Odour** pleasant, not distinct. **Taste** mild to slightly astringent. **Spores** cylindric to ellipsoid, 7–10 (-13) x 3.5–4.5 µm, strongly dextrinoid. **Spore deposit** white. **Cuticle** a tangle of smooth undifferentiated hyphae.

Uncommon, rarely reported, so the distribution is uncertain but known from southern England; found in wet, marshy habitats usually with *Sphagnum* moss and *Juncus*.

Hygrophoropsis fuscosquamula P.D. Orton
[fuscosquamula = with dark brown hairs or scales]
Cap whitish cream to pale yellowish cream, with dark olive-brown to sepia-brown hairs scattered over the cap surface. **Gills** are pure white to pale cream and finally ochraceous. **Spores** ellipsoid-cylindric, 6–8 x 3.5–4.5 µm, so significantly shorter than *H. macrospora*; strongly dextrinoid. **Spore deposit** white. **Cuticle** with tangled, filamentous hyphae, while the dark hairs consist of broad, cylindric-clavate end cells 12–16 µm across.

Very similar to *H. macrospora*, and perhaps even identical to that species. With equally pale colours *H. fuscosquamula* grows in exactly the same sort of habitat as *H. macrospora*, i.e. with *Juncus* in wet, boggy habitats, close to *Salix*, *Alnus*, etc. Should the two species prove synonymous then *H. fuscosquamula*, as the earlier described species, would have precedence.

Its distribution is uncertain but there are scattered records from Scotland down to the south of England, so it may actually be quite common but overlooked.

Genus *Aphroditeola*

Aphroditeola olida ** (Quél.) Redhead & M. Binder
[olida = scented]
Syn. *Hygrophoropsis olida* (Quél) Métrod
Cap 3–5 cm, strongly inrolled when young, then soon flattened, margin often wavy and irregular, pale flesh-pink to brick-red. **Gills** crowded, decurrent, white to pale pinkish white, forking. **Stem** rather short, 1–3 cm, tapered, white, slightly yellow below. **Flesh** whitish to dull yellowish below. **Odour** strong, sickly-sweet, almost floral, like that of *Hebeloma sacchariolens*. **Taste** mild. **Spores** broadly ellipsoid, 3.5–4.5 x 2.5–3 µm, not reacting to Melzer's iodine. **Spore deposit** white.

Uncommon, in coniferous woods and *Pinus* heaths on calcareous soils, mainly northern. Not known with certainty in Britain as previous records appear to be mis-identifications, but should occur here in Scotland.

This unusual species is very different from other species in the genus with much smaller, non-dextrinoid spores and with a striking odour. It was recently transferred to the new genus *Aphroditeola,* in the family Hygrophoraceae, named after the goddess Aphrodite.

Family *Tapinellaceae*
Fruitbody pleurotoid to resupinate (the resupinate species are not covered here), with decurrent gills bruising brown. Cap tomentose, usually eccentric. Spores pale brown, dextrinoid. Saprotrophic on conifer stumps and trunks, causing a brown rot of affected timber.

Genus *Tapinella*

Tapinella atrotomentosa (Batsch: Fr.) Šutara
[atrotomentosa = with blackish fur]
Syn. *Paxillus atrotomentosa*
Cap 10–25 cm, convex then expanding and becoming funnel-shaped, with the margin inrolled, cinnamon-brown, nut-brown to sienna, velvety-tomentose especially at the disc. **Gills** decurrent, branching and rejoining often near the stem, pale ochre to yellow-brown, darker rust-brown with age, bruising brown. **Stem** squat, robust, 3–10 x 4–5 cm, usually offset to one side of the cap, dark red-brown to bay, chestnut-brown, densely velvety. **Flesh** pale buff-ochre, paler in the cap, sometimes pale lavender when cut. **Odour** mild, not distinct. **Taste** bitter. **Spores** ellipsoid, 5–6 x 3–4.5 µm, dextrinoid. **Spore deposit** pale rust-brown.

Widespread and frequent on stumps or at the base of *Pinus sylvestris*.

Formerly placed in *Paxillus* the species of *Tapinella* cause a brown rot in the wood on which they grow and DNA analysis shows them as separated from *Paxillus* and deserving of a family and genus of their own.

Tapinella panuoides (Fr.: Fr.) E. -J. Gilbert
[panuoides = ear-like]
Syn. *Paxillus panuoides*
Cap 3–10 cm, spathulate to rounded, shell-shaped, without a true stem, surface dry, velvety or felty, yellow-ochre, sienna brown, darker at the base, margin at first inrolled. **Gills** crowded, forking, often very wrinkled and wavy-edged, with cross connectives, pale ochre to deeper orange-yellow when mature; separable from the flesh. **Odour** not distinct. **Taste** mild. **Spores** ellipsoid, 4.5–7 x 3.5–5.5 µm. **Spore deposit** yellowish rust.

Widespread but never very common, growing on rotted coniferous wood which turns vivid yellow where attacked as well as becoming soft and pulpy. It is a common cause of rotting of timbers in underground mines and even occasionally in houses.

A variety with bright violet tomentum at the base of the cap is var. ***ionipes*** Quél., while if the cap has prominent reddish scales it is var. ***rubrosquamulosus*** Svrcek & Kubicka.

Hygrophoropsis, Aphroditeola, Tapinella

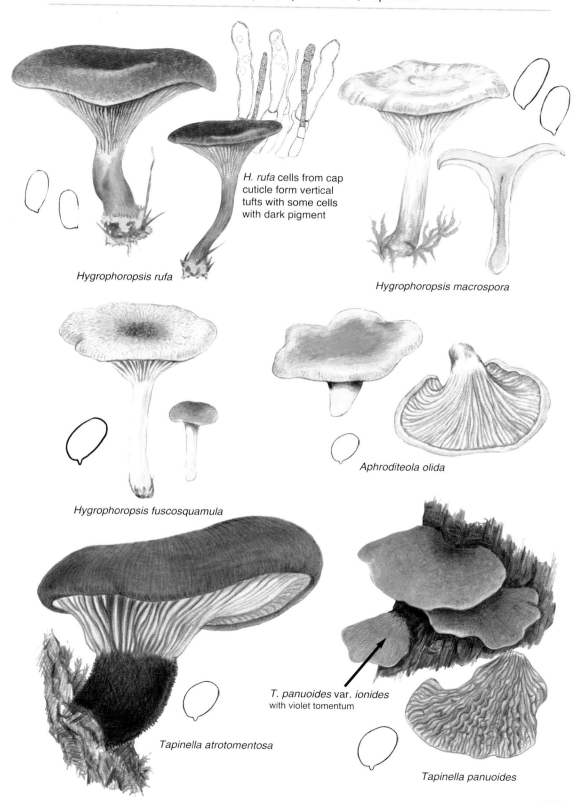

> **Family *Gomphidiaceae***
> Fruitbody agaricoid, with thick, waxy, decurrent gills. Cap tomentose-scaly to smooth and viscid. Spores dark brown to almost black, subfusiform. Mycorrhizal with conifers and many (all?) species forming a relationship with boletes of the genus *Suillus*.

Genus *Gomphidius*

Gomphidius glutinosus (Schaeff.: Fr.) Fr.
[glutinosus = sticky, slimy]
Cap 5–13 cm, umbonate, very glutinous-viscid when wet, at first whitish to grey then soon pale vinaceous pink, bruising blackish or fuscous. **Gills** deeply decurrent, thick, white then pale grey, darker as the spores mature, unchanging when bruised. **Stem** 5–10 × 1.5–2 cm, cylindric to spindle-shaped, white, whitish or sometimes brownish, bright yellow at the base, often spotted and stained blackish; with glutinous ring and veil in the upper part. **Flesh** white or whitish, yellow in the stem base, unchanging when exposed to air. **Odour** not distinct. **Taste** mild. **Spores** subfusiform, 17–20 × 5.5–6.5 μm. **Spore deposit** almost black.

Widespread although most abundant in Scotland, uncommon elsewhere, with various conifers including *Picea*, *Pinus* and *Pseudotsuga*.

Gomphidius maculatus Fr.
[maculatus = spotted]
Cap 3–7 cm, umbonate then flattened and finally depressed, viscid-glutinous when wet, pale buff, drab to hazel-brown, sometimes with a vinaceous tint, soon spotted and blotched with sepia to reddish brown, finally blackening. **Gills** decurrent, thick, whitish to vinaceous, soon spotting and bruising reddish black. **Stem** 5–8 × 0.7–1.2 cm, tapered below, slightly viscid, whitish to yellowish buff, streaked and spotted sepia to reddish black or black. **Flesh** white in the cap and stem, more yellow at the stem apex and slightly vinaceous at the base, blackening when old. **Odour** not distinct. **Taste** mild. **Spores** subfusiform, 16–23 × 6–8.6 μm. **Spore deposit** almost black.

Widespread in Britain but rare, commonest in the north, mycorrhizal with *Larix* and often close to *Suillus grevillei* with which it appears to form a close association.

Gomphidius gracilis Berk.
[gracilis = slender]
There is some dispute as to whether this second larch-associated species is really distinct from *G. maculatus*. Like that species it is associated with *Larix*, is rather slender and has rather similar colours. It differs by its non-blackening gills, presence of amber-coloured droplets at the margin of the cap and stem apex and shorter, broader spores. Until any DNA evidence is forthcoming to prove otherwise I regard it as separate.

Gomphidius roseus (Fr.) P. Karst.
[roseus = rose pink]
Cap 3–6 cm, rounded then soon flattened and then depressed, viscid to tacky, at first bright coral red then fading to pink or pinkish rust. **Gills** decurrent, thick, often forking, whitish then brownish black as the spores mature. **Stem** 2–4 × 0.5–1.5 cm, whitish, tinted pinkish or slightly yellowish at the base, with a glutinous white veil forming a ring-zone, often very poorly defined. **Flesh** white to faintly tinted pink below the cap cuticle. **Odour** not distinct. **Taste** mild. **Spores** subfusiform, 17–20 × 5–6.5 μm. **Spore deposit** fuscous-black.

Widespread and locally frequent, found under *Pinus* species, apparently only with 2-needled pines; sometimes in small groups, usually in close proximity to *Suillus bovinus* with which it has an association, most likely 'piggy-backing' on its mycorrhizal mycelium, perhaps stealing nutrients.

Genus *Chroogomphus*

Chroogomphus rutilus (Fr.) O.K. Miller
[rutilus = reddish orange]
Syn. *C. corallinus*
Cap 5–15 cm, umbonate, less commonly rounded, viscid at first, later dry and polished, brick-red, vinaceous-red, ochre to reddish brown or olivaceous-brown. **Gills** pale olive-buff darkening to fuscous-black, deeply decurrent, widely spaced. **Stem** 6–12 × 1–1.5 cm, tapered below, vinaceous at the apex, more yellow-ochre below, brighter chrome-yellow at the base, with pale yellowish mycelium; with a fine ring-zone above and several floccose zones below. **Flesh** vinaceous, yellower in the stem, bright chrome-yellow in extreme base. **Odour** not distinct. **Taste** slightly astringent. **Spores** subfusiform, 15–22 × 5.5–7 μm. **Spore deposit** fuscous-black. **Pleuro- and cheilocystidia** thick-walled, up to 2.7 μm thick. **Melzer's iodine** on the flesh = deep blue-violet.

Frequent and widespread in Britain, although commonest in Scotland, associated with 2-needled *Pinus* species, sometimes found fruiting in association with *Suillus* species.

C. britannicus, wrongly synonymised with *C. rutilus*, differs in its smaller, drier, fibrillose cap surface, slightly broader spores and thin-walled cystidia. Described from Hampshire under *Pinus sylvestris*.

Chroogomphus purpurascens (Lj.N. Vassiljeva) M.M. Nazarova
[purpurascens = becoming purple]
Cap 3–5 cm, usually distinctly umbonate, greyish red, reddish ochre flushed purplish or dark purple. **Gills** deeply decurrent, widely spaced, pale buff then soon deep fuscous. **Stem** cylindric-tapered below, ochraceous with reddish purple coating of fibrils, darker purplish below, with pinkish purple mycelium. **Flesh** ochraceous flushed vinaceous. **Odour** not distinct. **Taste** mild. **Spores** fusiform, 15.5–21 × 6.0–7 μm. **Spore deposit** fuscous-black. **Pleuro- and cheilocystidia** thin-walled, <1 μm thick. **Melzer's iodine** on the flesh = deep blue-violet.

Under *Pinus* spp. possibly widespread but confused with other species. Recently recorded from the island of Jersey.

Chroogomphus fulmineus ** (R. Heim.) Courtec.
[fulmineus = like lightning]
Cap 5–8 cm, rounded to slightly umbonate, not pointed, smooth, viscid, bright flame red to vermilion to copper-red or vinaceous red with age. **Gills** deeply decurrent, widely spaced, ochraceous-orange then fuscous-black. **Stem** 4–8 × 0.6–1.5 cm, tapering below, more orange above and often vivid vinaceous or flame-red below, with regular floccose bands scattered over the surface. **Flesh** orange-ochre to slightly olivaceous below. **Odour** not distinct. **Taste** slightly astringent. **Spores** 18–20 × 7–8 μm. **Spore deposit** fuscous-black. **Pleuro- and cheilocystidia** thin-walled, <1.2 μm thick. **Ammonia** on flesh = violet. **Melzer's iodine** on flesh = blue-black.

Under *Pinus* species in warm, Mediterranean areas especially along the Atlantic coast. Not known from Britain but might possibly occur in southern counties under imported pines.

Gomphidius, Chroogomphus

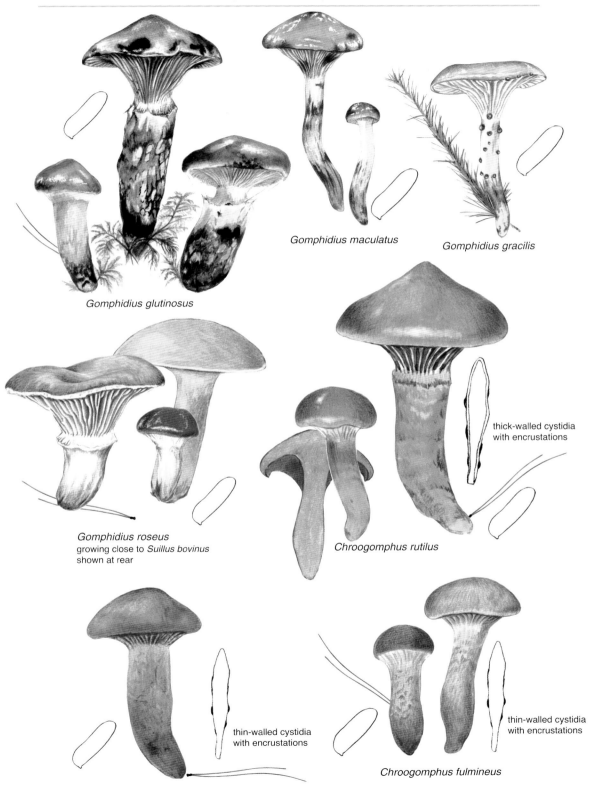

Gomphidius maculatus

Gomphidius gracilis

Gomphidius glutinosus

Gomphidius roseus growing close to Suillus bovinus shown at rear

Chroogomphus rutilus — thick-walled cystidia with encrustations

Chroogomphus purpurascens — thin-walled cystidia with encrustations

Chroogomphus fulmineus — thin-walled cystidia with encrustations

12. Russulas, Milkcaps & Their Relatives

The order *Russulales* includes fungi which do not conform to our usual ideas of what a mushroom or toadstool looks like. So there are truffle-like genera, resupinate genera, brackets and even toothed or spined forms. But like their gilled relatives they share microscopic, chemical or genetic characters which unite them into this one order.

Most genera within the *Russulales* have spores with some sort of ornamentation such as warts, spines or ridges and this ornamentation stains bluish black when treated with Melzer's reagent. This is referred to as an amyloid reaction. The active substance in Melzer's reagent is iodine which reacts with polysaccharides in the spore ornamentation to produce the very dark colour.

Many *Russulales* possess certain elements that stain in the presence of sulphuric benzaldehydes. The colour of this reaction varies from reddish to violaceous to black. Sulphovanillin (vanillin crystals mixed with sulphuric acid) is the most frequently used chemical to produce this reaction.

This book deals with the gilled or lamellate genera plus the toothed genera *Auriscalpium* and *Hericium* as well as the polypore-like *Albatrellus*.

The habitat and plant associations are important features to note when identifying species and should be recorded as accurately as possible. Many species in these genera are associated with a single tree species, others are less fussy and can be found in more varied habitats. Soil type is also important and it can be very useful for example, to know if you are on calcareous or acid soils.

Characters of particular use in *Russula*

Apart from colour, important macro-characters include the texture of the cap and stem surface, e.g. whether the surface is matt or viscid-sticky, pruinose or roughened. The cap cuticle may be peeled away from the underlying flesh in many species and the extent to which this is possible needs to be recorded. In some species the cuticle is completely adnate and not peelable.

Other characters include the attachment of the gills to the stem, the spacing of the gills and their colour when mature. Colour changes in the fruitbody from young to old may occur, often involving staining when bruised. Cap pigments are often water soluble to some extent so after heavy rains the cap colour may change completely, fading almost to white. Smell and taste are often extremely useful although these can be difficult and very subjective from person to person. Getting a second or even third opinion is always advisable. In *Russula* it is the taste of the flesh (usually a tiny piece of cap flesh and gill is sufficient) that is important and this can vary from mild, to bitter or acrid or more rarely an unusual taste such as cedar wood or menthol. Not everyone feels comfortable tasting fungi and of course if you are uncertain whether you have one of these genera then you absolutely **should not** taste them!

Spore deposits

The colour of the spore deposit is an essential character in *Russula* and a colour square representing the spore colour of each species is shown alongside each species illustrated. These colour squares are based on the now standard colour chart of the famous French *Russula* specialist Henri Romagnesi, and the chart is shown below and opposite the first *Russula* colour plate on page 172.

Spore deposits are best made on a glass microscope slide placing a cap gill-side down on the slide and covering it with a glass or plastic cup for a few hours. This should provide a sufficient deposit for examination. After allowing a few minutes drying time the deposit should be scraped together to make a thick deposit. This is then covered with a glass coverslip and the latter permanently fixed either with Sellotape or by sealing the edges with nail varnish (be careful that the varnish does not creep under the coverslip and wet the spores).

Spore deposits made in this way have the added advantage of allowing easy comparison with spore colour charts such as the one included here by placing the slide directly over the chart.

Ia   IIa   IIc   IId   IIIa   IIIc   IVc   IVe

Fig. 12.1. Russula spore deposit colour chart. 1a–IVe = Romagnesi colour system. Note that some intermediate shades were omitted.

Microscopy

Apart from their macroscopic characters the microscopy of these genera is often essential for accurate identification of many species. In *Russula* the following are the principal features to be examined under the microscope.

Spores

The size of spores (not including any warts, etc) is important so accurate measurement must be made, preferably under a x100 oil immersion lens, which, combined with standard x10 eyepieces will give a total magnification of x1000. When used with an eyepiece measuring graticule, calibrated against a micrometer slide you will be able to make measurements in microns (thousandths of a millimetre, shown as µm). Most *Russula* spores are typically in the range of 6–12 µm.

All *Russula* spores have some sort of ornamentation consisting of warts, spines or ridges (see Fig. 12.3) and the extent and type of this ornamentation is often critical to species identification. However, to see these warts and ridges they must first be stained using an iodine solution called Melzer's reagent. This may be obtained from some microscope dealers or via your local mycological society or club. Its formula is: 1.5 g iodine, 5.0 g potassium iodide, and 100 g chloral hydrate to 100 ml water.

A typical spore for each species is illustrated next to the fruitbody along with a colour square showing the spore deposit colour.

Cap cuticle

The outer cuticle of the cap, the pileipellis, is one of the critical features to study under the microscope. The lower layers of the cuticle will consist of swollen, balloon-like cells, but in the outer layers the cells are usually elongated and more typically hypha-like. The length, shape and in particular the way these cells terminate are of particular importance. Specialised cells are usually also present, either in the form of pileocystidia or of fuchsinophile hyphae (sometimes called primordial hyphae).

Pileocystidia

These can usually be easily seen standing out from the background cuticular hyphae, being more or less club-shaped or worm-like and with granular-appearing contents. When treated with sulphovanillin or Cresyl Blue they will stain blackish or blue and are more easily observed. Very often these cystidia may have septa dividing the cell and the number of septa is often a useful specific feature (Figs 12.5 to 12.8). More rarely pileocystidia may have encrustations at their base (Fig. 12.8). With sulphovanillin you may use either ready-made solution (crystals of pure vanillin dissolved in 80% sulphuric acid until golden yellow) or mix it fresh on the slide (just a few crystals in a drop of the acid, stirred until dissolved). The ready-made solution will blacken over time but will still be effective. Cresyl Blue is easier to obtain and use and may prove more practical for many.

Fuchsinophile hyphae

These are usually very long worm-like (vermiform) cells which, when stained with carbol fuchsin, may be seen to have blobs of encrusting material on their outer walls (Fig. 12.7). Although carbol fuchsin shows them most clearly they may also be seen less clearly in Cresyl Blue or even Congo Red and these are certainly easier to use. The technique for using carbol fuchsin is to use strong carbol fuchsin solution for at least 10-15 minutes then wash off in 10% hydrochloric acid, leaving it in the acid for exactly one minute, then mount in water.

Absence of specialised cells in the cuticle

More rarely neither pileocystidia or fuchsinophile hyphae are present (Fig. 12.5).

Preparing a specimen

Only a tiny fragment of cuticle is required for examination but it is important that this is with as little as possible underlying fleshy tissue and that the upper surface remains uppermost on the slide. An easy way to obtain cuticle is to take a sharp razor blade and make three cuts forming three sides of a small square, with the uncut side pointing towards the centre, then grab the leading edge of the cut with some fine forceps or your finger nails and gently peel the cuticle towards the centre until it breaks (Fig. 12.2). This broken edge will be only one or two cells in thickness—ideal for examination. If excess flesh has come away with the cuticle (the undersurface of the cuticle will appear white) then the fragment may be placed upside down on a glass microscope slide and gently scraped with the razor blade in order to remove the flesh. You should be left with just the translucent fragment of cap cuticle. Remember to turn the fragment back over to expose the upper surface.

Fig. 12.2. Cutting a piece of cap cuticle for microscopic examination.

Characters of particular use in *Lactarius* and *Lactifluus*

The cap cuticle in *Lactarius* species vary greatly from species to species and a number of types may be seen under the microscope in cross-section (Figs 12.9 a–f). Because of the latex present throughout the flesh the piece of tissue should be washed in water and blotted dry before attempting to stain (Congo Red is a useful basic stain). In many species the outer cells may be embedded in a layer of transparent slime and this layer may be of different thickness in different species.

In *Lactifluus* species distinctive thick-walled hairs are present (Fig. 12.9d) and their length and thickness should be carefully measured.

Like those of *Russula*, the spores may be extremely diagnostic, having similar ornamentation which stains blue-black in Melzer's reagent. Typical examples are shown in Fig. 12.10. Spore colour is much less variable than in *Russula* and hence is used less in identification but the colour of the deposit is nevertheless mentioned in each description but not illustrated.

The milk or latex of *Lactarius*/*Lactifluus* is usually white or clear but in many species will change colour after a few minutes and careful note should be made of any changes. In some sections the milk is already brightly coloured. The taste of the milk (apply a small drop to the tip of the tongue) is also important, varying from mild to bitter or very acrid.

Cap cuticle structures and spore ornamentation in *Russula*

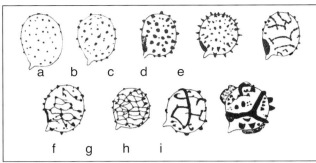

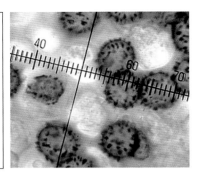

Fig. 12.3. Spore ornamentation.
a. Spore with simple, very small isolated warts.
b. Spore with larger, rounded warts.
c. Spore with large, isolated and rounded warts.
d. Spore with sharp, isolated warts or spines.
e. Spore with some connectives and some connate warts.
f. Spore with numerous connectives and some enclosed meshes.
g. Spore with complete reticulum of fine lines.
h. Spore with thick ridges forming partial net or in zebroid stripes.
i. Spore with enormous 'wings' or flanges as well as large, robust warts.

Fig. 12.4. Typical Russula spores stained with Melzer's reagent.

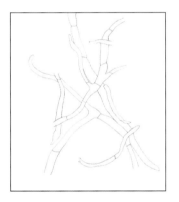

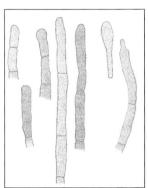

Fig. 12.5. Far left: **cap cuticle** without pileocystidia or fuchsinophile hyphae.

Fig. 12.6. Left: **typical pileocystidia** (note the cross-septa).

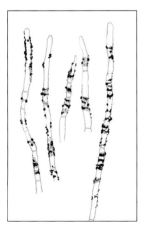

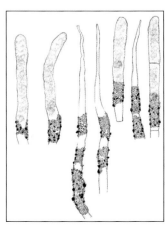

Fig. 12.7. Far left: **fuchsinophile hyphae** showing the external encrustations and droplets stained with carbol fuchsin.

Fig. 12.8. Left: **pileocystidia** with basal encrustation of fuchsinophile granules along with some pointed fuchsinophile hyphae.

Cap cuticle structures and spore ornamentation in *Lactarius*

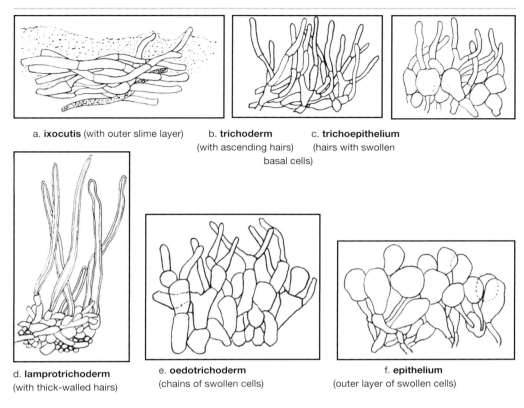

a. **ixocutis** (with outer slime layer)
b. **trichoderm** (with ascending hairs)
c. **trichoepithelium** (hairs with swollen basal cells)
d. **lamprotrichoderm** (with thick-walled hairs)
e. **oedotrichoderm** (chains of swollen cells)
f. **epithelium** (outer layer of swollen cells)

Fig. 12.9. **Common cap cuticle structures:** note that intermediates may occur, e.g. an ixo-trichoderm.

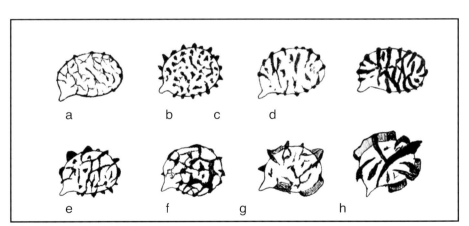

Fig. 12.10. **Spore ornamentation.** a. Low ornamentation, some connectives, b. Mainly isolated warts, some connate warts, c. Short to medium ridges, somewhat zebroid, d. Strongly zebroid ridges with partial reticulum, e. Large warts with some enclosed meshes, f. Almost complete reticulum, g. With tall warts and large flanges or wings, h. With encircling wings.

Family *Russulaceae*

In Europe the agaricoid members include three genera, *Lactarius*, *Lactifluus* (formerly part of *Lactarius*) and *Russula*. All are characterised by their mainly cellular, not filamentous, brittle tissues which lack the fibrous nature of the order *Agaricales*. *Lactarius* and *Lactifluus* are characterised by laticifers running through their flesh and gills, which 'bleed' when the flesh is broken. This latex varies from clear and watery to thick and cream-like. The colour of this milk and any change in colour is an important identification character. The flesh of European *Russula* species lacks these laticifers and hence remain dry when broken.

The spores of all three genera have warts or ridges of varying sizes and which stain blue-black in Melzer's solution. The patterns of the warts and ridges are important identification characters. The detailed structure of the cap cuticle and the presence or absence of specialised cystidia or unusual hyphae are also used in delimiting both sectional groupings and species in all three genera. All the European species form ectomycorrhiza with various tree or shrub species.

Genus *Lactarius*
Subgenus *Lactarius*
Section *Dapetes*: milk orange to red; with conifers

Lactarius semisanguifluus R. Heim & Leclair
[semisanguifluus = half blood-red fluid]
Cap 3–8 cm, at first rounded then soon flattened and finally depressed, funnel-shaped, smooth and viscid when wet, pale orange-buff to ochre-buff with darker concentric zones, very soon flushing entirely malachite green to bright blue-green. **Gills** moderately crowded, adnate to slightly decurrent, pale pinkish orange, bruising deep vinaceous red then slowly deep blue-green. **Milk** orange then soon vinaceous red in 5–8 minutes. **Stem** 3–5 x 0.8–2 cm, smooth, dry, pale pinkish orange to clay-pink, often with small orange scrobiculae (pits), bruising vinaceous red then dark green. **Odour** sweet, of carrots. **Taste** rather bitter. **Spores** pinkish ochre, broadly ellipsoid, 7.5–9 x 6–7 µm, with warts up to 0.5 µm high, ridges zebroid forming a very incomplete reticulum.

Locally common, mainly in southern England, associated with *Pinus*. Recognised by the extensive greening with vinaceous staining milk and flesh.

Lactarius sanguifluus (Paulet) Fr.
[sanguifluus = blood-red fluid]
Cap 3–8 cm, rounded then soon flattening and depressed at the centre, smooth, viscid, pinkish orange, pinkish buff, to clay, flushed pale vinaceous, faintly zonate, aging dull greyish green. **Gills** crowded, adnate to slightly decurrent, pale vinaceous, paler on the margin. **Milk** vinaceous or wine-red right from the start. **Stem** 2–4 x 1–2 cm, cylindric, smooth to slightly roughened, pale pinkish buff to greyish pink or flushed vinaceous, sometimes with darker spots or sunken scrobiculae (pits). **Flesh** brittle, more or less hollowed in the stem, pinkish buff to vinaceous or brownish red. **Odour** pleasant, not distinct. **Taste** mild to bitter and slightly acrid. **Spores** dark cream, 8–9.5 x 6.5–8.0 µm, warts and ridges up to 0.8 µm high, ridges forming a partial to almost complete reticulum.

Very rare in Britain, only one authenticated record from Hampshire, widespread on the Continent, associated with *Pinus* on calcareous soils in warm areas.

Lactarius deliciosus (L.: Fr.) Gray
[deliciosus = edible and delicious]
Cap 4–11 cm, rounded then expanded and depressed at the centre, smooth, viscid, bright salmon-orange, orange-buff, to yellowish brown, zonate especially at the margin, discolouring pale greyish green with age or where bruised. **Gills** crowded, adnate to slightly decurrent, pale peach-orange, bright carrot coloured then grey-green where bruised. **Milk** is rather scanty, carrot orange. **Stem** 3–9 x 1–2 cm, cylindric, concolorous with the cap or slightly paler, whitish at the apex, with darker spots or scobiculae scattered over its surface. **Odour** not distinct. **Taste** mild to slightly bitter. **Spores** pale buff, 7.5–9 x 5–7 µm, warts and ridges up to 0.8 µm high, ridges forming a well-developed but incomplete reticulum.

Widespread and frequent wherever *Pinus* is planted on neutral to calcareous soils, perhaps commonest further north. Best distinguished by its carrot-coloured and more or less unchanging milk and almost completely reticulate spores. Most likely to be confused with **L. deterrimus**.

Lactarius salmonicolor R. Heim & Leclair
[salmonicolor = salmon coloured]
Cap 4–11 cm, entirely bright salmon-orange to peach-orange and unchanging, with narrow darker zones at the cap margin. **Milk** is bright carrot-orange becoming very slowly reddish. **Odour** pleasant, not distinct. **Taste** mild then bitter. **Spores** pale buff, 8–10 x 6–8 µm, ellipsoid, warts and ridges up to 1 µm high, ridges forming a very incomplete reticulum, mostly zebroid.

Rather rare in Britain, with scattered records from England and Scotland, associated with *Abies*.

Lactarius deterrimus Gröger
[deterrimus = very poor taste]
Cap 4–11 cm, rounded then flattened and depressed, smooth, viscid, apricot-orange, dull brownish yellow, sometimes zonate at the margin, soon flushed or bruised green; sometimes entirely green even when young. **Gills** crowded, pale yellow-orange, bruising green. **Milk** scanty, carrot-orange, red in about 10 min. **Stem** cylindric, 2–9 x 1–2 cm, concolorous with cap, usually without scrobiculae. **Flesh** orange flushed green in the cap, slowly reddening in about 5–15 mins. **Odour** not distinct. **Taste** unpleasant, bitter. **Spores** dull buff, 7.5–10 x 6.5–8.5 µm, warts and ridges to 0.8 µm high, ridges with few connectives, rather sparse, not forming a reticulum.

Quite common and widespread, associated with *Picea* and occasionally *Pinus*.

Lactarius quieticolor Romagn.
[quieticolor = of calm colour]
Syn. *Lactarius hemicyaneus* Romagn.
Cap 4–11 cm, soon flattened and depressed, viscid when wet, greyish buff, greyish pink, yellowish cinnamon, distinctly zonate, often with a bluish flush, more rarely distinctly blue, staining green when bruised. **Gills** crowded, adnate-decurrent, pale orange. **Milk** scanty, orange then slowly reddish brown, finally green. **Stem** cylindric, smooth, concolorous with the cap. **Flesh** pale orange, whitish at the centre, often bluish green under the cap cuticle, slowly vinaceous-brown in the cap and stem. **Odour** not distinct. **Taste** mild to slightly bitter. **Spores** pinkish buff, 8–10 x 6.5–8 µm, warts and ridges up to 0.8 µm high, ridges forming thick zebroid bands, partial to almost complete reticulum.

Uncommon but widely distributed, associated with *Pinus* on acid soils.

Lactarius

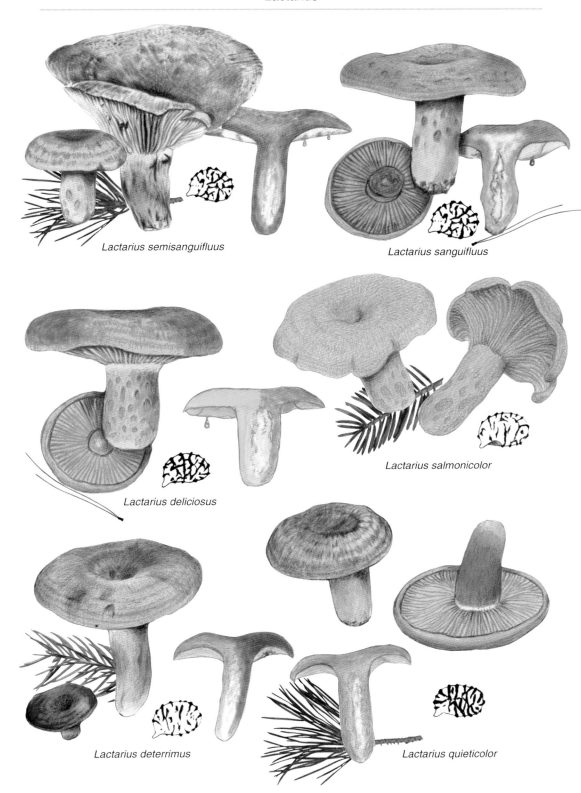

Lactarius semisanguifluus   *Lactarius sanguifluus*

Lactarius deliciosus   *Lactarius salmonicolor*

Lactarius deterrimus   *Lactarius quieticolor*

Section *Uvidi*: milk white, staining violaceous

Lactarius luridus (Pers.: Fr.) Gray
[luridus = lurid colour]
Cap 4–8 cm, rounded to depressed at centre, smooth and slightly viscid-sticky, vinaceous-grey, grey-buff, grey-brown, often with darker zones and spots, paler at margin. **Gills** crowded, adnate-decurrent, pale cream staining dark violet. **Milk** abundant, white, drying pale lilac, but not changing if isolated from the flesh. **Stem** 2–7 x 1–2 cm, cylindric-clavate, concolorous with the cap or paler, violet where bruised. **Flesh** whitish, hollow in the stem, deep lilac to greyish purple where cut. **Odour** not distinct. **Taste** mild then soon bitter. **Spores** pale cream, 8–11 x 6.5–9.5 μm, warts and ridges up to 1 μm high, ridges forming an almost complete, thick reticulum.

Very rare in Britain, known only from Scotland, growing in wet locations with broadleaf trees.

L. violascens (J. Otto: Fr.) Fr. has gills staining even darker purple but flesh pale lilac and spores with taller ridges and warts (up to 1.5 μm). A rare species known from scattered localities in Britain.

Lactarius uvidus (Fr.: Fr.) Fr.
[uvidus = moist or humid]
Cap 4–12 cm, soon flattened and depressed, almost funnel-shaped, smooth, viscid to glutinous, pale pinkish buff, grey-buff, lilac-grey or vinaceous-grey, staining lavender where bruised. **Gills** moderately crowded, adnate-decurrent, cream to pinkish buff, bruising lavender-violet. **Milk** abundant, white slowly staining lavender when in contact with the flesh. **Stem** 3–10 x 0.8–2.5 cm, cylindric, often tapered, white to pale cream or pinkish buff, turning pale lavender. **Flesh** whitish, staining pale lavender, hollow in the stem. **Odour** slightly fruity. **Taste** mild then soon bitter or even acrid. **Spores** pale cream, broadly ellipsoid, 8.5–11 x 6.5–8.5 μm, warts up to 0.8 μm high, ridges forming short crests.

Widespread but commonest in Scotland, associated with *Betula* and *Quercus*, possibly also *Salix*, usually on sandy or calcareous soils.

Lactarius flavidus Boud.
[flavidus = yellowish]
Cap 4–8 cm, soon expanded and depressed, smooth, viscid, pale yellow, yellowish cream to pale cream, usually slightly zonate with irregular spots, bruising livid reddish violet. **Gills** crowded, adnate-decurrent, pale cream bruising lavender-violet. **Milk** white, turning reddish violet both on and off the flesh. **Stem** cylindric, 4–6 x 1–2 cm, concolorous with cap, dry, bruising violet-lavender. **Flesh** whitish staining lavender-violet. **Odour** slightly fruity. **Taste** acrid. **Spores** pale cream, broadly ellipsoid, 8–11.5 x 6–8 μm, warts and ridges up to 1 μm high, with irregular, sometimes branched ridges but no reticulum.

Rather rare, with mixed broadleaved trees on calcareous soils, widespread in Britain. ***L. flavopalustris*** Kytöv. is yellower, has bitter milk which only changes colour on the flesh, not on paper and grows with *Betula*; not yet British.

Lactarius aspideus (Fr.: Fr.) Fr.
[aspideus = like a round shield]
Cap 3–6 cm, rounded and inrolled, then expanded-depressed, smooth, viscid but with the margin slightly velvety when young and becoming ribbed-crenulate when old, pale cream to cream-ochre, pale yellow. **Gills** crowded, adnate-decurrent, pale cream, lilac when bruised. **Milk** white, soon lavender-violet on the flesh but unchanging on paper. **Stem** 2–7 x 0.5–1.8 cm, cylindric-clavate, smooth, slightly viscid, concolorous with the cap, staining pale lavender. **Flesh** whitish, staining pale lavender. **Odour** slightly fruity. **Taste** mild then soon bitter. **Spores** pale cream, broadly ellipsoid, 6.5–9.5 x 5.5–7.5 μm, with warts and ridges up to 0.5 μm high, ridges forming a partial to almost complete reticulum, often with zebroid pattern.

Associated with *Salix* and possibly *Betula*, in wet areas, often by lakesides, widespread but uncommon. In highland areas of Scotland with dwarf *Salix* or *Dryas* is ***L. salicis-reticulatae*** which differs in its darker, pinkish buff gills and much larger spores (9–11.5 x 8–10 μm).

Lactarius repraesentaneus Britzelm.
[repraesentaneus = manifest]
Cap 5–15 cm, rounded with inrolled margin, then expanded and depressed, margin densely hairy-bearded, surface with concentric zones of hairy squamules, pale to bright yellow, viscid or glutinous but margin dry. **Gills** crowded, adnate-decurrent, cream, bruising lilac. **Milk** abundant, white, turning violet-lavender. **Stem** 4–12 x 1.5–4 cm, cylindric-clavate, becoming hollow, smooth, viscid, concolorous with the cap, usually with darker pits or spots. **Flesh** firm, whitish staining violet-lavender. **Odour** pleasant lilac hyacinth. **Taste** mild then bitter. **Spores** pale cream, 8–10.5 x 7–8.5 μm, broadly ellipsoid-subglobose, with warts and ridges up to 0.7 μm high, ridges forming an incomplete to almost complete reticulum.

Associated with *Betula*, mostly confined to Scotland.

Section *Zonarii*: cap ± viscid, often zonate, stem often pitted

Lactarius scrobiculatus (Scop.: Fr.) Fr.
[scrobiculatus = with small pits or depressions]
Cap 6–25 cm, soon flattened and depressed or even umbilicate at the centre, smooth to finely squamulose, margin minutely hairy, dry to viscid, pale ochre yellow to buff-yellow with darker, honey-coloured zones. **Gills** crowded, adnate-decurrent, cream to pale ochraceous. **Milk** quite abundant, white then slowly sulphur-yellow. **Stem** 3–8 x 1.5–4 cm, cylindric, pale cream to ochraceous with darker pits or scrobiculae. **Flesh** firm, hollow in the stem, white turning lemon-yellow when cut. **Odour** strongly acidic-fruity. **Taste** mild to bitter or acrid. **Spores** pale cream, 7–10 x 5.5–7 μm, with warts and ridges to 1 μm high; ridges thick and forming scattered meshes, no reticulum.

Associated with *Picea*, very rare in Britain, possibly absent, only a few very old records in herbaria.

Lactarius resimus (Scop.: Fr.) Fr.
[resimus = turned up or reflexed]
Cap 10–20 cm, soon flattened and depressed at centre, margin inrolled, minutely pubescent, viscid to glutinous, pure white then marked and stained with pale to darker yellow. **Gills** crowded, adnate-decurrent, pale cream. **Milk** scarce, white then turning lemon-yellow. **Stem** 4–5.5 x 2–3 cm, cylindric, concolorous with cap, sometimes with some ochre spots. **Flesh** firm, hollow in the stem, staining pale yellow. **Odour** slightly fruity. **Taste** mild then slowly acrid or bitter. **Spores** pale cream, broadly ellipsoid, 7–9.5 x 5.5–7 μm, with warts and ridges to 0.8 μm high, ridges forming a very incomplete reticulum, with some isolated warts.

Associated with *Betula*, very rare in Britain, known only from the Scottish Highlands.

Lactarius

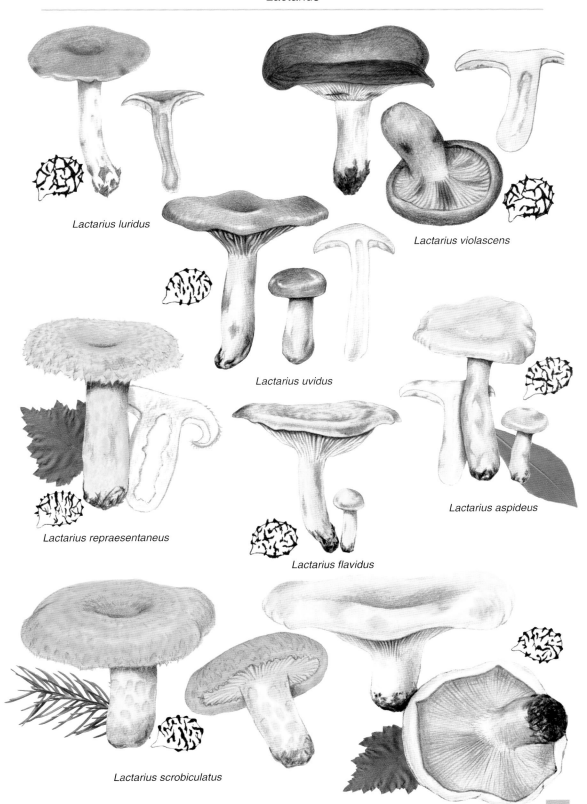

Lactarius citriolens Pouzar
[citriolens = citrus scented]
Cap 5–15 cm, funnel-shaped and strongly inrolled, margin prominently shaggy-hairy, surface viscid-glutinous (although the hairy margin is dry), pale cream to whitish yellow to pale yellow with age, faintly zonate, sometimes with watery spots. **Gills** moderately crowded, adnate-decurrent, pale cream to yellowish white. **Milk** white, quickly turning sulphur-yellow, lemon-yellow. **Stem** 4.5–6 x 2–2.5 cm, cylindric, pale cream, smooth, rather woolly-hairy at base. **Flesh** very firm, whitish becoming lemon-yellow when cut. **Odour** fruity-acidic, like lemons when old. **Taste** bitter then hot and acrid. **Spores** pale cream, ellipsoid, 6.5–8.5 x 5–6 µm, with warts and ridges to 0.8 µm high, ridges forming a few meshes but not really reticulate.

In mixed broadleaved woods, especially *Betula* on calcareous soils, fairly uncommon but widespread, especially in the south of England.

Subsection *Croceini*: cap dry to viscid, often zonate, margin pubescent; milk yellowing

Lactarius chrysorrheus Fr.
[chrysorrheus = gold flow]
Cap 4–8 cm, rounded then soon depressed, dish-shaped, pale salmon-pink, pinkish buff, ochraceous pink, with distinct zones and bands often of darker spots; surface smooth to greasy. **Gills** crowded, adnate, pinkish buff. **Milk** abundant, white then soon sulphur-yellow to chrome-yellow. **Stem** 2.5–4.5 x 1–2 cm, cylindric, smooth, slightly paler than the cap but otherwise concolorous. **Flesh** firm but often hollow in the stem, whitish, soon sulphur yellow especially in the cap. **Odour** not distinct. **Taste** mild then soon rather bitter and slightly acrid. **Spores** pale straw yellow, ellipsoid, 6–8.5 x 5–7 µm, with warts and ridges 0.5–1.0 µm high, ridges forming a ± well-developed reticulum.

Occasional and widespread in Britain, associated with *Quercus*. Easily recognised by the dry, zonate pinkish cap and strongly yellowing milk.

Subsection *Zonarii*: cap ± viscid, often zonate, milk white usually unchanging

Lactarius acerrimus Britzelm.
[acerrimus = very sharp pointed]
Cap 4–15 cm, soon flattened-convex then funnel-shaped, often irregular in outline, lobed and folded; pale yellow-buff, ochraceous cinnamon, pinkish buff, with darker concentric zones and spots, slightly viscid when moist. When young the cap margin is distinctly tomentose. **Gills** crowded, adnate, pale pinkish buff, anastomosing and with numerous cross-veins, especially near the stem. **Milk** fairly abundant, white, unchanging. **Stem** 2.5–5 x 2–4 cm, cylindric, pale buff to ochraceous. **Flesh** pale whitish buff, firm. **Odour** slightly fruity. **Taste** acrid to burning hot. **Spores** pinkish buff, 9–15 x 8–11 µm, with warts and ridges to 1.0 µm high, forming a partial network. **Basidia** are 2-spored.

Very uncommon, usually with *Quercus*, mainly in southern England but as far north as Yorkshire. The whole fungus often looks rather distorted or as if infected.

Lactarius zonarius (Bull.) Fr.
[zonarius = displaying zones]
Syn. *L. insulsus* (Fr.) Fr. *ss* Phillips (1981)
Cap 5–10 cm, soon flattened-convex then funnel-shaped, often irregular in outline, asymmetric; pale yellow-buff, pinkish buff, with numerous darker concentric zones of cinnamon-orange, slightly viscid when moist. **Gills** crowded, adnate, pale cream, bruising brownish. **Milk** fairly abundant, white, unchanging. **Stem** 2.5–5 x 2–4 cm, cylindric, pale buff to ochraceous, often with small, yellowish scrobiculae. **Flesh** pale whitish buff, firm, becoming pale vinaceous when cut. **Odour** slightly fruity. **Taste** mild then acrid to burning hot. **Spores** pale pinkish buff, 6–9 x 5–7 µm, with warts and ridges to 1.0 µm high, forming short, irregular ridges but never forming a reticulum.

Rare in Britain, possibly widespread but much confused with the similar *L. evosmus* and *L. acerrimus*; associated with *Quercus* on calcareous soils. From *L. acerrimus* and *L. evosmus* it differs in its paler spores and the clearly tomentose young margin. *L. acerrimus* also has much larger spores and 2-spored basidia, while *L. evosmus* starts much paler.

Lactarius evosmus Kühner & Romagn.
[evosmus = scented]
Cap 6–12 (15) cm, flattened-depressed with a strongly inrolled margin, gradually expanding to become broadly funnel-shaped but with margin remaining incurved, very pale cream at first, from almost azonate to quite densely zonate with yellowish brown to cinnamon-buff zones; bruising saffron-yellow to dark pinkish buff. **Gills** crowded, slightly decurrent or sinuate with a decurrent tooth, pale yellowish cream, pinker with age or brownish where bruised. **Milk** abundant, watery white, very quickly extremely acrid. **Stem** cylindric-clavate, dry, 3–6 x 2–3 cm, pale cream, bruising ochre to dark ochraceous-orange. **Flesh** firm, solid, whitish. **Odour** fruity-acidic, apple-like. **Taste** very acrid. **Spores** pinkish buff, broadly ellipsoid, 6–9.5 x 4.5–7.2 µm, with warts and ridges to 0.5 µm high, forming irregular ridges, sometimes in a zebroid pattern but no true reticulum (but see note below).

Uncommon in Britain, usually associated with *Quercus* or *Populus* in woodland or scrub on calcareous soils. Also known with *Helianthemum* on open, calcareous downland where it can grow in large rings. The latter collections can have rather more reticulate spores than normal and it would be interesting to see if a molecular study proved it to be an independent species, as with so many other of the numerous agarics now known to occur with that host.

Lactarius porninsis Rolland
[porninsis = after M. Pornin, French botanist]
Cap 5–12 cm, rounded with inrolled margin then soon flattened and depressed at centre, radially fibrillose, greasy to slightly viscid; bright yellowish orange with darker, cinnamon-orange narrow zones, especially at the margin. **Gills** crowded, adnate to slightly decurrent, pale buff to pinkish buff, yellowish buff. **Milk** quite abundant, white, unchanging. **Stem** cylindric-fusiform, 3–7 x 1–2.5 cm, smooth pale cream to yellowish or pinkish buff, with slightly darker bruising. **Flesh** firm, hollow in the stem, whitish to pale pinkish buff. **Odour** rather strong and fruity. **Taste** mild and pleasant then gradually bitter. **Spores** cream to pinkish buff, 6–9.5 x 5.5–7 µm, with warts and ridges to 0.5 µm high, forming irregular ridges connected by fine lines, sometimes forming an incomplete reticulum.

Very rare in Britain, only known from two localities in Scotland with introduced trees. Associated with *Larix*. Although traditionally placed with other members of subsection *Zonarii*, recent molecular studies indicate that this is actually an aberrant member of section *Dapetes* (which includes the familiar *L. deliciosus*). From these it differs in its white, unchanging milk and association with *Larix*. The latter association also separates it from other members of the subsection *Zonarii*.

Lactarius

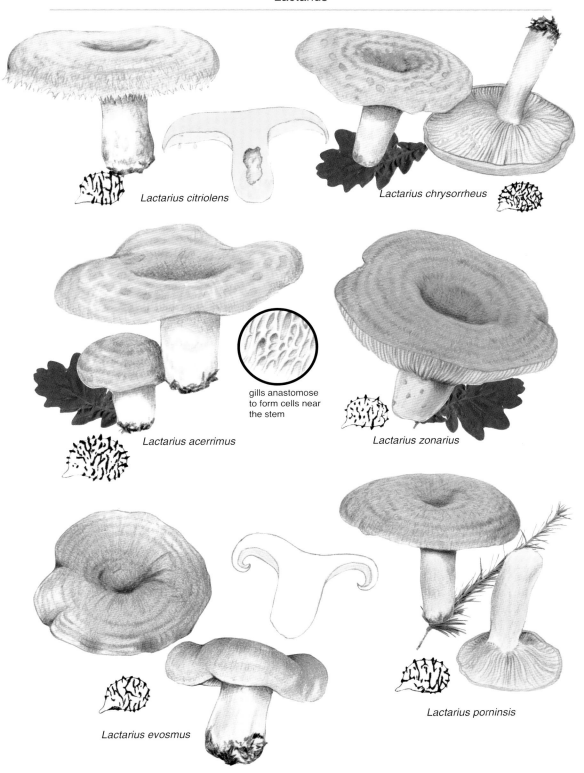

Lactarius citriolens

Lactarius chrysorrheus

gills anastomose to form cells near the stem

Lactarius acerrimus

Lactarius zonarius

Lactarius evosmus

Lactarius porninsis

Lactarius controversus Pers.: Fr.
[controversus = turned over or against]
Cap 6–30 cm, convex with strongly inrolled margin then soon flattened and depressed at centre, felty-tomentose, slightly viscid when wet; whitish becoming stained with reddish pink and finally entirely flushed pinkish buff. **Gills** very crowded, adnate-decurrent, distinctly pinkish as they mature. **Milk** sparse to moderately abundant, white, very acrid. **Stem** cylindric, tapered at base, 3–5 x 2–3 cm, smooth to tomentose, pale cream to yellowish or pinkish buff, with darker pink bruising. **Flesh** firm, solid in the stem, whitish to pale pinkish buff. **Odour** rather faint and fruity. **Taste** slowly becoming very acrid. **Spores** pale pinkish buff, ellipsoid to oblong, 6–8.5 x 4–6 µm, with warts and ridges up to 0.5 (0.8) µm high, ridges thick and well developed, forming a partial to almost complete reticulum.

Widespread and quite common in Britain wherever *Salix* and *Populus* grow, this is one of our largest species of *Lactarius*.

Section *Lactarius*: cap salmon, whitish, clay-buff, with hairy-bearded margin; milk white, unchanging

Lactarius torminosus (Schaeff.: Fr.) Pers.
[torminosus = griping]
Cap 4–12 cm, convex with strongly inrolled margin then soon flattened and depressed at centre, slightly to strongly felty-tomentose, surface and margin densely hairy-bearded, slightly viscid when wet; pale to deep flesh-pink, pink to brick with darker concentric zones. **Gills** very crowded, adnate-decurrent, distinctly pinkish as they mature. **Milk** abundant, white, very acrid. **Stem** cylindric, tapered, 4–7 x 1–2 cm, smooth to tomentose, pale cream to yellowish or pinkish salmon, with darker pink bruising. **Flesh** firm, hollow in the stem, whitish to pale pinkish buff. **Odour** rather faint and acidic-fruity. **Taste** quickly becoming very acrid. **Spores** pale cream, ellipsoid, 8–10 x 5.6–7.5 µm, with warts and ridges up to 0.75 µm high, ridges forming some closed meshes but no true reticulum.

Widespread and common with *Betula* everywhere.

Lactarius pubescens Fr.
[pubescens = minutely hairy or downy]
Cap 4–10 cm, convex with strongly inrolled margin, depressed at centre, slightly to strongly felty-tomentose, minutely hairy-woolly, slightly shaggier at the margin, slightly viscid when wet; pale cream to flesh-pink, becoming more ochraceous with age or when wet. **Gills** very crowded, adnate-decurrent, pale salmon-buff then pale pinkish as they mature. **Milk** abundant, white, very acrid. **Stem** cylindric, tapered, 4–7 x 1–2 cm, smooth to tomentose, cream to yellowish ochre when old. **Flesh** firm, hollow in the stem, pale pinkish buff. **Odour** rather faint and acidic-fruity. **Taste** quickly very acrid. **Spores** pale cream, ellipsoid 6–8.5 x 4–6 µm, with warts and ridges to 0.5 µm high, with some complete meshes, incompletely reticulate.

Common and widespread wherever *Betula* is found, often in open grassy areas, pathsides and lawns.

Lactarius scoticus Berk. & Broome
[scoticus = of Scotland]
Cap 2–6 cm, convex with strongly inrolled margin then flattened and deeply depressed at centre, dry and smooth to velvety, felty-tomentose, surface and margin minutely hairy-woolly, slightly shaggier at the margin, greasy to slightly viscid when wet; white to pale cream, becoming more ochraceous to yellow-brown at the centre. **Gills** very crowded, adnate-decurrent, pale cream then pale pinkish as they mature. **Milk** abundant, white, unchanging, very acrid. **Stem** cylindric, tapered at base, 4–7 x 1–2 cm, smooth to tomentose, pale cream to yellowish ochre when old. **Flesh** firm, hollow in the stem, whitish to pale pinkish buff. **Odour** rather faint and acidic-fruity. **Taste** quickly becoming acrid. **Spores** pale cream, ellipsoid, 5.5–7.7 x 4.5–5.5 µm, with warts and ridges to 0.5 µm high.

Rare, Scotland only, growing with *Betula*.

Lactarius mairei Malençon
[mairei = after René Maire, 1878–1948, mycologist]
Cap 3–12 cm, convex with strongly inrolled margin then soon flattened and depressed at centre, ochraceous buff, pinkish buff to cream-ochre, densely covered with large, tangled, radiating hairs, in pointed clumps at the margin. **Gills** crowded, adnate-decurrent, pale ochre-cream as they mature. **Milk** abundant, white, very acrid. **Stem** cylindric, tapered at base, 4–7 x 1–2 cm, smooth, pale cream to yellowish ochre when old. **Flesh** firm, hollow in the stem, whitish to pale pinkish buff. **Odour** acidic-fruity of *Pelargonium*, to slightly oily. **Taste** quickly becoming acrid. **Spores** cream, subglobose to broadly ellipsoid, 6–8.5 x 5–6.5 µm, with warts and ridges to 1 µm high, ridges thick, often branched, rarely forming closed meshes.

Rare in Britain, mainly southern, with *Quercus* on calcareous soils, it is considered endangered in several countries.

Lactarius spinosulus Quél.
[spinosulus = with small spines or hairs]
Cap 2–6 cm, convex with inrolled margin then soon flattened and slightly depressed at centre, flesh-pink, coral to pale vinaceous with darker vinaceous-pink zones and small dark squamules; the latter erect at the margin. **Gills** fairly crowded, adnate-decurrent, pale cream then pale ochre-cream as they mature. **Milk** moderately copious, white to cream, mild then slowly acrid and bitter. **Stem** cylindric, tapered at base, 4–7 x 0.5–1.2 cm, smooth to longitudinally wrinkled or furrowed, concolorous with the cap. **Flesh** firm, brittle, hollow in the stem, whitish to pale pinkish buff. **Odour** slightly fruity. **Taste** mild then slowly bitter-acrid. **Spores** yellowish buff, subglobose, 5–7 x 5–6 µm, with warts and ridges to 1 µm high, ridges forming zebroid stripes.

Very uncommon in Britain, mainly in Scotland, Wales and a few southern localities, it is associated with *Betula*.

Section *Atroviridi*: cap sticky, ± hairy at margin, olive-green turning purple with ammonia

Lactarius turpis (Wein.) Fr.
[turpis = ugly]
Cap 5–14 cm, convex with inrolled margin then soon flattened and slightly depressed at centre, yellowish olive to deep olive, sometimes almost greenish black when old, viscid when wet then dry and matt, often velutinous or squamulose near the margin. **Gills** fairly crowded, adnate-decurrent, pale cream then pale ochre-cream as they mature. **Milk** moderately copious, white, quickly very acrid and unpleasant. **Stem** cylindric, tapered, 4–6 x 1.5–2.5 cm, smooth to slightly wrinkled or furrowed, concolorous with the cap. **Flesh** brittle, hollow in the stem, whitish to pale pinkish buff. **Odour** not distinct. **Taste** very acrid. **Spores** pale cream, subglobose, 6.3–9 x 5–7 µm, with warts and ridges to 1 µm high, ridges forming a ± complete network. Ammonia on the cap = violet-purple.

Common everywhere with *Betula* on wet soils, occasionally with conifers also.

Lactarius

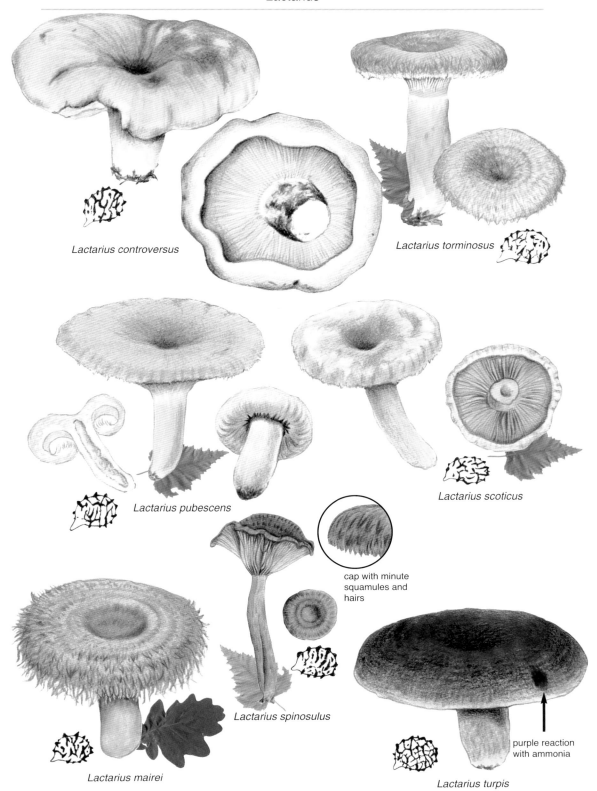

Section *Glutinosi*: cap sticky to slimy, ± smooth, milk white to greenish grey; spores often with zebroid ridges

Lactarius blennius (Fr.: Fr.) Fr.
[blennius = mucous or viscid]
Cap 5–10 cm, convex with inrolled margin then soon flattened and slightly depressed at centre, yellowish olive, grey-green to dull grey-buff, often with zones of darker droplet-like spots, viscid when wet. **Gills** fairly crowded, adnate to slightly decurrent, white then pale cream as they mature. **Milk** fairly copious, white to cream, drying on the gills pale olive-grey, quickly very acrid. **Stem** cylindric, tapered at base, 4–7 x 1.5–3 cm, smooth to slightly wrinkled or furrowed, concolorous with the cap, viscid. **Flesh** firm, brittle, hollow in the stem, whitish to pale greenish buff tint. **Odour** slightly acidic. **Taste** very acrid. **Spores** pale cream, broadly ellipsoid, 5.5–8 x 5–6.8 µm, with warts and ridges to 1 µm high, ridges loosely connected to form partial meshes, often in zebroid stripes.

Common everywhere with *Fagus*. The forma *virescens* differs in being much greener and with fewer spots on the cap.

Lactarius fluens Boud.
[fluens = flowing, i.e. exuding fluid]
Cap 5–10 cm, convex with inrolled margin then soon flattened and slightly depressed at centre, yellowish olive, grey-green to dull grey-buff, often with zones of darker droplet-like spots, viscid when wet then dry and shiny. **Gills** fairly crowded, adnate to slightly decurrent, white then pale cream as they mature. **Milk** moderately copious, white to cream, drying on the gills pale olive-grey, quickly very acrid and unpleasant. **Stem** cylindric, tapered at base, 4–7 x 1.5–3 cm, smooth to slightly wrinkled or furrowed, concolorous with the cap, viscid. **Flesh** firm, brittle, hollow in the stem, whitish to pale greenish buff tint. **Odour** slightly acidic. **Taste** very acrid. **Spores** pale cream, broadly ellipsoid, 5.5–8 x 5–6.8 µm, with warts and ridges to 1 µm high, ridges loosely connected to form partial meshes, often in zebroid stripes.

Rather uncommon, mainly on calcareous or clay soils with *Fagus*. The best field character is the noticeably pale, whitish margin of the cap.

Lactarius circellatus Fr.
[circellatus = with circular zones]
Cap 5–10 cm, convex with inrolled margin then soon flattened and slightly depressed at centre, grey, grey-green to dull grey-buff or greyish brown, with distinct narrow, darker zones and often with droplet-like spots, viscid when wet. **Gills** fairly crowded, adnate to slightly decurrent, white then pale cream to pinkish buff as they mature. **Milk** moderately copious, white to cream, drying on the gills pale olive-buff, quickly very acrid and unpleasant. **Stem** cylindric, tapered at base, 4–6 x 1.5–2.5 cm, smooth to slightly wrinkled or furrowed, concolorous with the cap, or paler, slightly viscid when wet but soon dry. **Flesh** firm, brittle, hollow in the stem, whitish to pale buff. **Odour** slightly acidic-fruity. **Taste** very acrid. **Spores** cream, subglobose to broadly ellipsoid, 5.5–8 x 4.5–6.5 µm, with warts and ridges to 1 µm high, ridges loosely connected to form several zebroid stripes, some isolated warts but no meshes.

Uncommon to rare but quite widespread wherever the host tree *Carpinus* is found.

Lactarius pyrogalus (Bull.: Fr.) Fr.
[pyrogalus = burning milk]
Cap 5–8 cm, convex with inrolled margin then soon flattened and slightly depressed at centre, often with wavy, irregular margin, surface smooth to slightly rugose, grey-green to dull grey-buff, pinkish grey or greyish brown, sometimes with indistinct narrow darker zones, viscid when wet. **Gills** noticeably widely spaced, adnate to slightly decurrent, cream then soon pale ochraceous as they mature. **Milk** copious, white to cream, drying on the gills pale olive-buff, quickly extremely acrid and unpleasant. **Stem** cylindric, tapered at base, 4–6 x 1.5–2 cm, smooth to slightly wrinkled or furrowed, concolorous with the cap, or paler, slightly viscid when wet but soon dry. **Flesh** firm, brittle, hollow in the stem, whitish to pale buff. **Odour** slightly acidic-fruity. **Taste** very acrid. **Spores** pale pinkish buff, subglobose to broadly ellipsoid, 5.0–7.5 x 4.5–6.2 µm, with warts and ridges to 1 µm high, ridges form partial zebroid stripes, some isolated warts and a few connectives.

Common everywhere under *Corylus*; the combination of wide-spaced, ochraceous gills, fiery milk and tree associate usually make it easy to identify.

Lactarius flexuosus (Pers.: Fr.) Gray
[flexuosus = flexing or wavy]
Cap 5–11 cm, convex with inrolled margin then soon flattened and slightly depressed at centre, fleshy, often with wavy, irregular margin, surface smooth to slightly rugose, variable in colour from grey to grey-buff, pinkish grey (strongly vinaceous-pink in the var. *roseozonatus*), usually with distinct narrow darker zones, viscid when wet then dry and shiny. **Gills** noticeably widely spaced, adnate to slightly decurrent, cream then soon pale pinkish buff as they mature. **Milk** copious, white to cream, quickly extremely acrid and unpleasant. **Stem** cylindric, tapered at base, 4–6 x 1.5–2 cm, smooth to slightly wrinkled or furrowed, concolorous with the cap, or paler, dry. **Flesh** firm, brittle, hollow in the stem, whitish to pale buff. **Odour** slightly acidic-fruity. **Taste** very acrid. **Spores** pale cream, subglobose to broadly ellipsoid, 5.0–7.5 x 4.5–6.2 µm, with warts and ridges to 1 µm high, ridges loosely connected to form partial zebroid stripes, some isolated warts and a few connectives.

In mixed woodlands. Rare, perhaps commonest in Scotland and northern England. The variety *roseozonatus* is sometimes treated as a separate species but seems to differ in little other than colour.

Lactarius vietus (Fr.: Fr.) Fr.
[vietus = shrunken]
Cap 5–8 cm, convex with rounded margin then soon flattened and slightly depressed at centre, fleshy, often with wavy, irregular margin, surface smooth to slightly rugose, variable in colour from grey to grey-buff, pinkish grey, clay-buff or even pale vinaceous grey, viscid when wet then dry and shiny; sometimes with faint zones at the margin. **Gills** fairly crowded, adnate to slightly decurrent, cream then soon pale pinkish buff as they mature. **Milk** fairly abundant, white to cream, drying dark olive-grey, quickly extremely acrid and unpleasant. **Stem** cylindric, tapered at base, 4–6 x 1–1.5 cm, smooth to slightly wrinkled or furrowed, concolorous with the cap, or paler, dry. **Flesh** fragile, brittle, hollow in the stem, whitish to pale buff. **Odour** faintly fruity. **Taste** very acrid. **Spores** pale cream, subglobose to ellipsoid, 7–9.5 x 5.5–7.5 µm, with warts to 0.8 µm high, ridges forming partial to entirely closed meshes.

Common and widespread wherever *Betula* is found in wetter areas, sometimes with *Sphagnum* but also in drier areas. **L. syringinus** Z. Schaef. differs in its larger fruitbodies with wavy, zonate caps with brighter, lavender-grey colours. It is possibly best regarded as a form of *L. vietus* pending any molecular work that might be done.

Lactarius

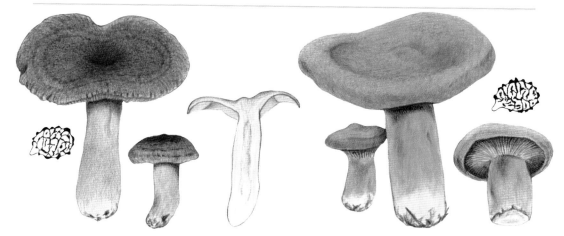

Lactarius mammosus

Lactarius helvus

Lactarius rufus

Lactarius quietus

Lactarius aurantiacus

Lactarius badiosanguineus

Lactarius sphagneti

Lactarius hepaticus Plowr.
[hepaticus = resembling liver, in colour]
Cap 2.5–7 cm, convex, soon expanded and depressed at the centre, usually with a blunt, rounded umbo, margin sometimes slightly crenulate, dull liver-brown, sepia to dark umber, paler at the margin. **Gills** medium crowded, slightly decurrent, pale pinkish buff to pale clay-buff. **Milk** white, turning yellow on white paper handkerchief, mild then distinctly bitter. **Stem** cylindric, clavate, 3.5–6 x 0.8–1 cm, dry and smooth, concolorous with the cap, or slightly paler. **Flesh** rather soft, brittle, often with a cavity in the stem, pale pinkish buff. **Odour** slightly 'buggy' like **L. quietus**. **Taste** mild to bitter. **Spores** pale cream, subglobose to ellipsoid, 6.5–9.5 x 5.5–7.5 μm, with warts and ridges to 1.3 μm high, ridges forming a more or less complete reticulum.

Associated with *Pinus* and usually fruiting late in the season, widespread and fairly common in Britain. Usually easily distinguished by its dull, liver-brown colours and yellowing, bitter milk.

Lactarius decipiens Quél.
[decipiens = deceptive, resembling another]
Cap 2–6 cm, convex, soon expanded and depressed at the centre, often with a blunt, rounded to acute umbo, smooth to slightly rugulose, pale pinkish buff, clay-pink to pinkish cinnamon. **Gills** medium crowded, slightly decurrent, pale pinkish buff to pale clay-buff. **Milk** copious, white, turning yellow on a white tissue, mild then distinctly bitter-acrid. **Stem** cylindric, clavate, 3.5–6 x 0.8–1 cm, dry and smooth, concolorous with the cap, or slightly paler, often furrowed or folded lengthwise. **Flesh** rather soft, brittle, often with a cavity in the stem, pale pinkish buff. **Odour** of *Pelargonium* or stewed fruits. **Taste** mild then acrid. **Spores** pale pinkish cream, subglobose to ellipsoid, 6.5–9.5 x 5–7.5 μm, with warts and ridges to 1 μm high, ridges forming a partial to complete reticulum.

Very uncommon in Britain, found with *Quercus* and *Carpinus*, mainly in the southern counties. Often found growing in small clumps with bases of stems together.

Lactarius britannicus Reid
[britannicus = from Britain]
Syn. *L. ichoratus*, *L. fulvissimus* of some authors
Cap 3–8 cm, convex, soon expanded and slightly depressed at the centre, smooth to slightly rugulose, at first dark brick-red, rust, then paler, more apricot, cinnamon-orange. **Gills** medium crowded, adnate-decurrent, pale ochraceous-cinnamon, cinnamon-buff. **Milk** fairly abundant, white then very slowly pale sulphur-yellow (10 mins), mild then slightly bitter-acrid. **Stem** cylindric, spindle-shaped, 3.5–10 x 1.5–2 cm, dry and smooth, concolorous with the cap, or slightly paler, darker at the base. **Flesh** rather firm, brittle, sometimes with a cavity in the stem, pale yellowish buff. **Odour** faint of bugs or of *Scleroderma*. **Taste** mild then bitter-acrid. **Spores** pale pinkish buff, subglobose to ellipsoid, 7–9 x 6–7.5 μm, with warts to 1.5 μm high, warts mostly isolated, occasionally in very short crests.

Associated with a variety of trees, *Fagus*, *Quercus*, etc and even conifers. Widespread in Britain especially in the southern counties.

Often synonymised with **L. fulvissimus**, I prefer to follow Basso in keeping them separate, at least until some DNA work is done. *L. fulvissimus* has non-yellowing milk and a different cap cuticle structure. It is likely that there is a species complex involved here, with several other names being available in the literature. They are badly in need of revision with molecular studies.

Lactarius fulvissimus Romagn.
[fulvissimus = very reddish brown]
Syn. *L. cremor* of some authors
Cap 3–7 cm, convex, then expanded and slightly depressed at the centre, smooth to slightly rugulose, at first dark brick-red, rust, then paler, more apricot, yellow-orange. **Gills** medium crowded, adnate-decurrent, pale ochraceous-cinnamon, cinnamon-buff. **Milk** moderately abundant, white unchanging, mild then slightly acrid. **Stem** cylindric, spindle-shaped, 3.5–6 x 1.0–1.2 cm, dry and smooth, concolorous with the cap, or slightly paler, darker at the base. **Flesh** rather firm, brittle, sometimes with a cavity in the stem, pale yellowish buff. **Odour** strong of bugs or of *Scleroderma*. **Taste** mild then slightly acrid. **Spores** pale pinkish buff, subglobose to ellipsoid, 7–9 x 6.5–7.5 μm, with warts to 1.5 μm high, warts sometimes isolated, usually forming short crests with some enclosed meshes.

Frequency and distribution in Britain is uncertain owing to confusion with **L. britannicus**. *L. fulvissimus* is best distinguished by its non-yellowing milk, spores with well-developed crests and cap cuticle formed of a loose palisade or cutis of slender cells embedded in a clear, slightly viscid outer layer. Associated with *Quercus*, *Tilia*, *Carpinus* and *Castanea*.

Lactarius subdulcis (Pers.: Fr.) Gray
[subdulcis = less than sweet]
Cap 2–5 cm, convex, soon expanded and depressed at the centre, smooth to slightly rugulose, pale pinkish buff, clay-pink to dull cinnamon. **Gills** medium crowded, slightly decurrent, pale pinkish buff to pale clay-buff, with ochre spots when old. **Milk** fairly abundant, white, mild then distinctly bitter. **Stem** cylindric, clavate, 3.5–6 x 0.8–1 cm, dry and smooth, concolorous with the cap, or slightly paler, darker at the base, often with white tomentose base. **Flesh** rather soft, brittle, often with a cavity in the stem, pale pinkish buff. **Odour** faint or slight of *Lepiota cristata*. **Taste** mild then bitter. **Spores** pale pinkish cream, subglobose to ellipsoid, 7–9.5 x 5.5–7 μm, with warts and ridges to 1.2 μm high, ridges forming irregular crests, no real reticulum.

Common and widespread everywhere, associated with *Fagus*. Most often confused with the equally common **L. tabidus**, which differs in its preference for *Betula* and slightly acrid milk yellowing on a white tissue.

Lactarius lacunarum Romagn. ex Hora
[lacunarum = hollow]
Cap 2–6 cm, convex, soon expanded and depressed at the centre, smooth to slightly rugulose, often crenulate at the margin, yellow-brown, rufous, dark cinnamon, slightly hygrophanous. **Gills** medium crowded, slightly decurrent, pale pinkish buff to pale clay-buff, with dark ochre spots when old. **Milk** fairly abundant, white, turning pale yellow on the flesh and on a white tissue, mild then slightly bitter-acrid. **Stem** cylindric, clavate, 3.5–6 x 0.8–1 cm, dry and smooth, concolorous with the cap, or slightly paler, darker at the base. **Flesh** rather soft, brittle, often with a cavity in the stem, pale pinkish buff. **Odour** faint or slight of *Lactarius quietus* or fruity. **Taste** mild then bitter. **Spores** pale cream, subglobose to ellipsoid, 6–8.0 x 5–6.5 μm, with warts and ridges to 1.2 μm high, ridges forming irregular crests, with partial reticulum.

Widespread and frequent in Britain although rarely reported. With *Alnus*, *Betula*, *Salix* or *Quercus* in wet, boggy areas, periodically flooded hollows, ruts, etc.

Lactarius

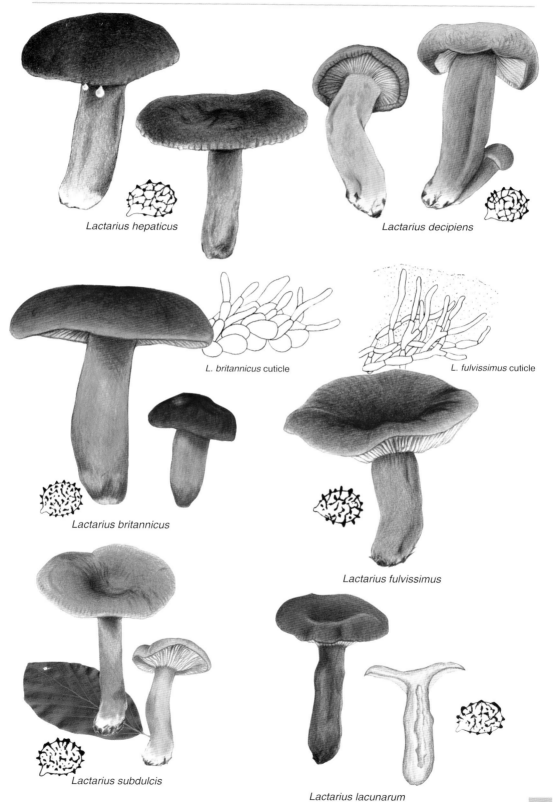

Lactarius hepaticus
Lactarius decipiens
L. britannicus cuticle
L. fulvissimus cuticle
Lactarius britannicus
Lactarius fulvissimus
Lactarius subdulcis
Lactarius lacunarum

Section *Tabidi*: cap cuticle complex, of swollen cells: i.e. epithelium, hyphoepithelium; no distinct odour

Lactarius tabidus Fr.
[tabidus = stunted]
Cap 2–4 (5) cm, convex, soon depressed at the centre, with a pointed umbo, often distinctly radially wrinkled around the umbo, crenulate at the margin when old; yellow-brown, orange-brown to pinkish ochre when old, ± hygrophanous. **Gills** medium crowded, pale pinkish buff. **Milk** fairly abundant, white, turning pale yellow on a white paper tissue, mild then slightly bitter-acrid. **Stem** cylindric-clavate, 3.5–7 x 0.5–1 cm, smooth, slightly paler than cap, darker at the base. **Flesh** brittle, hollow in the stem, pale pinkish buff. **Odour** faint or oily. **Taste** mild then bitter-acrid. **Spores** whitish, subglobose to ellipsoid, 6–8.6 x 5–7 µm, with warts and ridges to 1.2 µm high, with many isolated warts, ridges forming irregular fine crests, with only a partial reticulum.

Common and widespread wherever *Betula* occurs.

Lactarius rubrocinctus Fr.
[rubrocinctus = with a red band]
Cap 2–7 cm, soon expanded and depressed at the centre, distinctly radially wrinkled when old; yellow-brown, orange-brown, rust-brown to orange ochre. **Gills** medium crowded, ± decurrent, pale pinkish buff to pale ochre, with dark brown spots when old; edges bruising faintly violaceous. **Milk** fairly abundant, white, mild then slightly bitter. **Stem** cylindric, clavate, 3.5–7 x 1-2.5 cm, dry and smooth, concolorous with the cap, or slightly paler, with a darker, reddish brown zone immediately below the gills. **Flesh** rather brittle, often hollow in the stem, pale pinkish buff. **Odour** faint or slight of *Lactarius quietus*. **Taste** mild then bitter-acrid. **Spores** cream, subglobose to ellipsoid, 6.5–9 x 5–7.5 µm, with warts and ridges to 1 µm high, ridges forming irregular zebroid crests, with only a partial reticulum.

Uncommon to rare in Britain, associated with *Fagus* and also with *Helianthemum* on calcareous soils. The reddish zone just below the gills is not always prominent.

Lactarius obscuratus (Lasch: Fr.) Fr.
[obscuratus = darkly coloured]
Syn. *L. obnubilis*
Cap 0.5–3 cm, flattened and depressed at the centre, often with a small umbo at centre, often radially wrinkled when old; yellow-brown, sepia-brown, umber-brown to yellowish brown when old. **Gills** medium crowded, ± decurrent, pale pinkish buff to pale ochre. **Milk** rather sparse, white, mild. **Stem** cylindric, slender, 1–3 x 0.2-0.5 cm, smooth, slightly paler than cap. **Flesh** rather soft, brittle, often hollow in the stem, pale pinkish buff. **Odour** faint or sweetish. **Taste** mild. **Spores** cream, ellipsoid, 6.3–9 x 5–7 µm, with warts and ridges to 1.4 µm high, ridges forming irregular, partial reticulum.

Frequent and widespread in Britain, always associated with *Alnus*, often in small troops on very wet soils.

This species is quite difficult to separate from two other *Alnus* associates:
L. cyathuliformis usually has distinct olivaceous tints in the cap centre, often presenting a rather distinct 'eye-spot' and is sometimes more robust than *L. obscuratus* and with larger spores, 7–11 x 6–8.5 µm.
L. omphaliformis looks rather like a small *Laccaria* species with bright orange-brown colours and soon develops a concentrically scaly cap surface. Spore size is intermediate between the other two, 7–9.5 x 5.5–7.8 µm.

All three are equally frequent in Britain in suitable habitats.

Lactarius atlanticus** Bon
(atlanticus = originally described from Atlantic oak woods)
Cap 2–6 cm, smooth-granulose, deep reddish brown, garnet-red, margin often crenate. **Gills** adnate-decurrent, reddish ochre. **Stem** concolorous with cap. **Odour** of bugs-oily. **Taste** mild to bitter. **Spores** 7–9 x 6.5–8.5 µm with warts to 1 µm high, ridges forming a ± complete reticulum. **Cuticle** an epithelium of rounded cells.

With *Quercus ilex* in coastal woodlands, not yet British but might just occur here.

Lactarius camphoratus (Bull: Fr.) Fr.
[camphoratus = strong smelling]
Cap 2–4 cm, soon depressed at the centre, often with a small umbo, smooth to crenulate at the margin; red-brown, orange-brown, umber-brown to yellowish buff. **Gills** medium crowded, ± adnate-decurrent, pale pinkish buff to pale ochre. **Milk** rather sparse, watery white, mild to slightly astringent. **Stem** cylindric, slender, 2–6 x 0.5-1 cm, dry and smooth, dark brick to distinctly vinaceous brown. **Flesh** rather brittle, often hollow in the stem, pale pinkish buff. **Odour** weak then soon strong of curry as it dries. **Taste** mild to astringent. **Spores** yellowish cream, subglobose to broadly ellipsoid, 6–8.5 x 5–7 µm, with mostly isolated warts to 1.2 µm high, with some fine connectives but no reticulum.

Common and widespread in Britain with broadleaved trees. The vinaceous stem is a good field character, backed up by the distinct odour as it dries.

Lactarius rostratus Heilmann-Clausen
[rostratus = with a beak or rostrum]
Syn. *L. cremor* ss. Bon
Cap 1–3.5 cm, depressed at the centre, often with a small umbo at centre, irregularly rugulose, granular; red-brown, orange-brown, ochre-orange to yellowish buff when old. **Gills** medium crowded, ± adnate-decurrent, pale pinkish buff to pale ochre. **Milk** watery white, mild. **Stem** cylindric, slender, 1–2.5 x 0.5-1 cm, dry and smooth, dark brick to distinctly vinaceous brown. **Flesh** rather soft, brittle, often hollow in the stem, pale pinkish buff. **Odour** strong of bugs or *L. quietus*. **Taste** mild. **Spores** whitish, subglobose to broadly ellipsoid, 6–8 x 5.5–7 µm, with warts to 1.5 µm high, with ridges forming some closed meshes. **Gill cystidia** are characteristically rostrate with a long, slender appendage.

Rarely reported in Britain and probably uncommon, it is found among mosses under *Fagus*. There appears to be a larger species in Britain also with rostrate cystidia which might equate to the original **L. cremor** of Fries.

Lactarius serifluus (DC: Fr.) Fr.
[serifluus = watery milk]
Syn. *L. cimicarius, L. subumbonatus*
Cap 3–8 cm, flattened and depressed at the centre, sometimes with a small umbo at centre, surface radially rugulose-wrinkled; sepia-brown, umber-brown, fuscous to cinnamon when old. **Gills** medium crowded to distant, ± adnate-decurrent, pale pinkish buff to pale ochre. **Milk** watery white, mild. **Stem** cylindric, slender, 2.5–5 x 1–1.5 cm, dry and smooth, dark brick to distinctly vinaceous brown. **Flesh** rather soft, brittle, often hollow in the stem, pale pinkish buff. **Odour** strong, sweet, of bugs or *L. quietus*. **Taste** mild to unpleasant. **Spores** whitish, subglobose to broadly ellipsoid, 6–8.5 x 5.5–8 µm, with warts to 1.2 µm high, with ridges forming a ± complete reticulum.

Uncommon but widespread in Britain, in mixed woods, the very dull, dark brown colours, wrinkled and often crenulate caps are distinctive.

Lactarius

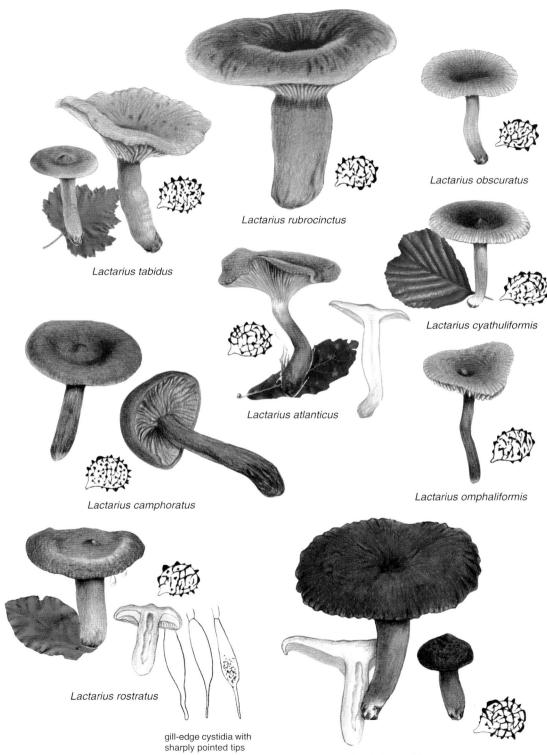

Subgenus *Plinthogali*
cap dry, often velvety; spores coarsely ornamented, often winged

Lactarius lignyotus ** Fr.
[lignyotus = smoky]
Cap 3–11 cm, convex-depressed, usually with a distinct pointed umbo at the centre, matt, velvety and often minutely wrinkled; margin crenate with age; fuscous black to very dark brown, paler fuscous with age. **Gills** rather widely spaced, decurrent, white to pale cream. **Milk** white, drying dull pink. **Stem** 4–10 x 0.5–1.5 cm, cylindric, often wrinkled grooved, concolorous with the cap, white tomentose at the base. **Flesh** thin, brittle, white then pale pinkish as the milk dries. **Odour** pleasant, not distinct. **Taste** mild. **Spores** pale yellowish cream, globose to broadly ellipsoid, 8–10 x 7.5–9.0 µm, with warts up to 1.8 µm high, with large, thick ridges or wings forming a partial network.

Not known in Britain, it might occur in Scotland associated with *Picea*.

Lactarius acris (Bolton: Fr.) Gray
[acris = with acrid taste]
Cap 4–9 cm, convex to funnel-shaped, often radially wrinkled, viscid-sticky, later dry, whitish when young then pinkish buff, clay-buff, often mottled with paler areas. **Gills** adnate to slightly decurrent, fairly crowded, pale buff to warm ochre-buff. **Milk** fairly copious, white slowly turning rose-pink, even when removed from the flesh. **Stem** 3–8 x 1–2.5 cm, irregularly furrowed, whitish cream, staining yellowish to clay-pink when bruised. **Flesh** firm, brittle, often hollow in the stem, white then quickly pinkish to brownish orange. **Odour** slightly aromatic, medicinal. **Taste** mild at first then quickly very acrid, finally of shellfish. **Spores** pinkish buff, globose to broadly ellipsoid, 7–9 x 6.5–8 µm, with warts and ridges to 1.8 µm high, ridges forming large, thick, often forked wings in a partial network.

Widespread in Britain although rarely reported, with *Fagus*, *Quercus*, *Corylus*, etc on calcareous soils.

Lactarius pterosporus Romagn.
[pterosporus = winged spores]
Cap 4–9 cm, convex to funnel-shaped, surface dry, matt and radially wrinkled, grey-brown, olive-buff, pale sepia-buff, often with a narrow paler zone at the margin. **Gills** adnate-decurrent, fairly crowded, pinkish buff to pale ochre. **Milk** white, pale grey-pink as it dries, bitter-acrid. **Stem** 3–8 x 1–1.5 cm, cylindric-clavate, often wrinkled-furrowed, whitish to pale pinkish buff. **Flesh** firm, brittle, often hollow in the stem, white then rose-pink when cut. **Odour** musty, medicinal. **Taste** astringent to very acrid. **Spores** dark pinkish buff, subglobose to broadly ellipsoid, 7–8.5 x 6–7.5 µm, with warts and ridges up to 2.5 µm high, ridges forming enormous wings encircling the spore, with some isolated warts.

Uncommon in Britain but widespread, associated with broadleaved trees, usually on calcareous soils.

Lactarius romagnesii Bon
[romagnesii = after H. Romagnesi, 1912–1999, mycologist]]
Cap 5–12 cm, convex to funnel-shaped, surface dry, pruinose-velvety, radially wrinkled around centre, dark fuscous brown, clay-brown, margin usually distinctly crenate when mature. **Gills** adnate-decurrent, distant, pale ochre when mature. **Milk** white, quite sparse, unchanging (flesh turns pink however). **Stem** 5–8 x 1–2.5 cm, cylindric-tapered, concolorous with cap or slightly paler. **Flesh** firm, brittle, often hollow in the stem, whitish then pink when cut. **Odour** faint, oily or of shellfish. **Taste** mild to slightly bitter. **Spores** pinkish buff, globose to broadly ellipsoid, 7–9.5 x 6.5–8 µm, with warts and ridges to 2.5 µm high, ridges forming large wings or flanges, often forked.

Rare and mainly southern in Britain, associated with *Fagus*, *Quercus* or *Corylus* on calcareous soils.

Lactarius ruginosus Romagn.
[ruginosus = wrinkled]
Cap 4–8 cm, convex, soon flattened and slightly depressed, sometimes with a small umbo, dry, minutely velvety, pale grey-brown, buff, margin crenate. **Gills** adnate, rather distant, pale cream to pale ochre. **Milk** fairly sparse, white, slowly reddish pink when in contact with the flesh, unchanging if isolated. **Stem** 3–5 x 1–1.5 cm, cylindric, concolorous with the cap but often much paler, whitish at the base. **Flesh** firm, whitish turning salmon-pink. **Odour** earthy, musty or spermatic. **Taste** very acrid. **Spores** dark pinkish buff, globose to broadly ellipsoid, 6.5–9.0 x 6–8 µm, with warts and ridges to 2 µm, with ridges forming large encircling wings.

Rare, mainly in southern England, with *Fagus*, *Quercus*, *Carpinus* or *Corylus* on rich soils.

Lactarius azonites (Bull.) Fr.
[azonites = without zones]
Cap 4–8 cm, convex-depressed, more or less smooth, minutely velvety, often slightly wrinkled, pale ochraceous, greyish brown to clay-buff, margin sometimes slightly crenate and often very pale. **Gills** adnate to slightly decurrent, crowded to moderately distant, rather irregular and frequently anastomosing, pale cream-ochre. **Milk** abundant, white then staining the flesh pink. **Stem** 3–5 x 1–1.5 cm, cylindric, smooth, white to pale greyish, staining pink. **Flesh** firm, solid, white then bright pink to coral when cut. **Odour** pleasant, fruity or medicinal. **Taste** mild to bitter or slightly acrid. **Spores** dark pinkish buff, globose to broadly ellipsoid, 7.5–9.0 x 7–8 µm, with warts and ridges to 1.5 µm high, ridges narrow, forming a partial reticulum.

Uncommon but widely distributed in Britain.

Lactarius fuliginosus (Fr.: Fr.) Fr.
[fuliginosus = sooty brown]
Cap 4–12 cm, convex to flattened, dry, smooth to slightly wrinkled at the centre, sooty brown, sepia to grey-brown or clay-buff. **Gills** fairly crowded, adnate-decurrent, cream to pale ochre-buff when mature. **Milk** white, fairly sparse, drying pink. **Stem** cylindric-clavate, 4–8 x 1–2 cm, concolorous with the cap. **Flesh** firm, solid, whitish then soon pink when cut. **Odour** nil to musty. **Taste** mild to slightly acrid. **Spores** pinkish buff, globose to broadly ellipsoid, 7.5–9.0 x 6.5–8.4 µm, with warts and ridges to 1 µm high, ridges forming an almost complete reticulum.

Fairly common and widespread in Britain, with broadleaved trees. If under *Picea* then see **L. picinus**.

Lactarius picinus Fr.
[picinus = pitch black]
Very rare in Britain, only one authentic modern record from Scotland, associated with *Picea*. Identical to **L. fuliginosus** except for its larger spore ridges.

Lactarius

Genus *Lactifluus*
Subgenus *Lactifluus*
cap dry, often velvety, frequently cracking

Lactifluus volemus (Fr.: Fr) Kuntze
[volemus = filling the palm of the hand]
Cap 5–15 cm, rounded, soon expanded, surface dry, smooth to minutely plush, often with fine cracks, bright cinnamon, orange-brown to yellowish orange, paler at the margin. **Gills** moderately distant, pale cream-buff, bruising brown to brick. **Milk** white, abundant, turning brown, mild. **Stem** 5–10 x 1.5–3 cm, cylindric-tapered, concolorous with the cap, dry, smooth. **Flesh** firm, pale cream bruising reddish brown, green with $FeSO_4$. **Odour** as it ages distinct of shellfish. **Taste** mild. **Spores** whitish, globose to subglobose, 9–11 x 8.5–10.5 µm with warts and ridges up to 0.5 (1.0) µm high, ridges forming thick lines connected into a complete reticulum.

Widespread in Britain but uncommon, commoner in parts of Scotland. With *Fagus*, *Quercus* or *Betula*.

Lactifluus subvolemus ** Van de Putte & Verbeken
[subvolemus = close to *L. volemus*]
Described in 2016 this species was previously considered as a simple variant of *L. volemus* but DNA analysis shows it to differ consistently from that species and hence worthy of separation. Although it is usually a pale yellowish buff it can occasionally be more orange-brown (especially when very young) and can then look more similar to *L. volemus*. Fully expanded specimens should be used to assess colour. With *L. volemus* it shares the long (15–125 µm), thick-walled hairs on the pileipellis.

Widespread although a little less common than **L. volemus** in mixed woodlands. Its presence and distribution in Britain is uncertain as yet.

Lactifluus oedematopus (Scop.) Kuntze
[oedematopus = swollen stem]
Syn. *L. volemus* var. *oedematopus*
Well separated from both *L. volemus* and *L. subvolemus* by its darker brown or orange-brown hues and particularly by the shorter (10–60 µm) thick-walled hairs on the pileipellis. It is found mainly under broadleaved trees and although widespread in Europe is distinctly uncommon to rare. It has been recorded in Britain under its synonym but its exact distribution is uncertain.

Lactifluus rugatus ** (Kühner & Romagn.) Verbeken
[rugatus = roughened]
Cap 5–7 cm, rounded then soon flattened and depressed, dry, velvety-matt, often with a concentrically rugose-wrinkled texture, bright rust-orange, red-brown. **Gills** widely spaced, pale cream-ochre. **Milk** copious, white, browning on the gills, mild. **Stem** 3–5 x 1.5–2.5 cm, cylindric, concolorous to slightly paler than the cap. **Flesh** firm, brittle, whitish, pink with $FeSO_4$. **Odour** indistinct. **Taste** mild. **Spores** 9–10 x 6–7 µm, with warts and ridges to 0.3 µm high, ridges forming an almost complete reticulum.

With broadleaved trees in the Mediterranean, often confused with **L. volemus**.

Lactifluus brunneoviolascens ** (Bon) Verbeken
[brunneoviolascens = brown and violaceous]
Syn. *L. luteolus* sensu auct non Peck
A white species whose milk and flesh stains violaceous brown. Distinguished by its extremely long, thick-walled hairs on the cap cuticle and its unusually isolated spore warts. Frequent under *Quercus* species in Mediterranean coastal woods.

Subgenus *Piperati*
cap dry, often velvety, pileipellis a lamprotrichoderm with thick-walled hairs

Lactifluus piperatus (L.: Fr.) Pers.
[piperatus = peppery]
Syn. *L. pergamenus* (Sw.: Fr.) Kuntze ss. Romagnesi
Cap 5–15 cm, rounded with inrolled margin, soon expanded and depressed, dry, finely velutinous, white to yellowish cream, sometimes with darker spots. **Gills** slightly decurrent, narrow, extremely crowded, whitish. **Milk** white, sometimes bluish green when dry, unchanging to yellow-orange with KOH, very acrid. **Stem** cylindric-tapered, 3–8 x 1–4 cm, white, minutely velvety. **Flesh** firm, solid, white then yellowish to bluish green after some hours. **Odour** faint of honey when drying. **Taste** very acrid. **Spores** white, ellipsoid, 7.0–10.0 x 5.5–7.5 µm, with warts and ridges up to 0.2 µm high, ridges forming short, irregular lines but no connectives.

Widespread but uncommon in Britain, associated with broadleaved trees.

Lactifluus glaucescens (Crossl.) Verbeken
[glaucescens = turning bluish green]
Usually distinguished from **L. piperatus** by the greenish colouration of the milk when drying and the yellow-orange reaction of the milk to KOH. This is variable however and these changes can occur in both species. The best characters for separating them are spore size and ornamentation—slightly smaller in *L. glaucescens* and with hardly any cross connections of the spore ridges—and the structure of the cap cuticle. In *L. glaucescens* the cuticle has an 80–120 µm layer of filamentous hyphae overlaying the ± completely rounded, cellular cells of the subpellis. In **L. piperatus** the cap cuticle has a thin, 10–30 µm layer of filamentous hyphae.

Lactifluus vellereus (Fr.: Fr) Kuntze Pl. 67
[vellereus = woolly or fleecy]
Cap 5–30 cm, with inrolled margin, soon broadly funnel-shaped, dry, velvety, white-cream or yellowish white, brown where damaged. **Gills** decurrent, crowded, narrow, often forked, white to pale cream, reddish brown when bruised. **Milk** abundant, white, unchanging with KOH, taste when isolated from the flesh mild. **Stem** cylindric, 3–7 x 2–6 cm, white, dry and velvety. **Flesh** firm, solid, whitish. **Odour** acidic. **Taste** of flesh soon very acrid. **Spores** white, globose to ellipsoid, 8–12 x 6.3–9.5 µm, with warts and ridges to 0.2 µm high, ridges forming fine, lines in a reticulum.

Widespread and common throughout Britain in broadleaved woods, more rarely with conifers.

Lactifluus bertillonii (Neuhoff ex Z. Schaef.) Verbeken
[bertillonii = after Prof. Louis Bertillon, 1821–1883]
Very similar to, and often indistinguishable macroscopically from, **L. vellereus** but easily distinguished by the orange KOH reaction on the milk and by the acrid taste of the milk when isolated from the flesh.

Associated with *Fagus*, *Quercus* and *Betula*, rather rare in Britain.

Lactifluus

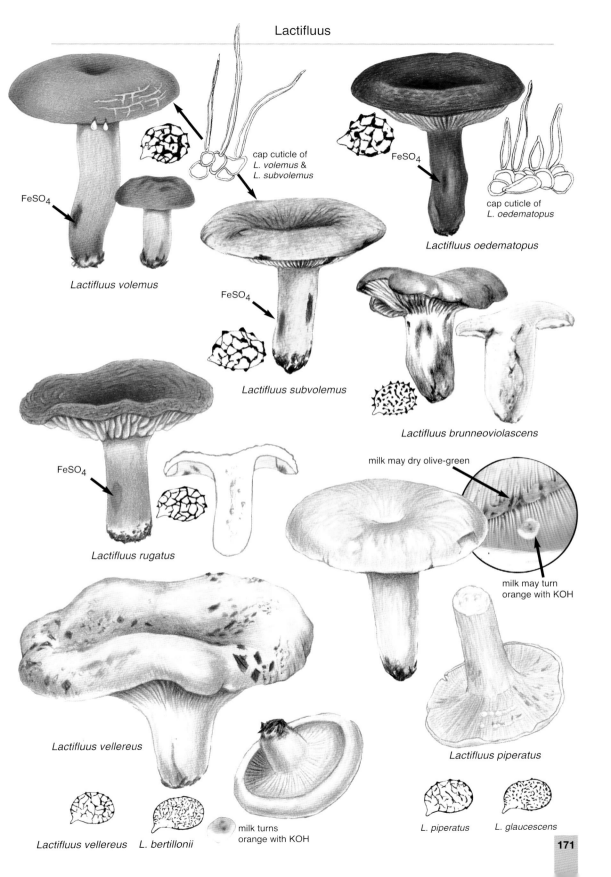

Genus *Russula*

Around 160 species in Britain and many more species in southern Europe. All share spores with amyloid ornamentation. Spore colour is an important feature.

Russula spore deposit colour chart
Ia–IVe = Romagnesi colour system

Ia   IIa   IIc   IId   IIIa   IIIc   IVc   IVe

Subgenus *Compactae*
Section *Compactae*
fruitbodies robust, very hard, brittle, flesh reddening and/or blackening; with numerous half-length gills

Russula nigricans (Bull.) Fr.
[nigricans = blackening]
Cap 10–20 cm, very firm, hard and brittle, soon depressed and crater-like with incurved margin, surface dry and smooth to slightly roughened, ivory-white then soon flushing brown and finally pitch black as if burned. **Cuticle** peeling 3/4. **Gills** very thick, brittle, widely to very widely spaced with numerous half or quarter-length gills present, ivory to pale buff bruising scarlet then brown. **Stem** stout, very firm and brittle, white bruising red and finally black. **Taste** mild then slowly acrid in the gills. **Odour** a mix of fruit and fungoid. **Macrochemicals**: $FeSO_4$ deep green, guaiac strongly positive. **Pileocystidia** difficult to separate from the cuticular hyphae. **Cuticular hyphae** long, 3–6 µm wide, hyaline, rather simple with few ramifications, end cells long and fusiform. **Spores** 7–8 x 6–7 µm, warts up to 0.25 (0.5) µm high, with very fine connectives and a partial reticulum. **Spore deposit** white (Ia).

Common everywhere under both broadleaf and coniferous trees on a wide range of soils.

Russula anthracina Romagn.
[anthracina = coal-black]
Cap 7–12 cm, rounded at first then soon expanded with depressed centre, dull white or cream soon flushing or staining sepia-brown to blackish brown, dry, matt. **Cuticle** adnate. **Gills** moderately crowded, cream and with a faint pinkish flush. **Stem** short and very firm, white bruising brownish (not bright red) when scratched, finally blackish. **Taste** mild in the flesh, slightly acrid in the gills. **Odour** with a faint fruity odour. **Macrochemicals**: $FeSO_4$ slowly grey-green, guaiac uncertain, probably positive. **Pileocystidia** absent. **Cuticular hyphae** and hyphae of stem cortex slender and when treated with SV with numerous black globules distributed through the cells. **Spores** 8–10 x 7–8 µm, warts tiny, 0.2 µm with very fine partial reticulum. **Spore deposit** white (Ia).

Apart from *R. nigricans* this is, in my experience, the commonest species in subgenus *Compactae* in southern UK, growing with a wide range of broadleaved trees.

Best distinguished by the flesh turning directly grey-brown and the cuticular characters.

Russula albonigra (Krombh.) Fr.
[albonigra = white and black]
Cap 4–10 cm, rounded then expanded with a depressed centre, dull white to cream staining blackish with age or injury. **Cuticle** adnate. **Gills** moderately crowded, narrow, white staining black. **Stem** robust, firm, white staining black. **Taste** mild but with an odd, slightly menthol, cooling or bitter flavour. **Odour** musty but not distinct. **Macrochemicals**: $FeSO_4$ pink-rust, guaiac strongly positive. **Pileocystidia** absent although some cystidia-like cells with oily contents may be visible but unstained in SV. **Spores** 7–9 x 5.5–7 µm, with very low warts, 0.2 µm and fine reticulum. **Spore deposit** white (Ia).

Growing with both conifers and broadleaf trees it is certainly rare in Britain and its distribution is uncertain due to confusion with *R. anthracina*.

All parts of this white species stain directly coal black without a red stage.

Russula adusta (Pers.) Fr.
[adusta = dark or swarthy]
Cap 5–12 cm, rounded then soon expanded and usually depressed at the centre, whitish to pale brown, does not darken as much as the more common *R. nigricans*. **Cuticle** ± adnate. **Gills** ivory white, intermediate in density. **Stem** robust, firm, often short and stocky, white to dirty cream. **Taste** mild. **Odour** musty, woody with wine overtones. **Macrochemicals**: $FeSO_4$ rose-rust then slowly grey-brown, guaiac strongly positive. **Pileocystidia** vermiform, often capitate or apically branched. **Spores** 7–9 x 6–8 µm, with low warts under 0.5 µm and with more or less complete fine reticulum. **Spore deposit** white (Ia).

Rather uncommon in Britain, it appears to be confined to *Pinus*. The flesh becomes slowly pinkish then grey and is mild to taste. The distinctive smell of old wine casks can be faint.

Russula densifolia Gillet
[densifolia = with densely crowded gills]
Cap 3–6(9) cm, broadly depressed, cream to ivory, becoming browner with age. When scratched turns red then blackish. **Cuticle** peeling about 1/4. **Gills** crowded (about 12–14 per cm at margin), adnate-decurrent, white to cream, browner with age. **Stem** firm, white, bruising blood-red then finally blackish. **Taste** moderately acrid in the gills after chewing for a short while. **Odour** slightly fruity or acidic. **Macrochemicals**: $FeSO_4$ pale rose then soon dark green (very distinctive!), guaiac rapidly and intensely blue. **Pileocystidia** sparse, narrowly fusiform and often capitate, weakly blackening in SV. **Spores** 6.5–8 x 5.5–6.7 µm, warts low and blunt, up to 0.6 µm, with a moderately well-developed reticulum (although often very fine). **Spore deposit** pure white (Ia).

Fairly common and widespread with both broadleaved and coniferous trees.

Russula acrifolia Romagn.
[acrifolia = acrid gills]
Cap 4–10 cm, broadly depressed, pale brown, distinctly viscid-tacky (look for bits of leaves stuck to the cap). **Cuticle** peeling about 1/4. **Gills** cream, rather crowded, narrow. **Stem** white to pale brown with age. **Taste** distinctly and strongly acrid, especially the gills. **Odour** not distinct. **Macrochemicals**: $FeSO_4$ green, guaiac rapid, intense blue. **Pileocystidia** vermiform, sometimes with apices bifurcating, grey-black in SV. **Spores** 7–9 x 6.5–7 µm, with partial to complete network. **Spore deposit** white (Ia).

Usually associated with *Quercus* on acid to neutral soils. Appears to be genuinely a rather rare species in Britain although possibly much confused with the other similar species in this group.

The distinguishing features are the viscid cap surface, the flesh which turns reddish then very slowly grey-brown and the extremely acrid taste of the lamellae.

Russula

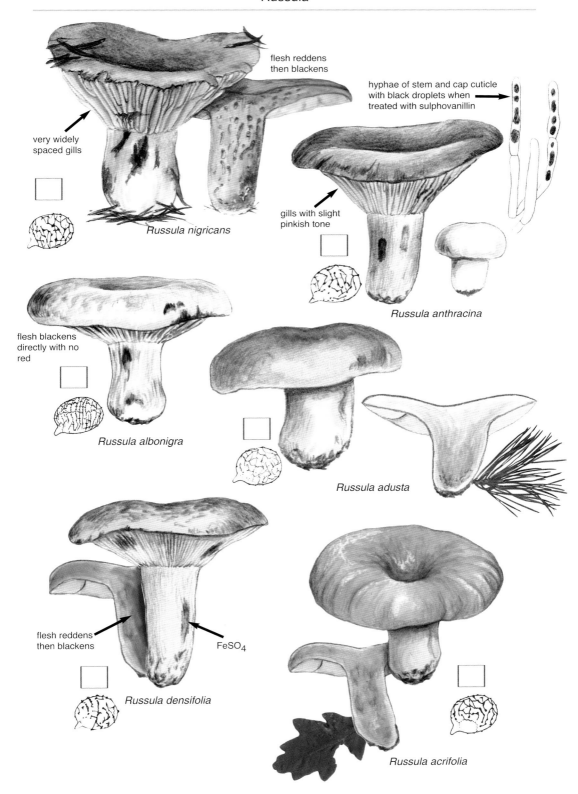

Section *Lactarioides*
fruitbodies robust, flesh unchanging in colour; gills often weeping clear droplets

Russula delica Fr.
[delica = without milk]
Cap 5–18 cm, very firm to hard, with inrolled margin when young, slowly unrolling with a depressed, centre, dry to minutely tomentose when young, ivory white, yellowish or brownish with age. **Cuticle** adnate, hardly peeling. **Gills** from moderately spaced to crowded, rather variable, narrow, decurrent, white to cream. **Stem** short, 2–5 x 1.5–4.5 cm, very firm, tomentose when young, white to staining brownish. **Taste** more or less acrid in the gills, mild in the stem flesh. **Odour** complex, fruity-fishy-buggy, not very pleasant. **Macrochemicals**: $FeSO_4$ pale orange, guaiac positive but often slow. **Pileocystidia** are little differentiated from background cuticular cells, vermiform, sometimes appendiculate, weakly staining. **Spores** 8–11 x 6.5–8.5 µm, with warts up to 0.75 µm, with numerous connectives and connate warts but only poorly developed reticulum. **Spore deposit** white to pale cream (Ib-IIa).

Fairly common and widespread, under both broadleaved trees and conifers although perhaps with a preference for alkaline soils.

There are several described variants and related species in the literature, none usually distinguished or recorded in the UK with the exception of **R. chloroides** which differs in its more constantly crowded gills, and bluish tints in both lamellae and stem apex, plus taller spore warts (1.5 µm); it is often the commoner of the two species in the UK.

Russula chloroides (Kromb.) Bres.
[chloroides = pale green]
Cap 5–13 c, white to yellowish ivory, centre depressed to funnel-shaped with margin slightly inrolled, surface dry and slightly roughened. **Cuticle** adnate. **Gills** very crowded (about 9–16 lamellae per cm at margin of adult cap), narrow, decurrent, white with a faint bluish green tint. **Stem** short, stout to attenuate at the base, white and with a narrow blue-green band at the apex just below the gills. **Taste** acrid and disagreeable, oily in the gills, mild in the stem flesh. **Odour** from unpleasant to slightly fruity-*Pelargonium*. **Macrochemicals**: $FeSO_4$ pale reddish pink, guaiac fairly intense blue. **Pileocystidia** poorly differentiated, narrow, irregular, with small blackening guttules when treated with SV. **Spores** 7–11 x 6–8.5 µm, with prominent warts up to 1.5 µm, some connectives but no true network. **Spore deposit** pale cream (Ib-IIa).

Common under mixed broadleaf trees but especially *Quercus*, preferring neutral to calcareous soils.

The glaucous tint to the lamellae varies in intensity but the blue-green ring at the stem apex is usually easy to see and is the best field character to distinguish this species from the closely related **R. delica**.

Russula pallidispora ** Blum ex Romagn.
[pallidispora = pale spores]
Cap 6–13 (15) cm, soon depressed and funnel-shaped, ivory-white to yellowish cream, discoloured yellowish brown when bruised or with age, felty-pubescent. **Cuticle** hardly peeling. **Gills** adnate-decurrent, crowded, with numerous intermediate gills, ivory white then finally ochre. **Stem** usually shorter than the cap diameter, cylindric, pointed at base, white to ochre-brown at base. **Taste** mild to slightly bitter-acrid in the gills. **Odour** complex, a mix of fruit and honey. **Macrochemicals**: $FeSO_4$ pinkish rust, guaiac dark blue. **Pileocystidia** cylindric with attenuate apices, 5–7 µm across, weakly colouring in SV. **Spores** 6.5–8 x 6–7 µm, with low warts up to 0.5 µm, with some connectives and crests forming partial reticulum. **Spore deposit** dark cream (IId).

In broadleaf woods, especially under *Quercus* but also occasionally under *Pinus*, rare.

The closely related species **R. flavispora** differs in its burning taste, darker yellow spores with large, isolated warts while the little-known and possibly doubtful northern taxon **R. pseudodelica** has large, isolated warts with only moderately acrid taste and spores dark cream. None of these species has been recorded from the UK with any certainty but might well be expected to occur here.

Russula flavispora ** Blum ex Romagn.
[flavispora = yellowish spores]
Cap 5–10 cm, broadly funnel-shaped when mature, surface dry, felty, dull white, ivory or cream. **Cuticle** adnate, hardly peeling. **Gills** moderately spaced (about 7–8 per cm at cap margin, with numerous intermediate gills, decurrent, ivory-cream to distinctly ochre-yellow. **Stem** short, stocky, firm, white, matt and pruinose when fresh then smooth and polished with age. **Taste** acrid to very acrid especially in the gills. **Odour** very distinctive but complex, sour to unpleasant (rather like *R. foetens*) but with fruit overtones. **Macrochemicals**: $FeSO_4$ rust-pink, guaiac strongly and rapidly blue. **Pileocystidia** long, filiform, difficult to distinguish from normal cuticular cells, only weakly staining with SV. **Spores** 6.5–8.5 x 5.5–7 µm, with isolated warts up to 0.75 µm (very different from *R. delica*). **Spore deposit** deep yellow (IVb).

Very rare, associated with *Fagus*, not yet recorded from the UK but should be here.

This species was split off from **R. pseudodelica** and described by Romagnesi (along with **R. pallidospora**) and although *R. pseudodelica* (a poorly defined and little understood species) has been recorded from the UK it is uncertain to which species in the narrow sense the record refers.

Section *Archaeinae*
fruitbodies fleshy; gills very widely spaced, rather waxy; spores with very tiny warts

Russula camarophylla ** Romagn.
[camarophylla = with arched gills]
Cap 4–7 cm, strongly convex, slowly expanding, fleshy, often undulating-lobed, surface scabrous-rough, pale ochre, pinkish brown to reddish. **Cuticle** ± adnate. **Gills** very widely spaced, adnate-decurrent, pale ochre, bruising brownish. **Stem** cylindric, fleshy, whitish, soon ochre to brown especially on handling, flesh tough, white then ochre-brown. **Taste** mild. **Odour** not distinct or of bread. **Macrochemicals**: $FeSO_4$ rust-pink, guaiac rapid and intense blue. **Pileocystidia** absent, cuticular hyphae often with swollen, capitate end cells. **Spores** ellipsoid, 5–5.7 x 3.8–4.4 µm, with a few tiny warts. **Spore deposit** pure white (1a).

Rare, with *Pinus*, *Fagus*, *Castanea*, etc in lowland to subalpine regions in France, Italy and Switzerland, not yet in Britain.

A member of a very primitive group closely resembling species of **Camarophyllus** (= *Hygrophorus*) with their waxy, tough texture, small, hardly ornamented spores and very long basidia.

Russula

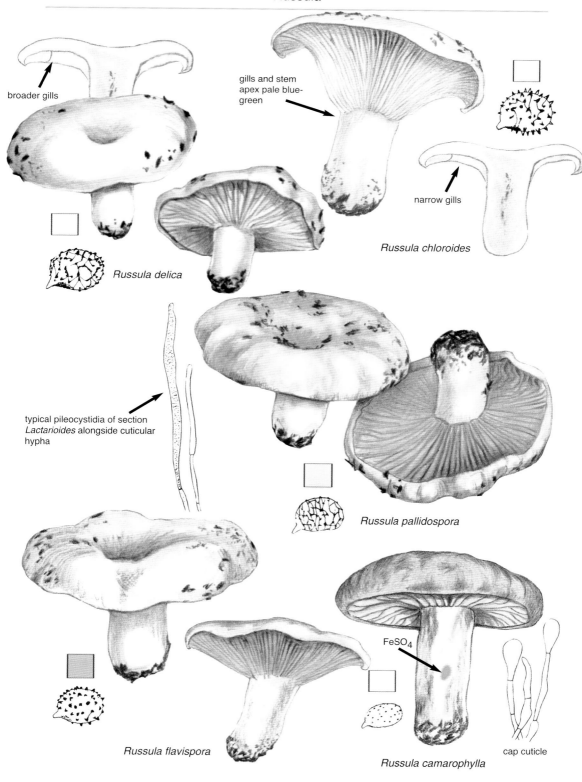

Subgenus *Ingratula*

species often have an unpleasant, sour odour, sometimes strong of aniseed or marzipan

Russula foetens Pers.
[foetens = foul smelling]
Cap 5–15cm, very fleshy, broadly convex, to slightly depressed, slimy-viscid when wet, smooth and glossy when dry, margin sulcate-pectinate, ochre, ochre-brown, orange-brown, paler at margin, often spotted with darker areas. **Cuticle** peels 1/4 –1/2. **Gills** distant, thick, cream often spotted with rust-brown spots, edge often dotted with drops of liquid. **Stem** robust, firm but finally cavernous when old, white staining yellow-brown when handled or with age. **Taste** very acrid in the gills but often almost mild in the stem flesh, also with a bitter or oily overtone. **Odour** strongly sour, rancid or oily. **Macrochemicals** $FeSO_4$ pink, guaiac blue-green, flesh with KOH reddish brown. **Pileocystidia** narrow (3–6 µm), fusiform to appendiculate, only slightly fatter than the surrounding cuticular cells, weakly staining in SV. **Spores** 7.5–10 x 6.5–9 µm, with large isolated warts to 1.5 µm. **Spore deposit** cream (IIb-c).

Uncommon in the south, commoner in northern England and Scotland, with broadleaf and coniferous trees.

R. subfoetens Wm.G. Sm.
Very similar in appearance to *R. foetens*, perhaps a little smaller in size on average, it differs in its flesh turning golden yellow with KOH and lower spore warts. It seems to be the commoner of the two species in the south of England.

Russula illota Romagn.
[illota = grubby]
Cap 5–16 cm, remaining globose-convex for some time before finally expanding, thick and fleshy, cuticle elastic, viscid, surface often pitted, margin sulcate-pectinate, tawny ochre, cinnamon, fawn to reddish brown, the mucus on the surface often tinted violaceous or with minute violet-brown spots. **Cuticle** peels 3/4. **Gills** crowded, pale yellowish cream with margins spotted with tiny purplish violet or dark brown droplets. **Stem** stout, firm, hard and brittle becoming hollow with cavernous chambers, white at first then soon marked or spotted with small brown flecks, soon reddish brown at base. **Taste** acrid and also nauseous, oily-sour. **Odour** strong of marzipan, bitter almonds with sour-foetid undertones. **Macrochemicals** $FeSO_4$ greyish rose, guaiac blue-green but not intense. **Pileocystidia** fusiform, obtuse, often in small bunches, yellowish, weakly staining in SV, non-septate. **Spores** 7–8.5 x 6–7 µm, warts up to 1.25 µm, often not completely staining in Melzer's iodine, with variable number of ridges, not forming a network, without the huge encircling ridges of the similar *R. grata*. **Spore deposit** pale cream (IIa-b).

Rare in the UK but widespread, under broadleaf trees especially *Quercus*, possibly preferring calcareous soils.

Similar to ***R. laurocerasi*** but lacking the extreme sporal ornamentation of that species.

Russula laurocerasi Melzer
[laurocerasi = after Cherry Laurel, *Prunus laurocerasus*]
Syn. *R. grata* Britzlmeyer?
Cap 5–8 cm, convex, soon broadly depressed, margin sulcate-pectinate, surface smooth, viscid when wet, with elastic cuticle, pale yellow-ochre, ochraceous, usually marbled with darker reddish yellow or reddish brown. **Cuticle** peels 1/2. **Gills** thick, moderately crowded, brittle, pale cream spotted and marked with ochre-brown. **Stem** stout, firm, thick-fleshed, becoming cavernous when old, white to yellowish cream, brownish at base. **Taste** very acrid in the cap and gills. **Odour** complex, strong of marzipan, bitter almonds with a slightly sour undertone. **Macrochemicals** $FeSO_4$ greyish rose, guaiac blue-green. **Pileocystidia** rare, poorly defined, cylindric, narrow and weakly staining. **Spores** 7–8.5 x 7–8 µm, almost globose with very tall warts up to 2 µm (often not staining in Melzer's iodine), often connected by enormous wing-like crests. **Spore deposit** pale cream (IIa-b).

Uncommon but widespread, with mixed broadleaf trees. One of a complex of closely related taxa.

Russula fragrantissima Romagn.
[fragrantissima = very fragrant]
This species is very similar to *R. laurocerasi*, differing in its often paler cap, and intense odour of aniseed. Its spores lack the tall crests and wings typical of *R. laurocerasi*, having instead an incomplete network of ridges to 1.4 µm high.

Russula pectinata Fr.
[pectinata = comb-like]
Cap 4–10 cm, margin strongly pectinate-sulcate, yellow-ochre to ochre-brown, viscid when wet, sometimes spotted or stained reddish brown. **Cuticle** peels 1/3–1/2. **Gills** rather widely spaced, attenuated-decurrent where they join the stem, interveined, pale cream to cream-ochre, often spotted brown. **Stem** robust, brittle, often cavernous internally, dirty white flushed with greyish brown, with ochre spots at the base. **Taste** acrid and unpleasant, oily in the gills. **Odour** sour, foetid as in *R. foetens*. **Macrochemicals** $FeSO_4$ greyish rose. guaiac rapidly azure blue. **Pileocystidia** slender, fusiform, often papilate at apex, non-septate. **Spores** 6.5–7.5 x 5–6 µm, warts up to 0.75 µm high, with a few connectives. **Spore deposit** cream (IIb-c).

An uncommon species in Britain, mainly associated with *Quercus* in the southern counties.

Russula praetervisa Sarnari
[praetervisa = overlooked]
Syn. *R. pectinatoides* sensu most European authors
Cap 3.5–7 cm, soon depressed at centre, smooth, viscid when moist, margin sulcate-pectinate, dull ochre-brown, yellowish brown, greyish yellow, sometimes olivaceous. **Cuticle** peels 1/3–2/3. **Gills** fairly crowded, narrow, whitish cream. **Stem** cylindric-clavate and slightly rooting at base, brittle, hollow when old, white, often with small purplish rust stains at base. **Taste** mild but oily. **Odour** sour, earthy, oily or even fishy. **Macrochemicals** $FeSO_4$ pale rose, guaiac positive, deep blue. **Pileocystidia** cylindric to tapered or attenuate, weakly staining in SV, 0-septate. **Spores** 7–8.5 x 5.6–7 µm, with warts up to 0.7 µm, with frequent connectives and crests to form a partial reticulum or zebroid pattern. **Spore deposit** dark cream (IIc-d).

Widespread under *Quercus*, *Tilia* and sometimes conifers, on sandy or loamy soils.

Widely recorded as *R. pectinatoides* Peck but that is an American species with isolated warts.

Russula

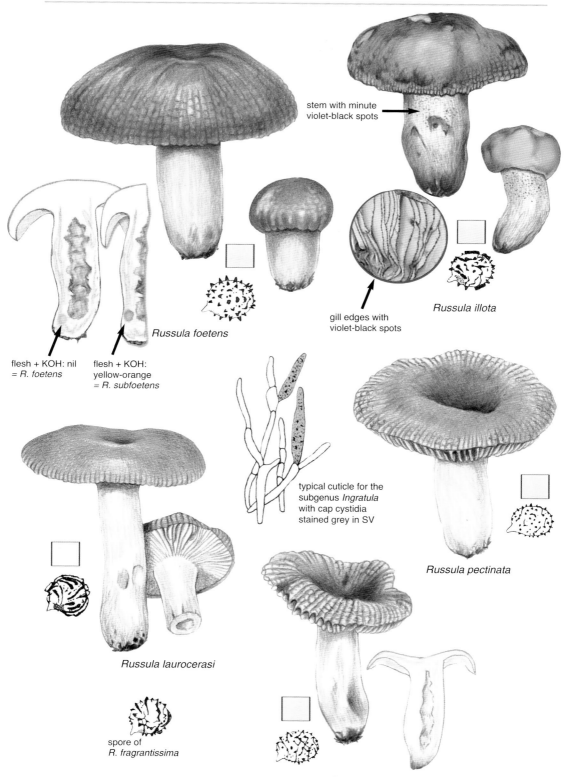

Russula amoenolens Romagn.
[amoenolens = pleasant smelling]
Cap 3–8 cm, dark brown, sepia grey to ochre-brown (a rare white form also occurs), with strongly pectinate-sulcate margin, often sticky and irregular or lobed in outline. **Cuticle** peels 1/2, elastic. **Gills** greyish cream, moderately crowded. **Stem** white or faintly flushed sepia. **Taste** flesh is slowly very hot and unpleasant. **Odour** a complex mixture of artichoke and Camembert cheese. **Macrochemicals:** $FeSO_4$ pink-rust, guaiac rapidly deep blue. **Pileocystidia** infrequent, difficult to differentiate from ordinary cuticular end cells, pointed-fusiform and weakly staining in SV. **Spores** 7–8 x 5–6 µm, with isolated, conical warts. **Spore deposit** cream (IIa-b).

Frequent under *Quercus* everywhere. It is most easily confused with **R. sororia**. The easiest distinguishing character to separate it from *R. sororia* seems to be the guaiac reaction: rapid and intense in *R. amoenolens*, weak or almost negative in *R. sororia*.

Russula sororia Fr.
[sororia = a sister]
Cap 5–9 (-12) cm, fleshy, rounded then depressed, margin sulcate-pectinate, viscid when wet, dull sepia brown, greyish to ochre-brown, sometimes with rusty spots. **Cuticle** peels 1/2, elastic. **Gills** crowded, narrow, dull cream to creamy-grey, bruising brownish, elastic. **Stem** cylindric, white to dull greyish, staining brown below. **Taste** oily and unpleasant, and also very acrid. **Odour** complex, spermatic-rancid with fruit components. **Macrochemicals:** $FeSO_4$ greyish yellow to greyish pink, guaiac slowly and only weakly blue or even nil. **Pileocystidia** sparse, little differentiated from surrounding cuticular hyphae, slender, fusiform-pointed, 4–5 µm across, only weakly staining in SV. **Spores** 7–8.2 (-8.8) x 5.7–6.4 (-7) µm, warts up to 0.4 µm, mostly isolated with a very few short connectives. **Spore deposit** cream (IIc).

Associated with mixed trees, although most commonly with *Quercus* in Britain.

Separated from **R. amoenolens** by the weak guaiac reaction, slightly more elongate spores and often larger fruitbodies with spermatic odour.

Russula pseudoaffinis Migl. & Nicolaj
[pseudoaffinis = false friend]
This species is very closely related to *R. sororia* and is considered by many as a form or variety of that species. It has similar colours to *R. sororia* but the cap surface is frequently cracked and disrupted and the surface has numerous small patches of greyish white veil. I feel that the microscopical character of the velar patches is distinct enough to maintain it as a separate species, at least until someone investigates their DNA. **Macrochemicals:** very weak or negative reaction with guaiac. **Pileocystidia** absent, the patches of white veil consist of very irregular, long hyphae with lots of vacuoles giving the appearance of being multi-septate. **Spores** 6.6–7.9 x 5–6 µm, warts up to 0.3 µm, mostly isolated. **Spore deposit** deep cream (IIa).

Described from the Mediterranean with *Quercus ilex* but in Britain associated with *Q. robur* and *Tilia*.

Russula insignis (Quél.) Quél.
[insignis = of striking appearance]
Syn. *R. livescens*, = *R. pectinatoides* ss. R. Rayner
Cap 3.5–7 cm, expanded with depressed centre, with margin irregular, split or lobed, moderately sulcate, pale sepia to deep umber-brown, sometimes paling to almost dirty white, surface ± viscid with minute crustose granules embedded in the surface. **Cuticle** peels 1/2, elastic 1/3–1/2. **Gills** rather crowded, cream to greyish cream, often spotted with reddish brown. **Stem** cylindric, white or flushed greyish, typically tinted yellowish in lower quarter and staining reddish brown. **Taste** mild, slightly oily. **Odour** fruity or a little like **R. fellea**. **Macrochemicals:** KOH bright red on stem base, $FeSO_4$ greyish rose, guaiac deep blue. **Pileocystidia** slender, poorly differentiated from background hyphae, filiform, papillate, weakly staining in SV. Also with finely encrusted hyphae from the very fine external velar tissue. **Spores** 6.5–8 x 5–6 µm, warts to 0.6 µm, with some connectives forming an irregular, partial network. **Spore deposit** pale cream (IIa-b).

Occasional but widespread under broadleaf trees, especially *Quercus*.

This is our only representative of a group of *Russula* species (section *Subvelatae*) united by the presence of an external veil, left usually as small granules on the cap surface and/or stem base.

Russula farinipes Romell in Britzelm.
[farinipes = floury stem]
Cap 3–8 cm, convex-globose with centre often depressed, margin irregular, pectinate-striate, viscid when wet, yellow-ochre, pale ochre to pale yellow. **Cuticle** tough, elastic, hardly peeling. **Gills** rather widely spaced, often strongly interveined, decurrent at junction with stem, pale cream-ochre. **Stem** cylindric, firm, white to palest ivory yellow, often with a white, floury scurf at the apex. **Taste** very acrid. **Odour** distinctive, fruity and agreeable. **Macrochemicals:** $FeSO_4$ greyish rose, guaiac very slow and very feeble. **Pileocystidia** long, fusiform-swollen, 5–9 µm across, with irregular blackening in SV although usually showing some small black guttules internally. **Spores** 6-8 x 5-6.5 µm, elliptic, with small isolated warts up to 0.6 µm high. **Spore deposit** white (Ia).

Not uncommon in some parts of the UK, under broadleaf trees.

Russula pallescens P. Karst.
[pallescens = becoming pale]
Very closely related to **R. farinipes** it is known in Britain only from the Cairngorm mountains in Scotland where it grows with dwarf *Salix*. It differs from *R. farinipes* in its smaller size (2–3 cm), often cracking stem and cap, slightly larger spores 7–9 x 6.5–7 µm, with small, rounded, isolated warts up to 0.4 µm and isolated habitat.

Russula fellea (Fr.) Fr.
[fellea = bitter]
Cap 4–10 cm, convex and often remaining so for a long period, only slowly flattening to broadly depressed, smooth, honey-yellow, ochre. **Cuticle** peels 1/3–1/2. **Gills** moderately spaced, broad, with occasional half-gills, pale cream-ochre. **Stem** robust, firm to spongy-cavernous, pale cream-ochre to honey-yellow. **Taste** often very acrid. **Odour** characteristic but often faint of *Pelargonium* or apple sauce, best detected on the gills. **Macrochemicals:** $FeSO_4$ almost negative, dull yellowish, guaiac slowly weakly greenish. **Pileocystidia** numerous, cylindric to irregular, swollen, 6–8 µm across, fusiform at the tips, sometimes appendiculate, 0–2 septate. **Spores** 7.5–9 x 6–7 µm, with warts to 0.75 µm and a well-developed partial network of connectives. **Spore deposit** white (Ia).

Growing with *Fagus*, very common. Often confused with **R. ochroleuca** which has brighter yellow-ochre, to more greenish yellow tones, stem often greying with age, no odour, less acrid taste and wider range of hosts as well as a rapid and strong guaiac reaction.

Russula

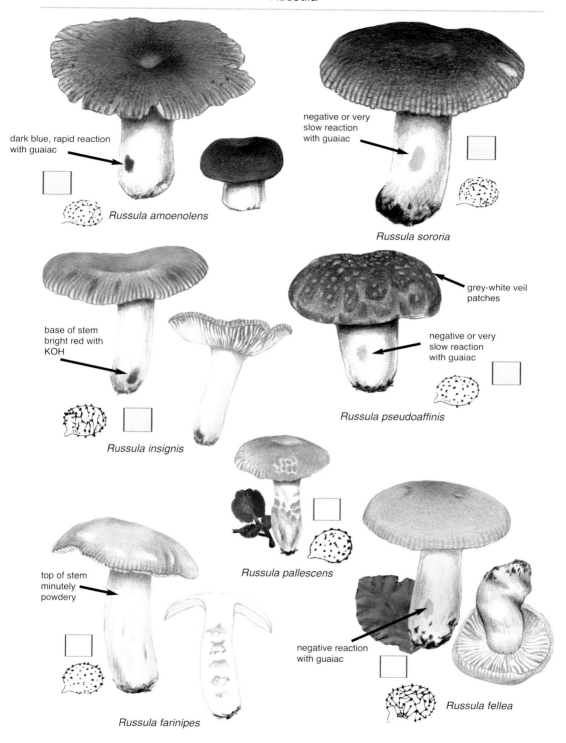

Subgenus *Heterophyllidia*
Section *Heterophyllae*
Subsection *Cyanoxanthinae*
spores white to cream, taste mainly mild, colours green to violaceous, blue never red, $FeSO_4$ to green

Russula cyanoxantha (Schaeff.) Fr.
[cyanoxantha = blue-green and yellow]
Cap 4–15 cm, smooth and shiny and slightly radially veined, violet, bluish violet to darker at centre but also greenish to entirely green (var. *peltereaui*). **Cuticle** peels 1/2. **Gills** crowded, white and flexible to touch, usually unforked but strongly and repeatedly forking in the var. *variata* (considered a species by many). **Stem** white to faintly flushed lavender. **Taste** mild to distinctly acrid in the var. *variata*. **Odour** not distinct. **Macrochemicals** $FeSO_4$ nil to slowly greyish green, guaiac azure blue. **Pileocystidia** infrequent, very slender (2–4 μm), narrowly fusiform and sometimes capitulate or appendiculate at apex. **Spores** 7–9.5 x 5–6 μm, with low, blunt isolated warts to 0.5 μm. **Spore deposit** white (Ia).

Common everywhere in broadleaf and conifer woods. The white, flexible-greasy gills, white spores and pale green reaction to $FeSO_4$ are good identification characters.

Russula cutefracta Cooke
[cutefracta = cracked skin]
Closely related to *R. cyanoxantha* and often considered a variety of that species but some preliminary molecular studies suggest it may be a good species. Usually greenish but occasionally purple; the cuticle cracks into a mosaic, particularly at the margin. $FeSO_4$ nil to pale greenish.

Rarely recorded in Britain but apparently widespread under broadleaved trees.

Subsection *Heterophyllae*
spores white, $FeSO_4$ bright salmon-orange, cuticle with long, narrow setae

Russula heterophylla (Fr.) Fr.
[heterophylla = mixed gills]
Cap 5–10 cm, convex, soon flattened, fleshy, smooth to finely radially 'veined', shades of green, yellowish green, blue-green, olive, to ochraceous or even brown. **Cuticle** peels 0–1/4. **Gills** very crowded, slightly decurrent, rather flexible and 'greasy' to fairly brittle with age, often forking and anastomosing near stem, white to cream. **Stem** cylindric, firm, white, browning slightly when bruised. **Taste** mild. **Odour** nil. **Macrochemicals** $FeSO_4$ strong salmon pink, guaiac rapidly blue. **Pileocystidia** clavate-fusiform, rather sparse and only weakly staining in SV. **Spores** 5–7 x 4–6 μm (smallest of all British species), warts extremely low, 0.2–0.6 μm, mostly isolated, some connate. **Spore deposit** white (Ia).

Frequent in the early part of the season (often one of the first species to appear), in mixed broadleaf woods.

Distinguished from the rather similar ***R. aeruginea*** by its white spores and from ***R. cyanoxantha*** var. *peltereaui* by its salmon $FeSO_4$ reaction.

Russula vesca Fr.
[vesca = edible]
Cap 5–10 cm, very firm and fleshy, surface smooth and slightly matt, colour like 'old ham' or bacon. Shades of pinkish brown, pale vinaceous to buff, usually rather pastel hues, margin slightly sulcate with age and cuticle retracting from the margin by a millimetre or so exposing the white flesh beneath. **Cuticle** peels 1/2. **Gills** rather crowded, narrow, slightly interveined, forking and anastomosing near their bases, pale cream. **Stem** cylindric-tapered, very firm and fleshy, white occasionally flushed pinkish vinaceous on one side. **Taste** mild and nutty. **Odour** nil. **Macrochemicals** $FeSO_4$ intense pinkish rust, very rapid, guaiac quickly blue. **Pileocystidia** slender, fusiform, 4–5.5 μm across, hardly staining in SV, cuticle also has long, pointed seta-like cells with thick walls, very striking. **Spores** 6.4–8 x 5.3–5.8 μm, warts up to 0.5 μm, mostly isolated with a few scattered connectives or catenulate crests. **Spore deposit** pure white (Ia).

Common everywhere, associated with broadleaf trees, especially *Quercus* but also on occasion with conifers.

The cap colours are subtle but very recognisable, as is the retraction of the cuticle when old. The unusual seta-like cells in the cap cuticle are shared with *R. heterophylla* (which can also be brown!) and *R. mustelina*.

Russula mustelina Fr.
[mustelina = pertaining to a weasel]
Cap 5–12 cm, fleshy, smooth and shining when wet, matt to finely granular when dry, uniformly ochre-brown to reddish brown. **Cuticle** peels 3/4. **Gills** adnate-sinuate, rather crowded, strongly forking and anastomosing near stem, cream to pale buff-straw, brittle. **Stem** robust, firm to hard, cylindric-ventricose, whitish becoming spotted or entirely brown, often hollow inside. **Taste** mild. **Odour** nil to slight of cheese. **Macrochemicals** $FeSO_4$ strong reddish brown, salmon (like *R. vesca*, guaiac rapid and intense. **Pileocystidia** sparse, cylindric-clavate, non-septate, poorly staining with SV and difficult to distinguish from background cap cells. **Spores** 8–12 x 6–8 μm, with warts up to 0.5 μm and numerous connectives forming a partial reticulum. **Spore deposit** cream (IIb).

Under conifers, especially *Abies* in far northern or subalpine areas. Although recorded by Rayner and others from both Scotland and England, voucher material does not appear to exist and fresh collections are required.

Subsection *Griseinae*
spores pale cream to ochre, $FeSO_4$ mainly pinkish salmon and cap colours cream, grey, green, brown, lilac to purple, never bright red

Russula parazurea Jul. Schäff.
[parazurea = more or less blue]
Syn. *R. plumbeobrunnea*
Cap 3–8 cm, soon flattened, fleshy to fragile, surface dry and often pruinose, colour very variable from dark grey-brown to glaucous, blue-grey, pale lilac or pallid. **Cuticle** peels 1/2–2/3. **Gills** pale cream to buff, quite crowded, brittle, often forked. **Stem** cylindric, white, firm then soon fragile. **Taste** mild to weakly acrid in the lamellae. **Odour** nil. **Macrochemicals** $FeSO_4$ pinkish brown, guaiac slowly pale blue. **Pileocystidia** cylindric-elongate, to fusiform-clavate or papillate, non-septate. **Spores** 6–8 x 5–6.5 μm, with tiny warts to 0.5 μm, numerous connectives and crests forming a moderate to fairly complete reticulum. **Spore deposit** cream (IIb-c).

Common to abundant in mixed woods everywhere, especially with *Quercus* or *Pinus*.

The ± reticulate spores, cream spore deposit and cap colours usually in tones of grey to bluish green, often with pruinose surface, are good determining characters. The grey-brown form (illustrated) was described as *R. plumbeobrunnea* but molecular studies suggest it is a synonym of *R. parazurea*.

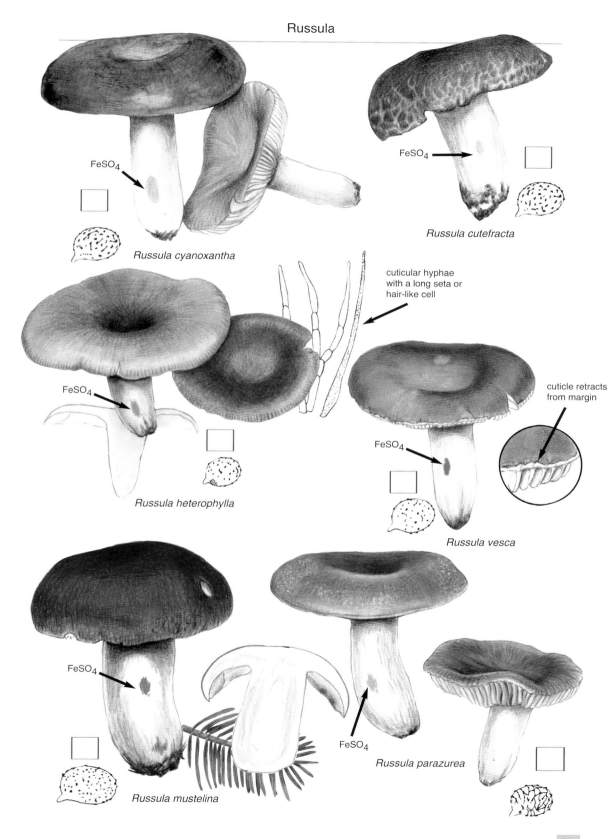

Russula ionochlora Romagn.
[ionochlora = violet-green]
Cap 4.5–8 cm, soon expanded and depressed at centre, rather fleshy but often fragile when mature, typically a mix of bright lilac at the margin and paler, yellowish green at the centre; dry, rather matt. **Cuticle** peels about 1/2. Flesh pale pink where eaten by slugs. **Gills** moderately crowded, fragile, forking at base, pale cream. **Stem** firm, cylindric, white to faintly flushed lilac. **Taste** mild to distinctly acrid in the young gills. **Odour** nil. **Macrochemicals**: $FeSO_4$ medium pinkish orange, guaiac slowly positive. **Pileocystidia** common, cylindric to subclavate, sometimes capitate, 4.5–7.5 µm across. **Spores** 6.5–7.5 x 4.7–6 µm, with low, isolated warts 0.2–0.5 µm high. **Spore deposit** pale cream (IIa).

Locally frequent in Britain, usually associated with *Fagus* or sometimes *Quercus*.

In its typical form quite easily recognised with its contrasting cap colours, occasionally however it is uniformly grey-lavender or violet-blue and then looks closer to **R. grisea** or even **R. parazurea**. The cap cuticle structure is decisive in their separation.

Russula grisea (Pers.) Fr.
[grisea = greyish]
Cap 5–11 cm, globose then convex, very soon depressed at centre, smooth and shiny, violet, violet-grey, blackish olive, bluish grey, discolouring at centre to pale ochre. **Cuticle** peels 1/2. **Gills** rather crowded, thin, very brittle, often forking, pale cream-ochre, sometimes tinted rose-violet nearest the cap margin. **Stem** firm, stocky and rather short, white or sometimes tinted lilac or rose-violet in part. **Taste** mild to very slightly acrid in the gills. **Odour** nil. **Macrochemicals**: $FeSO_4$ bright rust, guaiac deep blue. **Pileocystidia** cylindric or claviform, sometimes with terminal appendage. Cuticular hyphae cylindric, with terminal cells filiform. **Spores** 6.5–8 x 5.5–6.5 µm, warts up to 1–1.25 µm, with some few thick connectives or chains of warts. **Spore deposit** cream-ochre (IIc).

Uncommon, under broadleaf trees especially *Fagus* and *Quercus*.

Under *Populus* and with similar cap colours from greenish to greyish violet, is found **R. subterfurcata**, not yet British, which has slightly shorter spores (5.4–7.7 µm) and a darker spore deposit (IId–IIIa). Its cuticular hyphae are formed of much shorter, broader elements.

Russula pseudoaeruginea (Romagn.) Kuyper & Vuure
[pseudoaeruginea = false aeruginea]
Cap 5–9 cm, broadly rounded then depressed at centre, grey-green, glaucous, pale olive, often discoloured ochre at centre. **Cuticle** peels to 1/2. **Gills** fairly crowded, adnate to slightly decurrent, ivory-cream, slightly browning with age or bruising. **Stem** cylindric, rather short and often tapered at base, white, sometimes stained brown. **Taste** mild to very slightly acrid in young gills. **Odour** nil to distinctly fruity. **Macrochemicals**: $FeSO_4$ strongly pinkish rust, guaiac slowly pale blue. **Pileocystidia** cylindric-clavate often attenuate-appendiculate, 0-septate. Basal cells of the cuticular hyphae very broad, almost globular and in short chains with elongate terminal cell. **Spores** 6–8.5 x 5.5–6.5 µm, warts up to 0.7 µm, with partial connectives forming a partial reticulum. **Spore deposit** cream-ochre (IIc-d).

With *Quercus* on sandy soils, one collection known from W. Sussex but may be more common, it is easily confused with other species. The similar **R. aeruginea** differs in its narrower spores (7–8.2 x 4.8–6.2 µm), with mostly isolated tiny warts; its slender, filiform cuticular hyphae and association with *Betula*.

Russula galochroa ** (Fr.) Fr.
[galochroa = milk-coloured]
Cap 3–7 cm, soon flattened and depressed, rather firm when young then soon fragile, ivory-cream, pearl-grey, palest beige. **Cuticle** peels 1/3. **Gills** thin, rather crowded, adnate to very slightly decurrent, clear ivory, pale cream-ochre. **Stem** firm, fleshy, finally spongy-cavernous, white with brownish stains at base. **Taste** mild to a little acrid in the gills. **Odour** nil. **Macrochemicals**: $FeSO_4$ brownish pink, guaiac strong and rapid, blue. **Pileocystidia** clavate-fusiform, non-septate. Cuticular hyphae forming chains of swollen, short, rounded or oblong cells. **Spores** 6–8.5 x 5–6.5 µm, with isolated blunt warts to 0.9 µm, with the occasional connective. **Spore deposit** cream-ochre (IIc-d).

In broadleaf woods, especially *Quercus*, not known with certainty from the UK, previous records have proved to be misidentifications. The distinctive cap cuticle must be checked if this species is recorded, plus the isolated spore warts noted. The species should certainly occur here.

R. galochroides, not British but common in the Mediterranean is very similar but differs in its spore ornamentation (illustrated). It grows under *Quercus*, often with *Erica arborea* or *Arbutus unedo*.

Russula faustiana Sarnari
[faustiana = from 'faustini' the lucky one]
Cap 4.5–10 cm, soon flattened and broadly depressed, dry, pale grey-brown, ochre, beige, pale greenish, variable and normally rather pale. **Cuticle** mostly adnate or peeling to 1/4. **Gills** moderately crowded, rarely bifurcating and with some fine anastomosing at the base, pale cream. **Stem** cylindric with tapering base, fairly firm, white, rarely flushed greyish. **Taste** mild to slightly acrid in the lamellae. **Odour** nil. **Macrochemicals**: $FeSO_4$ pinkish rust, guaiac slowly pale blue. **Pileocystidia** rather short and broad (7–12 µm) clavate with slender, mucronate apical 'tail'. **Spores** 6.4–8 x 5.5–6.2 µm, warts up to 0.65 µm with several but irregular connectives or crests, no true reticulum. **Spore deposit** dark cream (IIc-d).

Associated with *Fagus*, more rarely under *Quercus*, prefering calcareous soils, usually early in the season. Rare but possibly under recorded.

Russula atroglauca Einhellinger
[atroglauca = blue-black]
Syn. *R. stenotricha*?
Cap 2–8 cm, rounded then flattened and depressed, smooth to pruinose, deep grey-green to very dark blackish green or flushed bluish, darker at centre. **Cuticle** peels 1/2. **Gills** crowded, cream, forking near the stem, fragile. **Stem** clavate, white or delicately tinted grey or pinkish, sometimes rust-brown at extreme base. **Taste** mild to slightly acrid when chewed for some time. **Odour** nil. **Macrochemicals**: $FeSO_4$ dull pink, guaiac blue-green. **Pileocystidia** clavate-fusiform, broad (6–12 µm), cells of cuticle cylindric to swollen with long, tapering end cells. **Spores** 6.5–8 x 6–6.5 µm, warts around 0.75 µm with numerous connectives but no true network. **Spore deposit** medium cream (IIc).

Found especially with *Populus* but also recorded in literature with *Betula* and *Picea*. Recently recorded in Britain and probably rare, although doubtless confused with other species.

Recent DNA studies indicate that *R. atroglauca* and **R. stenotricha** are synonymous. If confirmed then the name *R. stenotricha* will take precedence.

Russula

Russula sublevispora (Romagn.) Romagn.
[sublevispora = less than ornamented spores]
Cap 3–8 cm. greenish grey, blue-grey to greyish lilac, more ochraceous at centre, the margin sometimes minutely cracked. **Cuticle** peels up to 1/3. **Gills** fairly crowded, brittle, cream to palest ochre. **Stem** white, stocky and rather brittle, often eccentric. **Taste** mild. **Odour** nil or of cedarwood. **Macrochemicals**: $FeSO_4$ medium pink, guaiac very weak. **Pileocystidia** clavate-fusiform, grey-black in SV. **Cuticular cells** slender, on shorter, more swollen elements. **Spores** 6.5–8.5 x 5.5–6.4 µm, with isolated tiny warts 0.2–0.3 µm high. **Spore deposit** cream (IIc).

Associated with *Quercus*, extremely rare with only one British collection so far but possibly overlooked.

Russula anatina Romagn.
[anatina = inverted]
Cap 3–10 cm, grey-green to grey-bluish, more olivaceous at centre, the margin becoming minutely tessellated and cracked, slightly sulcate. **Cuticle** peels ± 1/2. **Gills** fairly crowded, brittle, cream to palest ochre. **Stem** white, stocky and rather brittle. **Taste** mild. **Odour** nil. **Macrochemicals**: $FeSO_4$ pale pink or almost nil, guaiac rapidly deep blue. **Pileocystidia** clavate-fusiform, grey-black in SV. **Cuticular cells** cylindric-swollen, end cells fusiform, 3-5 µm across. **Spores** 6–7(8) x 5–7 µm, with isolated blunt warts 0.5–0.7 µm high. **Spore deposit** deep cream (IIc-d).

Under broadleaf trees, especially *Quercus*, a rare species with only one British record but possibly confused with other taxa.

Note that other members of the section *Griseinae* such as the common **R. parazurea** can occasionally have a cracked cuticle.

Russula medullata Romagn.
[medullata = pithy]
Cap 4–12 cm, smooth, convex then soon flattened, usually with pale tones of grey, brown and lilac-pink to violet mixed with glaucous green (recalling *R. parazurea*). **Cuticle** peels 1/4–1/2. **Gills** moderately crowded, forking near the stem, very broad near cap margin, pale cream then buff-ochre, spotting rust-brown. **Stem** rather short, often swollen at base, white, rather spongy with age. **Taste** mild, or very rarely slightly acrid in the young gills. **Odour** nil. **Macrochemicals**: $FeSO_4$ rose-orange, guaiac feeble and slow to change. **Pileocystidia** rather uncommon and little apparent, cylindric-clavate, non-septate, weakly staining in SV but with greyish corpuscles often evident. **Cuticular cells** forming chains of short, cylindrical cells, about 3–5 µm across, end cells obtuse to slightly fusiform. **Spores** 6–8.5 x 5.5–6.5 µm, with low, isolated warts 0.2–0.5 µm high. **Spore deposit** ochre (IIIa-b).

Associated with *Populus* and possibly *Betula* also. Collections have been made in Shropshire and from Scotland.

Also with a dark spore deposit is **R. ochrospora** which looks rather like *R. medullata* but differs in its completely reticulate spores. It is a Mediterranean species not yet recorded from Britain.

Russula aeruginea Fr.
[aeruginea = verdigris green]
Cap 4–12 cm, very variable, grass green to yellow-green or pale greyish green or brownish olive, the umbo usually with small spots or areas of rust-brown. **Cuticle** peels easily. **Gills** fairly crowded, white to pale cream, fragile. **Stem** rather fragile, white to slightly rusty-spotted at base. **Taste** mild or often a little acrid in the gills. **Odour** nil.

Macrochemicals: $FeSO_4$ slowly pinkish rust, guaiac weakly to strongly positive. **Pileocystidia** are clavate-fusiform, 1-septate, 4.5–7 µm wide staining well in SV. **Cuticular cells** very long and narrow. **Spores** 6–10 x 5–7 µm, with low (0.5 µm) warts and partial connectives and meshes. **Spore deposit** cream (IIb-IIc).

Common under *Betula* everywhere. The $FeSO_4$ reaction is particularly strong pink and the small rust coloured spots are very characteristic.

Section *Virescentinae*
spore deposit white, cap cuticle strongly cracking

Russula virescens (Schaeff.) Fr.
[virescens = becoming green]
Cap 4–10 cm, subglobose then expanding, margin usually remaining rather incurved, dry, matt with the outer cuticle finely cracking to form a delicate mosaic of plaques, a beautiful bluish green, verdigris, paling to a dull yellowish ochre when old or exposed to the sun. **Cuticle** peels 1/2. **Gills** crowded, intervened, cream. **Stem** cylindric-tapered, robust, white. **Taste** mild, nutty. **Odour** nil. **Macrochemicals**: $FeSO_4$ pink, guaiac variable, slowly to rapidly positive. **Pileocystidia** and fuchsinophile hyphae both absent, terminal cells of cuticular hyphae slender, filiform, emerging from basal cells which are grossly swollen, globose, barrel-shaped and which form a loose, basal cellular layer. **Spores** 7–9 x 5.7–7, with tiny warts up to 0.5 µm, mostly isolated with a few scattered connectives. **Spore deposit** white to pale cream (Ib).

Fairly common and widespread, especially in the south associated with *Fagus*, *Quercus* and *Castanea*.

Perhaps the only possibility of confusion would be with **R. cutefracta** but that species has slender dermatocystidia and filiform cuticular hyphae.

Subgenus *Amoenula*
cap cuticle with broad cells with very long, pointed end cells; spores pale cream-ochre; spores ± reticulate

Russula violeipes Quél.
[violeipes = violet stem]
Cap 5–10 cm, globose then slowly expanding to broadly flattened, dry and matt, pruinose, very variable in colour from deep purple to lemon yellow, greenish yellow, bronze, sometimes a mixture of these. **Cuticle** adnate. **Gills** fairly crowded, slightly sinuate to sub-decurrent, greasy to touch, pale cream-yellow to buff. **Stem** cylindric-fusiform, tapered at base, very firm or even hard when young, white often tinted with purple or pinkish red, sometimes on one side only. **Taste** mild. **Odour** nil or slight of shrimps. **Macrochemicals**: $FeSO_4$ dull pink, guaiac extremely slow and weak blue. **Pileocystidia** and fuchsinophile hyphae both absent, terminal cells of cuticular hyphae very long, attenuated, seta-like, the preceding cells are short, swollen, often cuboid to globose. **Spores** almost globose, 6.7–9 x 6.4–8 µm, warts up to 0.8 µm, with a thick, almost complete reticulum. **Spore deposit** pale cream (IIa).

Frequent and widespread, mainly southern, associated with broadleaf trees, especially *Quercus* and *Fagus*, also with conifers.

In Britain *R. violeipes* is most commonly found with a lemon-yellow cap and violet tinted stem (forma *citrina*), in its purple form it is easily confused with **R. amoena** and **R. amoenicolor**. The latter species is not yet recorded from Britain.

Russula

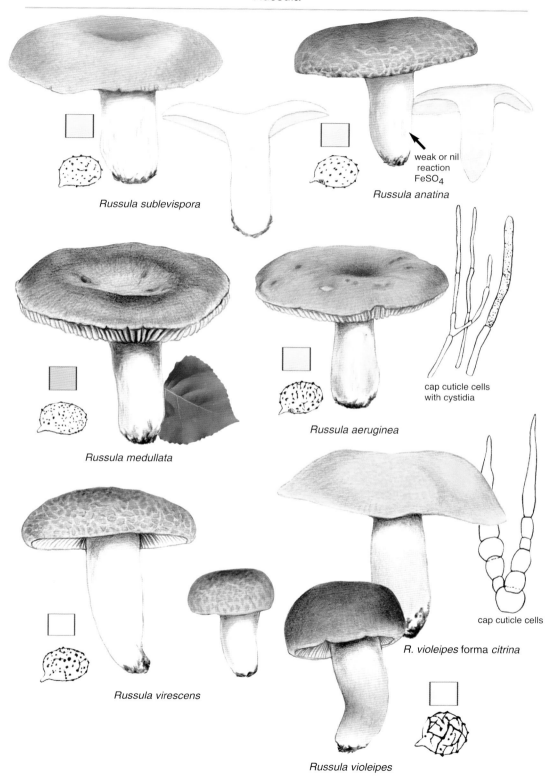

Russula amoena Quél.
[amoena = beautiful]
Cap 3–8 cm, dark purple-red to vinaceous-violet, velvety-pruinose. **Cuticle** peels 1/2. **Gills** white to cream. **Stem** white flushed more or less violaceous-lilac. **Taste** mild. **Odour** of the crushed flesh with a smell of artichokes. **Macrochemicals**: $FeSO_4$ deep rust, guaiac negative. **Pileocystidia** are absent, end cells of the cap cuticle are long and pointed and supported by chains of variously shaped, globose, brick-shaped to cylindric cells. **Spores** 6–8 x 5.5–6.5 μm, with low warts and strong connectives forming a partial reticulum. **Spore deposit** cream (IIb). **Phenol** = purplish.

Grows with both broadleaf and coniferous trees on acid, sandy soils, mainly in the south.

Appears to be genuinely rare in Britain with only two valid collections known from southwest England; most herbarium material under this name has been redetermined as purple forms of **R. violeipes**.

Russula amoenicolor Romagn.
[amoenicolor = beautiful colour]
Cap 3–6 cm, olive green, citrine-green colours, often mixed with a flush of purple, velvety. **Cuticle** peels 1/2. **Gills** white to cream. **Stem** white flushed partially or entirely violet or purple. **Taste** mild. **Odour** of crushed flesh with a smell of artichokes. **Macrochemicals**: $FeSO_4$ pale rust, guaiac slowly blue. **Pileocystidia** absent, the cells supporting the terminal pointed cells of the cap cuticle are uniformly long and rectangular, without any inflated or rounded cells. **Spores** 6–8 x 5.5–6.5 μm, with large warts and partial crests or reticulum. **Spore deposit** darker than *R. amoena* (IIc–IId). **Phenol** = brownish.

Under hardwoods, especially *Quercus*, mainly in southern Europe. Doubtfully British.

Difficult to separate from **R. amoena** and **R. violeipes** without microscopic examination. Differing from both in its darker spores and different cuticle structure.

Subgenus *Russula*
Section *Russula*
Subsection *Russula*
spores white to palest cream, taste ± acrid, odour often fruity of apples or coconut

R. atropurpurea (Krombh.) Britzelm.
[atropurpurea = blackish purple]
Syn. *R. krombholzii*
Cap 5–12 cm, rounded to depressed, smooth, shiny, from deep blood red to purple or even blackish at centre, often with decoloured cream spots or entirely cream at centre. An entirely yellow form, ***dissidens*** has not yet been recorded in Britain. **Cuticle** peels 1/3–1/2. **Gills** white, crowded, margin entire, not serrated. **Stem** white, quite stocky and firm when fresh, later soft and fragile and often flushed greyish with age. **Taste** mild to distinctly acrid in the gills. **Odour** fruity, of apples. **Macrochemicals**: guaiac positive from slow to rapid. **Pileocystidia** cylindrical to subclavate, often capitate, 4–6 μm across, 0–1 septate. **Spores** 7–9 x 5.5–7 μm, warts blunt, 0.5–0.7 μm high, forming short chains with scattered fine connectives. **Spore deposit** white (Ia).

A very common species everywhere, under both broadleaved and coniferous trees.

The epithet *atropurpurea* was used by Peck in 1888 for an American species in the *R. xerampelina* complex and predates our familiar taxon, so another name is needed. Many alternative names have been suggested but it seems that **R. bresadolae** is perhaps the best candidate.

Russula aquosa Leclair
[aquosa = watery]
Cap 3–8 cm, rounded, very fragile, smooth and shiny, deep cherry red, brownish rose, lilac, sometimes very washed out. **Cuticle** peels 3/4. **Gills** white, distant, fragile, margin even, not serrated. **Stem** white, clavate and often narrowed at the apex, fragile and sometimes rather translucent-waterlogged. **Taste** very slightly hot in the gills, otherwise mild. **Odour** weak fruity or slightly coconut. **Macrochemicals**: $FeSO_4$ slowly brownish pink, guaiac slowly blue. **Pileocystidia** slender clavate, strongly staining in SV, non-septate. **Spores** 7–8.5 x 6–7 μm, with conical warts up to 0.75 μm and a partial to almost complete network. **Spore deposit** white (Ib).

Uncommon in the UK, commonest in Scotland, it is found mainly in marshy, mossy places near water.

Russula fragilis (Pers.) Fr.
[fragilis = fragile]
Cap 2–6 cm, fragile, smooth and dry, viscid when wet, very variable in colour, usually purplish lilac, violet-black with greenish areas or even entirely so with blackish centre (the rare var. ***gilva*** is entirely yellow), margin slightly sulcate. **Cuticle** peels 2/3–3/4. **Gills** moderately to widely spaced, pure white to palest cream, fragile, margin often minutely serrate-crenulate. **Stem** cylindric, very fragile, white slightly yellowing at base. **Taste** very acrid. **Odour** pleasant of coconut mixed with fruit. **Macrochemicals**: $FeSO_4$ flesh-pink, guaiac almost nil, only very slowly pale blue. **Pileocystidia** abundant, clavate or cylindric, 0–2 septate, strongly staining in SV. **Spores** 7.5–10 x 6–8 μm, with warts to 0.5 μm high and a well-developed complete reticulum. **Spore deposit** white (Ib).

Common everywhere under both coniferous and broadleaved trees.

The serrated gill edge is best seen using a hand lens. The negative guaiac reaction is a good character.

Russula poikilochroa Sarnari
[poikilochroa = many colours]
Very similar to **R. fragilis** even to having an all-yellow form, it differs in its more matt, shagreened cap surface, positive (although fairly slow) guaiac reaction and appears restricted to *Quercus* spp. Its spores are distinctly smaller, more rounded and even more reticulate, 6.6–8.8 x 5.5–7 μm with warts to 0.5 μm high.

Very few British records but probably overlooked. This is probably identical to **R. rubrocarminea** Romagnesi with that name taking precedence.

Russula atrorubens Quél.
[atrorubens = blackish red]
Syn. *R. olivaceoviolascens, R. knauthii*
Cap 3–8 cm, rounded then depressed, smooth and shiny to viscid, deep blood-red to purple with a blackish centre, sometimes becoming paler and more greenish in the outer third. **Cuticle** peels 3/4. **Gills** white, medium spaced, with even edges. **Stem** cylindric-clavate, pure white to slightly yellowish at base. **Taste** strongly acrid. **Odour** fruity of amyl acetate or slightly coconut. **Macrochemicals**: $FeSO_4$ flesh-pink, guaiac rapidly and strongly positive. **Pileocystidia** abundant, swollen-clavate, to 200 μm long, 5–8 μm across, 2–3 septate SV. **Spores** 7–8 x 5–6.5 μm, with low warts 0.3–0.5 μm and numerous connectives forming a good network. **Spore deposit** white (Ia).

Found with conifers and possibly also with *Salix*, its distribution in the UK is uncertain but probably commoner in the north.

Russula

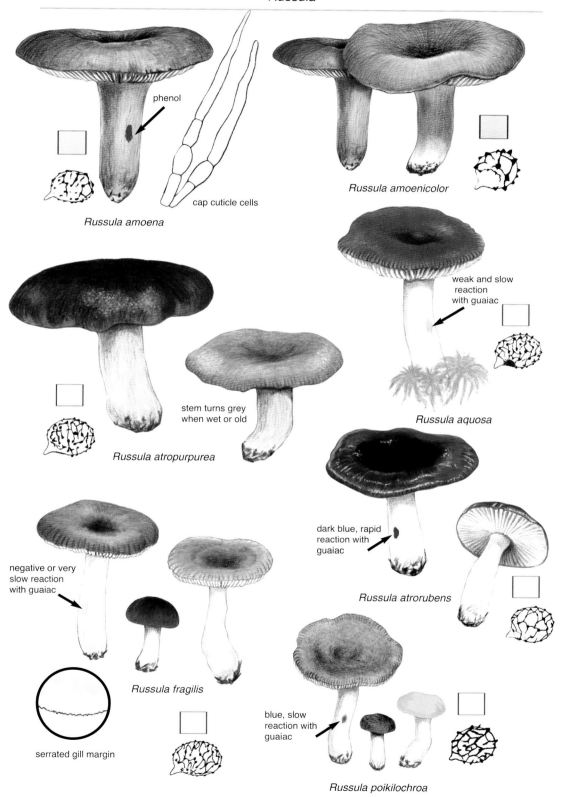

Russula laccata Huijsman
[laccata = lacquered]
Syn. *R. norvegica*
Cap 2.5–5 cm, rounded then soon flattened, very smooth and usually extremely glossy as if lacquered, deep blood-red, red-brown to purplish and darker, almost black at centre. **Cuticle** peels 3/4 or more. **Gills** rather widely spaced, pure white, brittle. **Stem** pure white to pale cream at base, then pinkish in very mature waterlogged specimens, clavate, fragile. **Taste** very acrid. **Odour** of stewed fruits or slightly of dessicated coconut. **Macrochemicals**: $FeSO_4$ pale pink, guaiac positive but not rapid. **Pileocystidia** moderately abundant, slender to clavate, non- to rarely 1-septate. **Spores** 7–9 x 5.5–7.2 μm, with warts up to 0.5 μm and numerous connectives forming a partial to almost complete network. **Spore deposit** white (1b-IIa).

Associated with *Salix* both in the lowlands and in the alpine regions.

Russula betularum Hora
[betularum = growing with *Betula*]
Cap 2–5 cm, bright rose-pink, pale pink to almost white, very fragile, often washed out at centre, margin somewhat tuberculate-sulcate. **Cuticle** peels almost entirely, flesh white below cuticle. **Gills** pure white, rather distant, margin even to slightly crenulate. **Stem** slender, clavate, very fragile, white. **Taste** hot to very hot. **Odour** nil to slightly coconut. **Macrochemicals**: $FeSO_4$ medium pink, guaiac positive but slow and weakly blue. **Pileocystidia** numerous, elongate-clavate (5-10 μm wide), 0–2 septate, strongly staining in SV. **Spores** 8.5–10.5 x 7–8.5 μm, warts 0.5–0.8 μm, with strong, well-developed network. **Spore deposit** pure white (Ia).

Often abundant under *Betula* in the UK and very widespread. Its small size, pale pink cap and hot taste are all distinctive.

Russula raoultii Quél.
[raoultii = after the original collector, Dr Raoult]
Cap 2–6 cm, pale lemon yellow, yellow-buff to almost white with the faintest tint of yellow, smooth and shining, margin slightly sulcate. **Cuticle** peels 3/4. **Gills** white to pale cream, brittle, with some intervening, edges entire, not serrated. **Stem** cylindric, rather slender, fragile, white to slightly grey-brown with age or bruising. **Taste** distinctly acrid. **Odour** pleasant, slightly fruity. **Macrochemicals**: $FeSO_4$ pale pinkish rust, guaiac rapid and intense blue. **Pileocystidia** plentiful, clavate, 0-septate, very dark with SV. **Spores** 7–8.5 x 5.5–6.5 μm, with warts to 0.5 μm, with a prominent and complete network of connectives. **Spore deposit** white (Ia).

Uncommon but widespread, with various broadleaf trees; like a yellowish, washed out **R. fragilis**. Might also be confused with **R. solaris**, which has cream spores and isolated warts.

Russula solaris Ferdinandsen & Winge
[solaris = belonging to the sun]
Cap 2–7 cm, soon flattened and depressed, surface smooth, dry, bright lemon yellow, egg-yellow, may be slightly darker at centre, margin smooth to slightly sulcate. **Cuticle** peels 2/3. **Gills** moderately spaced, broad, pale cream. **Stem** cylindric-clavate, rather short, soft and fragile, white bruising yellowish brown. **Taste** acrid to strongly so. **Odour** fruity or vinegary. **Macrochemicals**: $FeSO_4$ pinkish rust, guaiac rapid but not very strong. **Pileocystidia** numerous, elongate, cylindric-clavate, 6–8 μm across, 0–2 septate. **Spores** 7–9 x 5.5–7.2 μm, warts up to 1 μm, mostly isolated with a few short connectives. **Spore deposit** cream (IIb-c).

Very uncommon but widely distributed in Britain, associated with *Fagus*. Similar in appearance to **R. raoultii** but with strongly acrid taste and isolated spore warts.

Russula silvestris (Singer) Reumaux
[silvestris = pertaining to woodlands]
Syn. *R. emeticella* sensu Rayner
Cap 2.5–4.5 (-5) cm, smooth and shiny, bright pinkish red, rose (but not strong scarlet as in *R. emetica*), often whitish cream at centre, margin slightly sulcate. **Cuticle** peels 3/4 to completely. **Gills** moderately spaced, narrow, pure white. **Stem** slender, cylindric-clavate, pure white, very soft and fragile. **Taste** very acrid. **Odour** pleasant, faintly of dessicated coconut. **Macrochemicals**: $FeSO_4$ pink, guaiac almost nil to slight blue. **Pileocystidia** numerous, elongate-clavate, 0–1 septate, reacting strongly with SV, 7–9 μm across. **Spores** 7.2–9.8 x 6.2–7.8 μm, warts up to 1 or 1.2 μm, with more or less complete reticulum. **Spore deposit** white (1a).

Under broadleaf trees especially *Quercus* and *Castanea* but also sometimes under *Pinus*, on sandy, ± acid soils. Widespread and not uncommon in Britain.

Russula emetica (Schaeff.) Pers.
[emetica = causing vomiting]
Cap 5–10 cm, smooth and glossy, viscid when wet, bright scarlet, to blood red, sometimes paling with age. **Cuticle** peels almost completely. **Gills** rather widely spaced when mature, broad, obtuse, white with cream-yellow tint at certain angles. **Stem** tall, clavate-cylindric, spongy and fragile when mature, pure white, yellowing slightly with age. **Taste** very hot. **Odour** clearly fruity-coconut. **Macrochemicals**: $FeSO_4$ flesh-pink, guaiac almost nil. **Pileocystidia** numerous, clavate-fusiform, mostly strongly clavate, 1–3 septate, strongly blackening with SV. **Spores** 8–11 x 7.5–8.5 μm, with conical warts up to 1.2 μm, numerous connectives forming a well-developed reticulum. **Spore deposit** white (Ia-b).

In wet, mossy areas usually in *Sphagnum* under conifers, also under *Betula nana* and dwarf *Salix* in northern highland areas; can be quite common in suitable habitats.

R. grisescens (Bon & Gaugué) Marti, is very similar to *R. emetica* and is found in upland *Sphagnum* bogs; its stem flushes grey with age and it has smaller spores, 7–9 x 5.5–7 μm plus a stronger guaiac reaction.

One British record under mixed trees in Surrey.

Russula mairei Singer
[mairei = after René Maire, 1878–1949, mycologist]
Syn. *R. nobilis* of some authors, *R. fageticola*
Cap 3–6 cm, firm-fleshed, convex, broadly expanded, smooth and dry, bright pinkish scarlet to rose, often decolouring in part to pale cream or even entirely white. **Cuticle** peels 1/3 and exposed flesh is pinkish. **Gills** rather crowded, fleshy but brittle, white at first then soon cream with a distinct glaucous tint viewed at certain angles. **Stem** usually equal to or slightly shorter than cap diameter, stout and firm, white. **Taste** rapidly very acrid. **Odour** pleasant, faint of coconut mixed with fruit. **Macrochemicals**: $FeSO_4$ yellowish pink, guaiac rapidly intense blue (compare with *R. emetica* which is almost nil). **Pileocystidia** numerous, clavate to fusiform, 0–2 septate, 6–8.5 μm across, strongly staining in SV. **Spores** 7–8 x 5.5–6.5 μm, with quite prominent warts up to 0.5 μm high, with numerous connectives forming a nearly complete reticulum. **Spore deposit** pure white (Ia).

Common everywhere with *Fagus* on clay to calcareous soils.

Russula

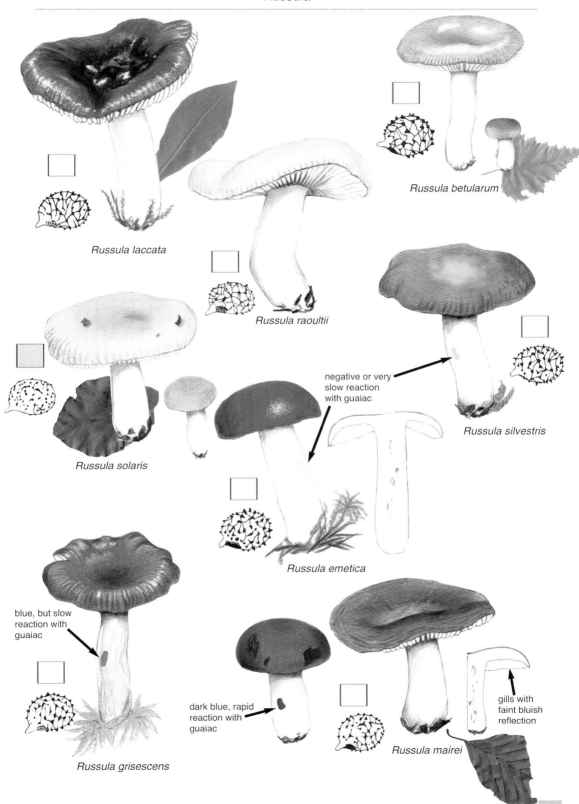

Russula rhodomelanea Sarnari
[rhodomelanea = rose-red and black]
Syn. *R. pulchrae-uxoris* Reumaux
Cap 2–4(-5) cm, convex then slowly flattened, smooth, shiny, blood red, crimson, sometimes with darker, almost black centre. **Cuticle** peels 1/2–3/4. **Gills** white, crowded, slightly arched, sometimes with a faint glaucous reflection (like *R. mairei*), edges turning grey-brown when bruised or with age. **Stem** cylindric, slender (0.5-0.7 cm), white but very often with grey-black stains at base or even overall (see note below). When scratched the outer cuticle turns dull red before blackening. **Taste** very acrid. **Odour** of *R. fragilis* i.e. fruity-coconut. **Macrochemicals**: $FeSO_4$ pale pink, guaiac rather slow and weak. **Pileocystidia** cylindric-clavate, (0)1–4 septate, 6-12 µm across. **Spores** 7.4–9.5 x (6-)6.4–7.8 µm, with warts to 0.8 µm, with numerous meshes to form a good reticulum. **Spore deposit** pure white (Ia).

Rare, found with *Quercus* in several sites in Surrey, and Kent, preferring dry, often barren soils.

The species was described from thermophilic sites in the Mediterranean growing on dry soils with *Quercus cerris*; it has been found here with *Q. ilex* and *Q. petraea*. The blackening seems to be very variable and occurs particularly as the specimens start to collapse and begin to decay.

Russula consobrina* (Fr.)Fr.
[consobrina = a cousin]
Cap 5–10 cm, convex then depressed when expanded, smooth and shiny, uniformly dull brown, sepia, hazel brown, grey-brown. **Cuticle** peels almost 4/5. **Gills** crowded, with numerous half gills, with some forking at the base, pale cream. **Stem** robust (2–3 cm across), white, generally flushed with grey especially on the fine longitudinal ridges, often reddening slightly when bruised. **Taste** soon extremely acrid, painful to taste! **Odour** nil or weakly of apples. **Macrochemicals**: $FeSO_4$ greyish, guaiac slowly azure blue, formalin rapidly red. **Pileocystidia** numerous, clavate, 1–3 septate, sometimes irregular in outline. **Spores** 8.5–10 x 7.5–9 µm, with blunt warts up to 0.8 µm, forming a partial to almost complete network. **Spore deposit** white-cream (IIa-b).

With *Picea excelsa*, *P. abies*, widely distributed in northern Europe, especially common in Scandinavia, not yet known in the UK, a few old records (doubtful?) in Scotland. This species is so distinctive in its colour, size and extremely acrid taste it is unlikely to be confused with anything else, but the name was formerly widely used for **R. sororia**.

Russula pumila Rouzeau & Massart.
[pumila = dwarfish]
Cap 2–4 cm, deep purplish violet, often blackish at centre, margin rather coarsely sulcate when old. **Cuticle** peels 1/2. **Gills** fairly widely spaced, broad, white. **Stem** white but soon flushed yellow-grey to ochre as it ages, looking very water-logged and fragile. **Taste** slowly and moderately acrid. **Odour** nil. **Macrochemicals**: $FeSO_4$ greyish yellow, guaiac slowly blue. **Pileocystidia** are clavate-fusiform, 1-septate, reacting strongly with SV. **Spores** 8–11 x 6–8 µm with small warts and partial connectives to a well-developed network. **Spore deposit** white (Ib).

Found only in *Alnus glutinosa* thickets. Rather rare in the UK although this may reflect the paucity of mycologists searching the specialised habitat!

Very similar is **R. alnetorum**, found with Green Alders (*A. alnobetula*) in subalpine zones; it has a brighter violet cap, less discoloured stem, and larger spores.

Subsection *Violaceinae*
spore deposit cream, cap colours variable, taste ± acrid, odour of *Pelargonium*

Russula pelargonia Niolle
[pelargonia = smelling of *Pelargonium*]
Cap 2–4 cm, fragile, sulcate at margin, pale purplish red, greyish rose, sometimes with paler, ochre areas, finely subvelutinous-chagrined. **Cuticle** peels 1/2. **Gills** pale cream, fragile, rather widely spaced, slightly interviened. **Stem** slender, fragile, cylindric-clavate, white, greying slightly with age. **Taste** moderately but clearly acrid. **Odour** distinct of *Pelargonium* or stewed fruits. **Macrochemicals**: $FeSO_4$ greyish rose, guaiac slowly dark blue. **Pileocystidia** cylindric-clavate, 6-10 µm across, 1–2 septate, reacting strongly with SV. **Spores** 7–9 x 6–8.5 µm, with warts up to 1 µm high, with numerous crests and connectives in a partial reticulum. **Spore deposit** pale cream (IIa-b).

Under *Populus* but also apparently under *Quercus*, *Ulmus*, *Carpinus*, etc, rather uncommon but widespread.

Russula clariana* Heim ex Kuyper & Vuure
[clariana = clear]
Very similar to *R. pelargonia* but more robust, cap less sulcate, and with lower spore warts in a more complete reticulation. Associated with *Populus*. Not yet British.

Russula violacea Quél.
[violacea = violet coloured]
Cap 3–5 cm, subglobose, sometimes slightly umbonate, smooth to slightly matt, purplish lavender, pale violet sometimes darker violet-black at centre, sometimes paling to greenish cream or even almost entirely so. **Cuticle** peels 1/2. **Gills** fairly widely spaced, arched, fragile, forking a little near the stem, pale cream. **Stem** clavate, fragile, white, often yellowing slightly at the base. **Taste** clearly acrid, with cedarwood component. **Odour** complex, of fruit, menthol or *Pelargonium*. **Macrochemicals**: $FeSO_4$ greyish rose, guaiac rapidly azure blue. **Pileocystidia** clavate, up to 11 µm across, 0–4 septate, strongly staining in SV. **Spores** 7–8.5 x 6–7 µm, with isolated spiny warts up to 1 µm high. **Spore deposit** pale cream (IIa-b).

Rare, associated with *Populus tremula*. Very few British records.

Subsection *Sardoninae*
taste acrid, spores cream-ochre, flesh often pink with ammonia

Russula cavipes Britzelm.
[cavipes = with cavities in the stem]
Cap 3–6 cm, broadly rounded, smooth to viscid when wet, purplish brown, greyish rose, lilac, greenish, or mixed, margin slightly sulcate. **Cuticle** peels 1/2. **Gills** fairly widely spaced, whitish cream. **Stem** tall and cylindric, firm to fragile, white. **Taste** acrid. **Odour** distinct of *Pelargonium*, fruit or honey. **Macrochemicals**: $FeSO_4$ pale pink, guaiac almost nil, ammonia (on gills) bright pink. **Pileocystidia** long and fusiform, 6-11 µm across, non-septate. **Spores** 7.2–9.5 x 6.5–8 µm, warts up to 1.2 (1.6) µm high, with some connectives or even a few meshes. **Spore deposit** pale cream (IIa-b).

Associated with *Abies* or rarely *Picea*, rare in Britain, only two confirmed records.

Looks a little like a small, delicate **R. sardonia** and has a similar KOH or ammonia reaction but differs in its negative guaiac reaction and different spore ornamentation.

Russula

Russula sanguinaria (Schum.) S. Rauschert
[sanguinaria = blood red]
Syn. *R. sanguinea* Fries
Cap 4–8 (-10) cm, fleshy, rounded, soon flattened, smooth and shiny, margin slightly sulcate with age, bright scarlet, pinkish red, often with white areas, sometimes completely so. **Cuticle** adnate, barely peeling. **Gills** moderately spaced, arched-subdecurrent near stem, pale cream-yellow, forking or anastomosing near stem, margin sometimes pink near cap. **Stem** firm and fleshy then brittle, white or variably flushed pinkish red, sometimes entirely so, very often (and characteristically) stained yellow with age. **Taste** moderately acrid and sometimes also bitter. **Odour** nil or fruity. **Macrochemicals:** $FeSO_4$ pale pink, guaiac slowly to rapidly dark blue. **Pileocystidia** numerous, slender, attenuate to capitate, 4–7 μm across, 0–2 septate (Sarnari states that they are unicellular but illustrates them with 2 septa!). **Spores** 7.2–9.6 x 6.3–7.4 μm, with warts to 0.8 μm high, with a few short crests and connectives joining 3 or 4 warts. **Spore deposit** deep cream (IId-IIIa).

Abundant everywhere that pines are planted, often in very large numbers.

The all white or yellowish white forms can be very deceptive but the acrid taste, habitat and tendency to yellow are good clues to identity.

Russula rhodopus Zvara
[rhodopus = red-stemmed]
Cap 4–10 cm, broadly convex, smooth, glossy, viscid, often appearing lacquered, bright scarlet to blood red. **Cuticle** peeling 1/3. **Gills** crowded, with frequent half gills, interveined, arched-sinuate, ivory-white. **Stem** cylindric-clavate, stout, firm, white and more or less flushed with red, especially around the middle. **Taste** mild at first then soon acrid in the lamellae. **Odour** distinct, fruity. **Macrochemicals:** $FeSO_4$ pinkish rust, guaiac instantly and strongly blue. **Pileocystidia** fusiform-clavate, rather long and slender, 5–9 μm across, 0–1 septate, sometimes capitulate at the apex. **Spores** 7.2–9 x 6.3–7.2 μm, with warts to 0.6 μm with numerous connectives and crests to form a partial network. **Spore deposit** deep cream to ochre (IIc-IIIb).

Associated with *Picea* in northern Europe, known from one site at the Linn of Dee in Scotland.

A striking species common in montane habitats, it is most likely to be confused with the pine associate *R. sanguinaria* which differs both in its host and mostly isolated spore warts.

Russula helodes Melzer
[helodes = of marshy places]
Cap 5–13 cm, rapidly flattening and becoming dished, often with margin very flexuose-undulate, surface matt, glossy only when wet, bright blood-red, occasionally dark vinaceous-red, often mottled with paler areas. **Cuticle** adnate, hardly peeling. **Gills** very crowded, often forking, adnate to slightly decurrent, pale ochre. **Stem** robust, rather short, white but often flushed partially pink. **Taste** acrid. **Odour** nil. **Macrochemicals:** $FeSO_4$ greyish rose, guaiac slowly blue-green. **Pileocystidia** abundant, cylindric-fusiform to vermiform, often with teat apex, 0-septate. **Spores** 8.5–10.3 x 7.5–8 μm, warts up to 0.7 μm, with numerous connectives to form a complete network. **Spore deposit** dark cream (IIIa).

Often found in *Sphagnum* swamps under conifers. Rare in the UK although good recent records have been made under conifers in wet mossy areas especially in upland regions.

Russula sardonia Fr.
[sardonia = pertaining to Sardinia]
Cap 4–12 cm, broadly convex, fleshy, robust, smooth, shiny, variable in colour from blood-purple, violet, vinaceous, livid violet, greenish (**forma** *viridis*) to partly or entirely citrine yellow or ochre yellow (**forma mellina**). **Cuticle** adnate, hardly peeling. **Gills** fairly crowded, very slightly decurrent, narrow, arched, forking at base, palest primrose yellow when mature (important field character). **Stem** firm, white, sometimes flushed partly or entirely lilac to pinkish red. **Taste** intensely acrid. **Odour** fruity, stewed fruits. **Macrochemicals:** $FeSO_4$ pink, guaiac quickly positive, deep blue, ammonia on gills or flesh slowly (5 mins) bright pink. **Pileocystidia** numerous, slender, pointed, flexuose, non-septate 3.3–6 μm across. **Spores** 7–9 x 5.8–7.6 μm, warts up to 0.6 μm, with numerous crests and connectives forming a partial reticulum. **Spore deposit** deep cream (IIc-d).

Common under pines everywhere, some of the colour forms are difficult to recognise as the same species, however the pale lemon tinted gills and pink reaction with ammonia are key field characters. Often accompanies *R. sanguinaria*.

Russula torulosa Bres.
[torulosa = cylindrical with bulges]
Cap 5–10 (-12) cm, broadly convex, fleshy, smooth and glossy, deep purple, violet-purple, reddish purple, sometimes almost blackish purple. **Cuticle** adnate. **Gills** crowded, finely anastomosed at the base, cream-white without a lemon-yellow hue (unlike *R. sardonia*). **Stem** cylindric-clavate, stout, firm and fleshy, flushed reddish purple. **Taste** very acrid. **Odour** pleasant of fresh fruit or honey. **Macrochemicals:** $FeSO_4$ pale pink, guaiac rapid and intensely blue, ammonia on flesh = negative. **Pileocystidia** long and slender, cylindric to slightly clavate, non-septate. **Spores** 7.2–9 x 5.6–7 μm, warts to 0.5 μm, with numerous thick meshes, connate warts, forming a partial reticulum. **Spore deposit** deep cream (IId).

Nationally rare but locally very abundant under *Pinus nigra* var. *maritima*, recorded from Wales, Scotland and West Sussex. Confused with **R. sardonia**, but the usually coastal location under Corsican Pine, the cream—not yellowish—lamellae and the negative reaction to ammonia are characteristic.

Russula queletii Fr.
[queletii = after Lucian Quélet, 1832–1899, mycologist]
Cap 4–7 cm, broadly convex then expanded, often with a broad umbo, fleshy, smooth and shiny, deep purple-magenta, vinaceous, dark violet to brownish purple. **Cuticle** peeling 1/2–1/3. **Gills** pale cream, never with a yellow hue (see *R. sardonia*), fairly crowded, subdecurrent. **Stem** cylindric-clavate, narrowed at apex, often slender, white flushed with carmine, vinaceous-red, occasionally pale. **Taste** strongly acrid. **Odour** intense, fruity, of apples. **Macrochemicals:** $FeSO_4$ pale pinkish, guaiac positive but not rapid and rather weak, ammonia negative on gills. **Pileocystidia** clavate, 0–1 septate, rather broad 5–9 (12) μm across. **Spores** 7.3–9 (-9.8) x 6–7.3 (-8) μm, with large isolated warts to 1.2 μm high. **Spore deposit** cream (IIc-d).

Associated primarily with *Picea* but also *Pinus*, on alkaline soils, uncommon everywhere.

Easily confused with **R. sardonia** but that species has pale, lemon-tinted gills which turn bright pink with ammonia in 5 minutes. **R. fuscorubroides** found with *Picea* on acid soils is extremely similar; it has uniformly darker colours and slightly darker spores.

Russula

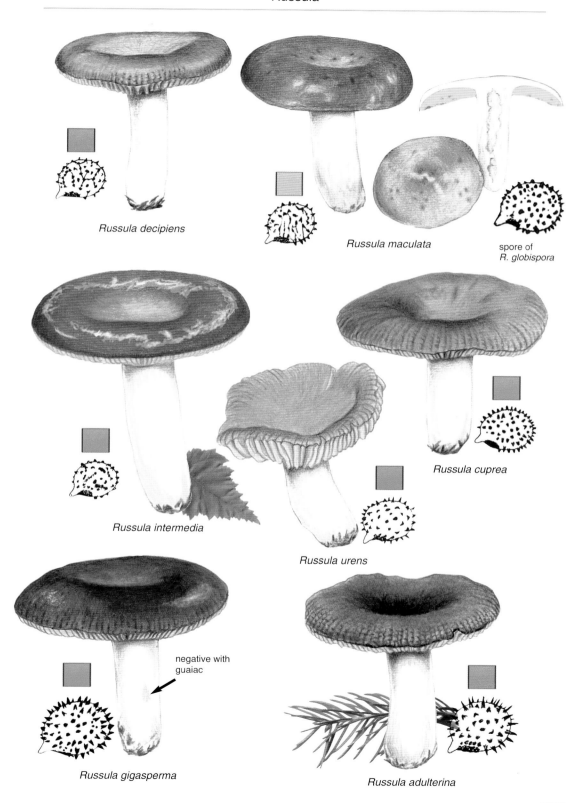

Russula firmula** Jul. Schaeff.
[firmula = fairly firm]
(non sensu R. Rayner which = *R. cuprea*)
Cap 4–8 cm, soon flattened, smooth, glossy, typically violet-lilac, bluish violet, brown-vinaceous, sometimes with greenish areas or at centre. **Cuticle** peels 1/2. **Gills** rather crowded, bright ochre-yellow to yellow-orange when mature. **Stem** often rather tall, cylindric, white. **Taste** quite acrid. **Odour** strong of apples. **Macrochemicals**: $FeSO_4$ dull reddish pink, guaiac rather slowly blue. **Pileocystidia** numerous, cylindric-clavate, 2–3 septate, 5–9 µm across. **Spores** 8–10.5 x 7–8.5 µm, with isolated warts up to 1 µm high. **Spore deposit** yellow-ochre (IVd).

Rather uncommon to rare, in subalpine woods, especially *Picea* on calcareous soils, not yet recorded for certain in the UK, but perhaps in Scotland.

This species has been much confused and the name has been applied to the very similar broadleaf species *R. cuprea* by Rayner among others; the latter does occur in the UK.

Russula veternosa Fr.
[veternosa = aged]
Cap 4–10 cm, broadly rounded, rather firm and fleshy, surface almost matt when dry, pale pink, coral, sometimes vinaceous, often with centre paler, ochre yellow, sometimes entirely so. **Cuticle** peels 1/2–2/3. **Gills** crowded, thin, brittle, broad, pale to medium ochre. **Stem** cylindric, firm then soon soft and spongy, white. **Taste** slightly to distinctly acrid. **Odour** nil at first then distinctly of honey as it ages (more noticeable when kept in a container). **Macrochemicals**: $FeSO_4$ greyish pink, guaiac slowly blue-green. **Pileocystidia** numerous, elongate-clavate, 2–3 septate, strongly staining in SV, 4–10 µm across. **Spores** 6.4–8 (8.5) x 5.8–6.5 µm, warts very large and conical, 1–1.5 µm tall, isolated. **Spore deposit** deep ochre-yellow (IVb).

Uncommon but widespread, associated with *Fagus*, especially in the south.

Russula vinosopurpurea (Lam.:Fr.) Fr.
[vinosopurpurea = wine-purple]
Cap 4–9 cm, expanded and depressed, margin obtuse, surface shiny, reddish purple, vinaceous, brownish purple, often decoloured at centre to pale yellow-cream or with yellowish spots and patches. **Cuticle** peels 1/2. **Gills** fairly crowded, broad, ventricose, interveined, medium ochre-yellow when mature. **Stem** cylindric-clavate, white flushed greyish brown on handling or with age especially at the base. **Taste** a little acrid in the gills, mild elsewhere. **Odour** a little fruity. **Macrochemicals**: $FeSO_4$ pinkish rust, guaiac positive but often weak. **Pileocystidia** long, filiform, 4.5–8.5 across, 1–2 septate. **Spores** (7.5) 8–10 x 6.4–8 µm, densely echinulate with isolated warts to 0.8–1.3 µm high. **Spore deposit** deep yellow-ochre (IVd).

Very rare, known from a few localities in England, Scotland and Northern Ireland, associated with *Fagus*.

Russula rubra Jul. Schäff.
[rubra = red]
Syn. *R. pungens* Beardslee
Cap 4–8 cm, expanded and depressed, margin obtuse, surface matt, pruinose, scarlet, carmine, darker at centre, often whitish at margin. **Cuticle** peels nil to 1/4. **Gills** fairly crowded, broad, ventricose, interveined, medium ochre-yellow when mature. **Stem** cylindric-clavate, fleshy, white flushed greyish brown when wet or with age. **Taste** distinctly acrid. **Odour** of honey. **Macrochemicals**: $FeSO_4$ pinkish rust, guaiac positive but often very weak. **Pileocystidia** long, filiform-clavate, 5–10 across, 1–4 septate, with fuchsinophile granules over the lower half. **Spores** (6.5) 7–9 x 6.5–7.5 µm, sparsely warted with isolated warts to 0.5 µm high and some short connectives. **Spore deposit** yellow-ochre (IIIb-c).

Recorded in Northern Ireland and England, rare but may well be under-recorded, associated with *Quercus*, *Castanea* and *Fagus*.

This is one of a small number of species whose pileocystidia also have fuchsinophile granules.

Russula rutila Romagn.
Cap 3–6 cm, fleshy, smooth and shiny to slightly chagrined or even matt, bright red to darker brownish red, often paler at centre, cream-yellow. **Cuticle** peels 1/3-1/2. **Gills** pale yellow-ochre, moderately crowded, brittle. **Stem** cylindric-clavate, white, fleshy. **Taste** distinctly acrid, especially in the lamellae. **Odour** slightly fruity. **Macrochemicals**: $FeSO_4$ greyish pink, guaiac pale blue. **Pileocystidia** with fuchsinophile granules at base, quite broad, 7–10 µm across, cylindric-clavate, 2–4 septate. **Spores** 7.5–9.5 x 6.4–8 µm, with warts to 1 µm high and numerous short crests or connectives but without a true reticulum. **Spore deposit** deep ochre-yellow (IVc).

Very rare in Britain and few authentic records exist, associated with *Quercus* and *Carpinus*.

Russula quercilicis** Sarnari
[quercilicis = associated with *Quercus ilex*]
Cap 4–9 cm, often curiously mottled, buff, vinaceous, livid greyish ochre, lilac, etc. **Cuticle** ± adnate. **Gills** strongly interveined, ochre. **Stem** cylindric, white, often faintly flushed pink at the apex. **Taste** acrid. **Odour** slight of cedarwood. **Macrochemicals** $FeSO_4$ pale rust-pink. **Pileocystidia** swollen, clavate, with fuchsinophile encrustations, also accompanied by long hyphae with similar encrustations. **Spores** ellipsoid-ovoid, 8–10.4 x 6.8–8 µm, with warts to 0.65 µm high, and a few short connectives. **Spore deposit** deep yellow (IVc).

With *Quercus ilex* on calcareous soils, in the Mediterranean, not yet British.

Section *Viscidinae*
with encrusting veil on stem base and reddish KOH reaction; pileocystidia with very fine encrusting pigment

Russula viscida Kudřna
[viscida = sticky or viscid]
Syn. *R. artesiana*
Cap 5–14 cm, convex then slowly flattening, very firm, fleshy, smooth and shiny, viscid when wet, minutely rugulose, deep vinaceous-red, purple-brown to ochraceous, occasionally entirely so. **Cuticle** adnate. **Gills** fairly widely spaced, arched, anastomosing at their base, ivory white to cream, discolouring brownish. **Stem** often tall and usually robust, firm and fleshy, white with a fine crust of yellowish brown, granulose veil at the base. **Taste** mild to slightly acrid in the lamellae. **Odour** pleasant of honey or fruit. **Macrochemicals**: $FeSO_4$ pinkish rust, guaiac rapidly blue. **Pileocystidia** cylindric-clavate, 5–7 µm across, 0–1 septate. **Spores** 8–10.5 (-11) x 6.5–8.5 (-9.5) µm, with tiny warts up to 0.4 µm and a more or less complete reticulum. **Spore deposit** pale cream (IIa-b).

Rare in Britain, known from a very few sites in southern England with both broadleaved and coniferous trees. The all yellow form has not yet been recorded in Britain.

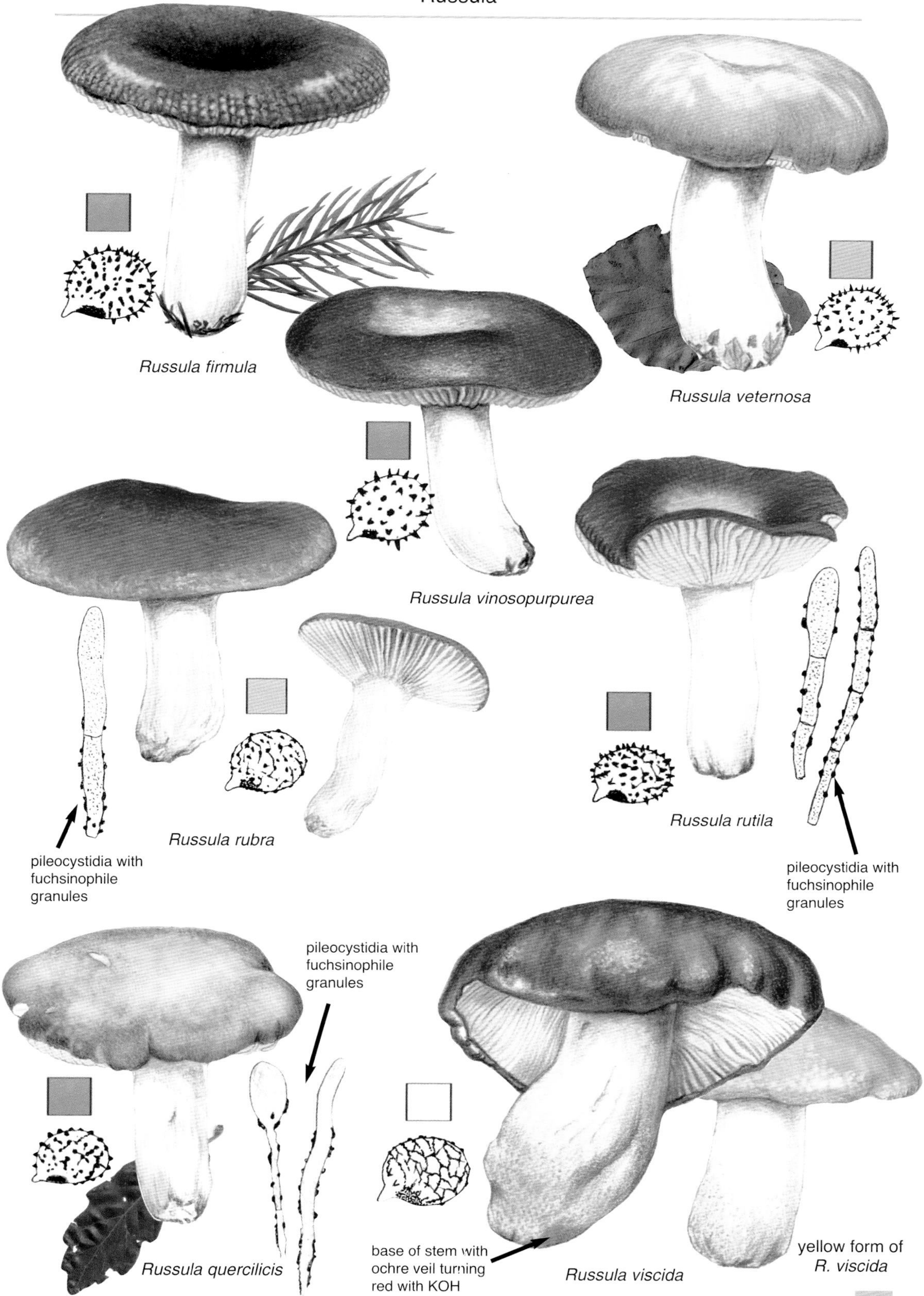

Russula

Russula firmula

Russula veternosa

Russula vinosopurpurea

Russula rubra

pileocystidia with fuchsinophile granules

Russula rutila

pileocystidia with fuchsinophile granules

pileocystidia with fuchsinophile granules

Russula quercilicis

base of stem with ochre veil turning red with KOH

Russula viscida

yellow form of *R. viscida*

Russula ochroleuca Pers.
[ochroleuca = pale ochre]
Cap 5–10 cm, fleshy and firm, soon expanded and flattened although margin often incurved, smooth and dry, uniformly ochre-yellow, citrine to distinctly greenish yellow. **Cuticle** peels 2/3. **Gills** cream, crowded to moderately spaced when mature, brittle. **Stem** cylindric, firm then spongy and brittle, white to cream often with ochre, cracking surface at base, becoming grey with age or when waterlogged. **Taste** from mild to distinctly slightly acrid in the gills. **Odour** nil. **Macrochemicals**: $FeSO_4$ pale pinkish rust, guaiac positive, more or less rapid and intense. **Pileocystidia** absent but with odd cells, variously interpreted by different authors as pileocystidia or fuchsinophile laticifers which are moderately thick-walled and which are visible when treated with Carbol-fuchsin or KOH and show external incrustations. **Spores** 8–10 x 7–8 µm, with large warts up to 1.25 µm high and numerous meshes forming ± complete reticulum. **Spore deposit** white to palest cream (Ia-IIc) very variable.

Common to abundant everywhere under both broadleaved and coniferous trees.

Often confused with **R. fellea** which is mainly with *Fagus*, has a more honey-ochre hue in all parts without any greenish tints, and a faint smell of *Pelargonium*.

Section *Polychromae*
Subsection *Xerampelinae*
with greenish reaction to $FeSO_4$, odour of shellfish, often bruising brown

Russula xerampelina (Schaeff.) Fr.
[xerampelina = colour of old leaves]
Syn. *R. erythropoda*
Cap 5–10 cm, soon expanded, viscid when wet but soon dry and shiny to slightly matt, firm and fleshy when young, vivid blood-red, purplish red, vinaceous to blackish red at centre. **Cuticle** adnate. **Gills** somewhat distant, broad, ventricose, pale ochre-yellow, brittle, bruising brown. **Stem** cylindric-clavate, firm and fleshy, white flushed more or less reddish, sometimes entirely and vividly so, staining yellowish brown on ageing or handling. **Taste** mild, pleasant. **Odour** nil at first then soon strong of old shellfish, very strong on drying. **Macrochemicals**: $FeSO_4$ deep blackish green, guaiac blue-green. **Pileocystidia** slender, cylindric, 5–6 µm across, 0–2 septate, cuticular hyphal endings slender, not swollen, 3–5 µm across. **Spores** 8–10.8 x 6.5–8.2 µm, with warts up to 0.75 µm and scattered short connectives but no real reticulum. **Spore deposit** ochre (IIIb).

Common and widespread throughout Britain, associated with *Pinus* species.

Russula favrei **M.M. Moser
[favrei = after J. Favre, 1882–1959, Swiss mycologist]
Closely related to **R. xerampelina**, differing in its browner cap, isolated spore warts and broader cap cuticle hyphae. It is an alpine species growing mainly with *Picea*. Not British.

Russula subrubens (J.E. Lange) Bon
[subrubens = somewhat rust red]
Syn. *R. chamiteae* ss. auct.
Cap 4–9 cm, broadly convex to flattened, smooth and shining, bright crimson, brown-red or even reddish purple, margin slightly sulcate. **Cuticle** more or less adnate. **Gills** white to pale ochre when mature, moderately spaced, bruising brownish. **Stem** cylindric-clavate, firm, white or flushed reddish, bruising brown on handling. **Taste** mild. **Odour** weak at first then soon strong of shellfish as it ages. **Macrochemicals**: $FeSO_4$ deep blackish green, guaiac positive but slow and weak. **Pileocystidia** slender, 5–8 µm across, 0–2 septate. **Spores** 7.8–9.8 x 6–7.2 µm, warts up to 0.6-0.8 µm, with many isolated warts plus a few to numerous connectives and some meshes in a very partial reticulum. **Spore deposit** pale ochre (IId-IIIb).

An uncommon to rare species although perhaps locally more common, associated with *Salix* both in lowland areas and in subalpine regions (with *Salix herbacea*).

The alpine form is squatter, with more rapidly washed out cap colours and has normally been called **R. chamiteae** but in its microscopic characters matches the lowland form; I follow other authors in synonymising the two.

Russula amoenoides Romagn.
[amoenoides = resembling *R. amoena*]
Cap 3-6 cm, deep purple, reddish purple, purple-brown to almost black at centre, frequently mottled with paler cream areas; surface plush, tomentose. **Cuticle** peels about 1/3. **Gills** deep cream-ochre, moderately crowded. **Stem** white, faintly flushed pink or red especially at the base, bruising brownish on handling. **Taste** mild. **Odour** strong of shellfish, shrimps and crab as the mushroom ages. **Macrochemicals**: $FeSO_4$ deep green to almost black, guaiac deep blue. **Pileocystidia** frequent, narrowly fusiform, 0–1 septate. **Spores** 7.7–8.5 x 7–8 µm, with isolated, conical warts to 0.7 µm high and a few connectives. **Spore deposit** ochre (IIIa).

Rare under *Quercus* in southern England. One of a group of species around **R. xerampelina** this species is distinctive by its beautifully tomentose cap surface, deep purple colours and isolated spore warts. It is perhaps most closely related to **R. graveolens** which differs in its smoother cap and more reticulate spore ornamentation.

Russula graveolens Romell
[graveolens = with strong odour]
Cap: 5-8 cm, broadly convex, smooth to rugulose, very variable in colour, purple-vinaceous, brown-purple, brown, often with centre olivaceous or buff. **Cuticle** adnate. **Gills** moderately crowded, broad, ivory-cream to yellow-ochre when mature, browning when bruised. **Stem** cylindric, firm, white, staining brown on handling. **Taste** mild. **Odour** at first nil then soon becoming strong of old shellfish. **Macrochemicals**: $FeSO_4$ deep green, guaiac rapidly dark blue. **Pileocystidia** narrow, fusiform (4–9 µm), weakly staining in SV. Terminal cells of cuticular hyphae never short and inflated at the cap centre, often very narrow and acute, almost hair-like. **Spores** 8–10 x 6.5–7.5 µm, warts up to 0.75 µm, with numerous connectives forming a partial network although a proportion of warts may be mostly isolated. **Spore deposit** ochre-yellow (IIIa-b).

Common and widespread under broadleaf trees, especially *Quercus*. Formerly a part of the much broader species concept that made up the old **R. xerampelina**.

Russula

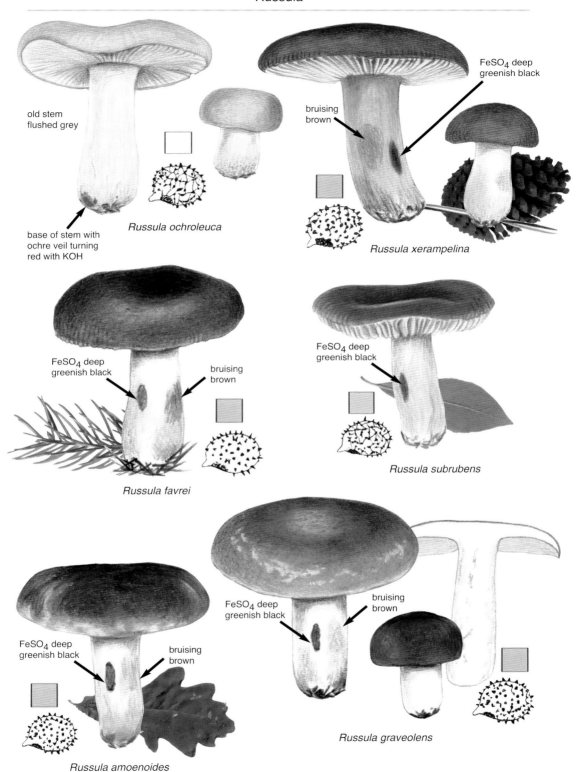

Russula faginea Romagn.
[faginea = growing with *Fagus*]
Cap 6–14 cm, broadly convex, then flattened, fleshy, often depressed at centre, smooth to slightly rugulose, rose-red, copper, vinaceous, ochre-orange, yellowish to cream-greenish at centre. **Cuticle** very adnate, peeling only at extreme margin. **Gills** becoming fairly widely spaced, cream-ivory, pale ochre. **Stem** cylindric, firm, white staining reddish brown or ochre when bruised. **Taste** mild. **Odour** soon like old shellfish when mature. **Macrochemicals**: $FeSO_4$ dark green, guaiac slowly dark blue. **Pileocystidia** very scarce, narrow, fusiform, weakly staining in SV. Cuticular cells narrow, filiform, with numerous swollen hyphal terminations (3–6.5 µm). **Spores** 8.5–10 x 7.5–8.5 µm, with large isolated warts up to 1.37 µm high. **Spore deposit** dark ochre (IIIb-IVa).

With mature *Fagus* and more rarely *Quercus*, widespread but not common in the UK.

The large, robust stature, swollen end cells of the cap cuticle and habitat are all diagnostic.

Russula cicatricata Bon
[cicatricata = scarred]
Cap 3–9 cm, rounded then depressed, firm, fleshy, surface matt to minutely wrinkled concentrically, dull olive-green, ochre-green, ochre to brownish or copper. **Cuticle** peels 1/2. **Gills** crowded, arched to broad, pale cream-ochre bruising ochre-brown. **Stem** cylindric, firm and fleshy, white bruising brown on handling. **Taste** mild. **Odour** nil at first then soon strong of crab or shrimp. **Macrochemicals**: $FeSO_4$ pinkish at first then dark green at the edges, guaiac rapidly blue. **Pileocystidia** long, slender, fusiform, 4–7.5 µm across, non-septate. Cuticular hyphal endings often swollen, vesiculose. **Spores** 8–9 (10) x 6.5–7.5 (8) µm, warts up to 0.75 (1) µm, mostly isolated with a very few connectives. **Spore deposit** deep ochre (IIIa-b).

Uncommon but widespread in Britain, usually with broadleaf trees, especially *Quercus*.

Russula clavipes Velen.
[clavipes = swollen, club-shaped stem]
Syn. *R. elaeodes*, *R. pascua*?
Cap 5–9 cm, convex to depressed, mainly olivaceous, to olivaceous brown, often pinkish at margin, with centre more tobacco-brown. Almost matt to slightly shining at centre. **Cuticle** peels 1/3–2/3. **Gills** moderately crowded, light cream to cream yellow. **Stem** robust, usually distinctly clavate (1.5–3 cm at base), white often tinted slightly pink on one side staining dull rust-brown when bruised. **Taste** mild. **Odour** nil when fresh then soon crab-like. **Macrochemicals**: $FeSO_4$ blue-green, guaiac blue-green. **Pileocystidia** cylindric-fusoid, 0–2 septate. Terminal elements of pileipellis hyphae long and attenuate at cap margin (35–56 x 9–12.5 µm), terminal elements at cap centre with large number of inflated, pear-shaped, ovate or ampulliform cells. **Spores** 7–9 (10) x 5.5–7 µm, with tall acute warts to 1 µm, mostly isolated with a very few connectives. **Spore deposit** ochraceous (IIIb).

In lowland to submontane forests, mainly associated with *Betula*, but also *Picea*, *Pinus* and perhaps also *Alnus* on moist, almost boggy soils. Recorded in the north and west of Britain.

Subsection *Melliolentinae*
cream spores, flesh browning, $FeSO_4$ pink, spores round with very low ornamentation

Russula melliolens Quél.
[melliolens = smelling of honey]
Cap 4–12 cm, fleshy, convex then depressed, smooth to matt, dark red, carmine to copper-red, brownish rose, vinaceous-rust or ochraceous. **Cuticle** peels 1/2. **Gills** broad, brittle, pale cream spotting rust-brown or ochre. **Stem** cylindric-clavate, white, rarely flushed pinkish, spotting ochre-brown at base. **Taste** mild. **Odour** nil when young but then distinct of honey as it becomes very mature, also in drying material. **Macrochemicals**: $FeSO_4$ rose-pink, guaiac rapidly intense blue. **Pileocystidia** cylindric, composed of numerous short, cylindric elements (6–10+! septate), 3–7 µm wide, numerous and very distinctive. **Spores** 8–11 x 9–11, almost spherical, with minute warts 0.1–0.25 µm high and extremely fine connectives forming almost complete reticulum. The roundest spores with the lowest ornamentation of any British *Russula*. **Spore deposit** cream (IIa-b).

Uncommon under broadleaf trees. Variable in colour, but the microscopic characters make this one of the easiest species to identify.

Subsection *Integriforminae*
taste mild; often bright colours, spores pale to dark ochre

Russula decolorans (Fr.) Fr.
[decolorans = losing its colour]
Cap 4–12 cm, fleshy and robust, globose to broadly expanded, smooth and shiny, bright copper-orange, yellow-orange, staining blackish. **Cuticle** peels at margin only. **Gills** pale ochre, broad, strongly interveined, blackening with age or bruising. **Stem** robust, white staining grey-black with age or when injured. **Taste** mild. **Odour** nil to fruity. **Macrochemicals**: $FeSO_4$ grey-rose, guaiac rapidly azure blue. **Pileocystidia** abundant, large, cylindric-clavate, 0–1 septate, rather weakly staining with SV. **Spores** large, 9–14 x 7–9 µm, with strongly spiny warts to 1.5 µm, with some few connectives or connate ridges. **Spore deposit** pale ochre (IIIa).

Under conifers in the north (perhaps confined to Scotland), uncommon.

Most likely to be confused with **R. paludosa** which is usually redder, with reddish tints on the stem, unchanging flesh and spores with a well-developed partial network.

Russula rivulicola Ruots. & Vauras
[rivulicola = growing in a riverine habitat]
Cap 3–7 cm, fairly fragile, margin sulcate with age, deep red but soon discolouring, often with centre paler yellow, often entirely copper-yellowish. **Cuticle** peels 1/2 or a little more. **Gills** moderately spaced, broad, fragile, deep cream-buff. **Stem** cylindric-clavate, often rather slender, fragile, white soon flushing grey to grey-brown with age or when wet. **Taste** mild to very slightly acrid in the lamellae. **Odour** nil. **Macrochemicals**: $FeSO_4$ pale pink, guaiac quickly blue. **Pileocystidia** clavate, occasionally with attenuate apex, broad, 6–10 µm, 0–1 septate. **Spores** 7–8.5 x 6–7 µm, with warts up to 0.8 µm high, with numerous meshes and connate ridges forming a partial reticulum. **Spore deposit** pale ochre-yellow (IIIb)

In wet, swampy woods with *Betula*, one record from Western Scotland.

Described from Finland this species looks a little like **R. nitida** or **R. robertii** but has more copper-red colours and distinctly greys with age.

Russula

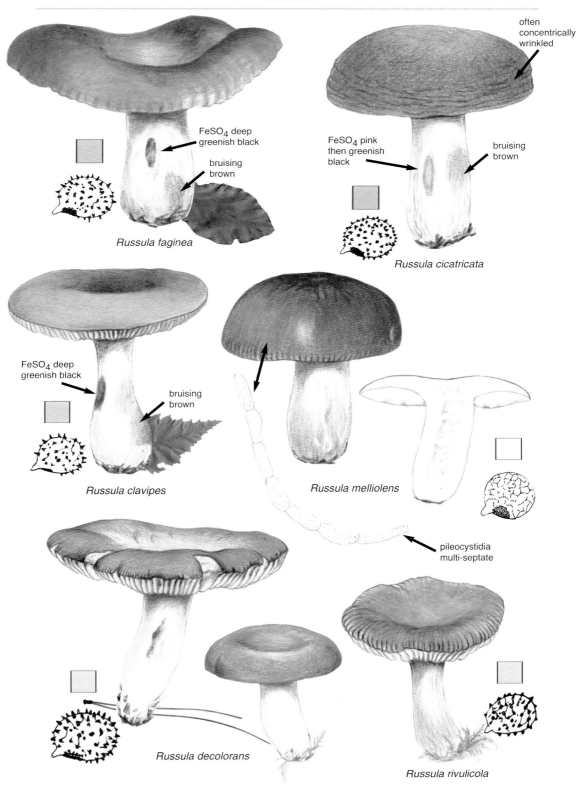

Russula vinososordida Ruots. & Vauras
[vinososordida = wine-red discolouring blackish]
Cap 6–10 cm, soon flattened and depressed at centre, smooth and shiny, bright blood-red, vinaceous-red, darker at centre, sometimes fading to brownish bronze-red, slightly sulcate at margin. **Cuticle** peels 1/2. **Gills** fairly crowded, brittle, interveined, cream to medium ochre. **Stem** tall, cylindric-tapered, white then slowly bruising or ageing grey, a cut through the base of the stem usually reveals grey-black flesh. **Taste** mild to slightly and briefly acrid in the young gills. **Odour** fruity or slightly of resin or cedarwood. **Macrochemicals**: $FeSO_4$ greyish rose, guaiac rapidly blue. **Pileocystidia** cylindric-clavate, with occasional diverticulae, 0–2 septate. **Spores** 6.5–8 x 5.6–6.8 µm, warts up to 0.8 µm, with a dense and partial to almost complete reticulum. **Spore deposit** clear ochre yellow (IIIa-b).

Recently recorded from three localities in Scotland, it is associated with *Betula* and *Picea*.

Easily confused with **R. vinosa** but that species has more purple cap colours, isolated warts and fuchsinophile hyphae rather than dermatocystidia.

Russula paludosa Britzelm.
[paludosa = associated with marshes]
Cap 6–15 (-22) cm, a long time convex for a long time before finally expanding and flattening, very fleshy, firm, smooth and shiny to viscid, bright blood-red, orange-red, apricot to rust, often with decoloured spots or areas of cream. **Cuticle** peels 1/2–3/4. **Gills** pale creamy-yellow, often with edges reddish near cap margin. **Stem** large and robust, clavate, firm, often flushed pink in part or overall, especially at base. Flesh slightly greyish when cut or bruised. **Taste** mild. **Odour** nil or weak of fruit. **Macrochemicals**: $FeSO_4$ greyish rose, guaiac dark blue-green. **Pileocystidia** elongate, weakly clavate, 0–1 septate, 5–7 µm across, only weakly colouring in SV. **Spores** 8–10.5 x 7–8 µm, with large warts up to 1 (-1.25) µm plus some connectives or catenulate warts, occasionally forming a very partial reticulum. **Spore deposit** dark ochre (IIIa-b).

Under conifers, often in moss, myrtle, or other heathland plants; only common in Scotland, rare elsewhere.

Russula carpini R. Girard & Heinem.
[carpini = associated with *Carpinus*]
Cap 5–10 cm, convex and usually remaining so, usually a mix of colours, from purple, brownish purple to greenish, often with small spots of red or purple. **Cuticle** peels about 1/2. **Gills** pale cream-ochre, only very slowly darkening to deep yellow. **Stem** clavate, white, sometimes stained yellowish at base or where handled. **Taste** mild. **Odour** nil to slightly fruity. **Macrochemicals**: $FeSO_4$ pale rose, guaiac more or less strongly positive, azure blue. **Pileocystidia** scarce, poorly staining in SV, clavate-cylindric, non-septate to 2–3 septate. **Spores** 8–10 x 7–8.5 µm, with isolated warts up to 1.25 µm. Spore-print dark yellow (IVd).

Restricted to *Carpinus* and *Ostrya carpinifolia*. Known from only a few localities in southern England (all with *Carpinus*) but should be more widespread.

Russula romellii Maire
[romellii = after L. Romell, 1854–1927, Swedish mycologist]
Cap 5–12 (-14) cm, large and fleshy, soon flattened and depressed at centre, smooth and shiny, shades of reddish violet, purple, or lilac often marked with yellowish areas or at centre. **Cuticle** peels 1/2. **Gills** fairly crowded, broad, interveined, pale to deep ochre-yellow. **Stem** cylindric-clavate, fleshy, white, spongy with age and often marked with ochre spots at base. **Taste** mild. **Odour** nil to slightly fruity when old. **Macrochemicals**: $FeSO_4$ pale pinkish rust, guaiac usually rather weak and slow, exceptionally rapidly blue. **Pileocystidia** slender, 3–5 µm across, 1–2 septate, only weakly staining in SV. **Spores** 7.2–9 (9.5) x 5.6–7.2 µm with warts to 1 µm, with many crests and connectives in a partial to complete reticulum. **Spore deposit** deep yellow (IVd).

Widely distributed but rather uncommon to rare, usually with *Fagus* but also with *Castanea* and *Carpinus*.

Best distinguished microscopically, the slender, weakly staining dermatocystidia and reticulate spores are characteristic. Macroscopically it can resemble other species such as **R. carminipes**, **R. melitodes** or **R. integra**.

Russula rubroalba (Singer) Romagn.
[rubroalba = red and white]
Cap 6.5–12 cm, fleshy, rounded and then soon flattened and depressed, smooth and shiny, bright scarlet, rust-red to even purplish red, often with centre paler, ochre-yellow or whitish. **Cuticle** adnate, hardly peeling. **Gills** crowded, attenuate-subdecurrent where they join the stem, brittle, anastomosing, deep ochre-yellow when mature. **Stem** cylindric to tapered at base, firm and fleshy, white to slightly yellow-brown at base. **Taste** mild. **Odour** nil to very slight of honey when old. **Macrochemicals**: $FeSO_4$ brownish pink, guaiac almost nil. **Pileocystidia** sparse, very slender, 3–5 µm across and attenuate, 0–1 septate, weakly staining in SV. Cuticular hyphal endings also very slender and attenuate. **Spores** 6.5–8.3 x 6–7 µm, with warts to 0.5 µm and with partial to complete reticulum of meshes and crests. **Spore deposit** deep yellow-ochre (IVd).

A rare species in Britain with only a couple of authentic records from the southern counties, associated with *Quercus* on slightly calcareous soils.

Closely related to **R. romelli** which differs chiefly in its cap colours and more elliptic spores with taller warts. It seems that this widespread interpretation of **R. rubroalba** is not the same as that noted by Romagnesi, which has a more purplish red cap and yellow tinted stem, resembling **R. aurea**. More work is needed on this species.

Russula curtipes F.H. Møller & Jul. Schaeff.
[curtipes = short stem]
Cap 6–12 cm, convex and soon flattened to slightly depressed, dry and matt to shiny and chagrined, very firm, flesh-pink to pinkish brown, vinaceous red, reddish brown, often paler cream at centre. **Cuticle** extremely adnate, non-peeling. **Gills** moderately spaced, thick and brittle, cream to ochre when mature. **Stem** noticeably short (3–7 cm), shorter than the cap diameter, often pointed at base, white, firm. **Taste** mild. **Odour** nil or weakly fruity. **Macrochemicals**: $FeSO_4$ greyish pink, guaiac rapidly blue. **Pileocystidia** scarce, clavate, non-septate. Cuticular hyphae ending in very long, very attenuate 'hairs' (similar structures may be seen in the common *R. virescens* or *R. violeipes*). **Spores** 7–9.5 x 6–7.5 µm, with warts up to 0.75 µm with a partial reticulum or at least numerous connectives. **Spore deposit** dark ochre (IIIc-IVa).

Under *Fagus*, mainly southern in the UK, rare. Very distinctive with its short stem and adnate cuticle.

Russula

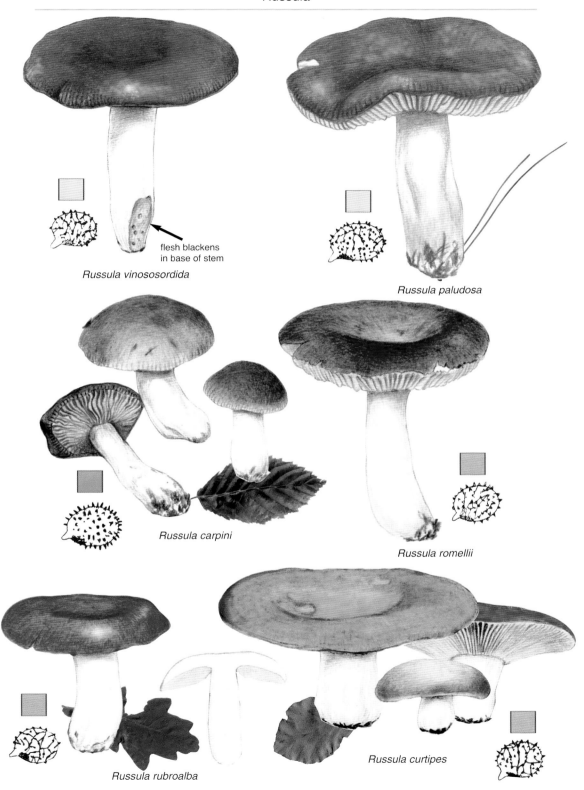

Russula velenovskyi Melzer & Zvára
[velenovskyi = after Josef Velenovsky, 1858–1949, Czech mycologist]
Cap 3.5–8.5 cm, often with a low umbo, smooth and shiny, deep red to brick-red, rust to almost orange, often more ochre-yellow at centre, sometimes darker. **Cuticle** peels 1/2–2/3. **Gills** fairly crowded, broad, brittle, pale cream to medium ochre. **Stem** cylindric-clavate, white, or flushed a little pinkish at base. **Taste** mild. **Odour** nil. **Macrochemicals**: $FeSO_4$ pale pink, guaiac deep blue-green. **Pileocystidia** long, filamentous, irregular and flexuose, 4–6.5 µm across, with distinct fuchsinophile encrustations, 0–1 septate. **Spores** 6.7–8.8 x 5.6–6.5 µm, warts up to 0.75 µm, isolated with scattered connectives but no meshes. **Spore deposit** deep ochre (IIIa-b).

Common and widespread in Britain, especially in the south, with broadleaf trees, especially *Betula*, *Quercus*, *Carpinus* and *Castanea*, more rarely with conifers.

Although variable in colour this species has a certain 'look'; the broad umbo, brick-red cap with paler centre and the medium ochre gills are all good field characters.

Russula tinctipes J. Blum ex Bon
[tinctipes = with a tinted or coloured stem]
Cap 6–12 cm, strongly convex then flattening, fleshy, bright scarlet, recalling *R. pseudointegra* but usually mottled or with paler ochraceous. **Cuticle** peels 1/4–1/3. **Gills** moderately crowded, yellow-ochre when mature. **Stem** clavate-cylindric, firm, white to faintly tinted pink, becoming slightly greyish when wet. **Taste** mild to very slightly acrid in young gills. **Odour** fruity-Pelargonium to honey-like. **Macrochemicals**: FeSO4 greyish pink, guaiac rapidly deep blue. **Pileocystidia** elongate filamentous to clavate, 1–3 septate. **Spores** 8–10.5 x 6.5–8.8 µm, with ± isolated warts to 0.3 µm high. **Spore deposit** yellow-ochre (IVb).

With *Quercus*, rare in Britain with only one record so far.

Russula seperina ** Dupain
[seperina = associated with the river Sèvre, France]
Cap 4–10 cm, fleshy, smooth and shiny, from vinaceous-purple, vinaceous-brown to partly or entirely yellow-ochre or even greenish. **Cuticle** peels 1/2. **Gills** moderately crowded, arched, deep yellow-ochre, becoming grey-black with bruising or age. **Stem** cylindric-clavate, firm to spongy, white staining grey-black with age, when scratched becoming red then black. **Taste** mild. **Odour** nil. **Macrochemicals**: $FeSO_4$ pinkish rust, guaiac quite rapid and intense blue. **Pileocystidia** sparse, cylindric filiform. 3.5–6 µm across, non-septate and sometimes encrusted with fuchsinophile material. **Spores** 7.5–10 x 6.5–8 µm, warts up to 0.8 µm, with some short crests or connectives but rarely any partial network. **Spore deposit** deep yellow (IVc-d).

Associated with *Quercus* in the Mediterranean region usually on calcareous soils.

Although recorded from Britain with hesitation by Rayner, there is no herbarium material, but as with other Mediterranean species it might just occur in the south.

Russula aurea Pers.
[aurea = golden]
Syn. *R. aurata*
Cap 5–10 cm, fleshy, rich copper-red to bright rust-red or almost blood, often bright golden yellow. **Cuticle** adnate. **Gills** crowded, obtusely rounded at cap margin, deep yellow-ochre when mature and the margins are typically (but not always) flushed with bright golden yellow. **Stem** clavate, rather fragile, white but often with a fine, bright yellow flush over part or all of the surface. **Taste** mild. **Odour** not distinct. **Macrochemicals**: $FeSO_4$ pale pink, guaiac deep blue. **Cuticle** without dermatocystidia or fuchsinophile hyphae. **Spores** 7–10 x 8.5–9 µm, with warts 0.5–0.7 µm high, numerous connectives and a partial network. **Spore deposit** deep ochre (IIIb-IVb).

Uncommon in the UK as a whole, but locally frequent in some areas under broadleaf trees.

Where it has bright yellow gill edges and yellow-flushed stem it is unmistakable, in cases where the yellow is reduced then the unusual cuticle is diagnostic.

Section *Paraincrustatae*
Subsection *Integrae*
caps variably coloured; spores yellow; with both fuchsinophile hyphae and encrusted cystidia

Russula melitodes Romagn.
[melitodes = dark toothed]
Cap 5–8 cm, fleshy, convex to flattened-depressed, with obtuse margin, slightly sulcate, brownish red, sienna, vinaceous-brown-red, chestnut. **Cuticle** peels 1/3. **Gills** fairly crowded, broad, pale cream then deep ochre-yellow. **Stem** cylindric-clavate, white, rather soft. **Taste** mild. **Odour** fruity-acidic or of honey as it dries. **Macrochemicals**: $FeSO_4$ rose-brown, guaiac bright blue. **Pileocystidia** cylindric-clavate, 5–10 µm across, 0–2 septate, often with fuchsinophile encrustations, also accompanied by filiform **fuchsinophile hyphae** with encrustations mostly only over the lower portions of the cells. **Spores** 7.4–10.5 x 6.8–6.9 µm, with large, isolated conical warts 1–1.4 µm high. **Spore deposit** ochre (IVb).

Locally common in southern Britain under *Quercus*, *Populus*, *Betula*, etc on diverse soils.

Russula integra (L.) Fr.
[integra = entire]
Syn. *R. polychroma* ss. Rayner
Cap 5–10 (12) cm, fleshy, expanding to become broadly depressed, smooth and dry to slightly viscid when wet, blood-red, reddish brown, chocolate-brown, sometimes olive. **Cuticle** peels 1/3–1/2. **Gills** crowded, broad, yellow to deep ochre, strongly intervened. **Stem** robust, cylindric-clavate, white, flecked with reddish brown at base. **Taste** mild. **Odour** very slight of iodoform or fruit. **Macrochemicals**: $FeSO_4$ greyish rose, guaiac slowly blue. **Pileocystidia** numerous, cylindric-clavate, with sparse fuchsinophile encrustations, also with **fuchsinophile hyphae**, narrow, attenuate. **Spores** 8–10.5 x 7–9 µm, with isolated sharp warts or spines up to 1.75 µm. **Spore deposit** yellow-ochre (IVc).

Frequent under conifers, mainly in Scotland.

Russula carminipes J. Blum
[carminipes = carmine stem]
Cap 5-10 cm, dark brownish purple, purple-violet, sometimes ochre at centre, smooth and matt to slightly pruinose. **Cuticle** peels 1/2. **Gills** yellow when mature, broad, obtuse at outer margin of cap, moderately spaced, clearly forked-anastomosing at base. **Stem** stout, white usually with a pale carmine or pinkish flush at the base. **Taste** mild. **Odour** weakly but distinctly fruity or honey-like. **Macrochemicals**: $FeSO_4$ pale rose-rust, guaiac rapidly azure blue. **Pileocystidia** clavate and numerous with incrustations, **fuchsinophile hyphae** narrow, fusiform and rather scarce. **Spores** 7.5–9.5 x 7–8.5 µm, almost spherical with very low warts and a very fine partial network. **Spore deposit** yellow (IVb)

This is quite a common species in the south growing with *Quercus*. Easily recognised by the faint red flush on the stem and very low spore ornamentation.

Russula

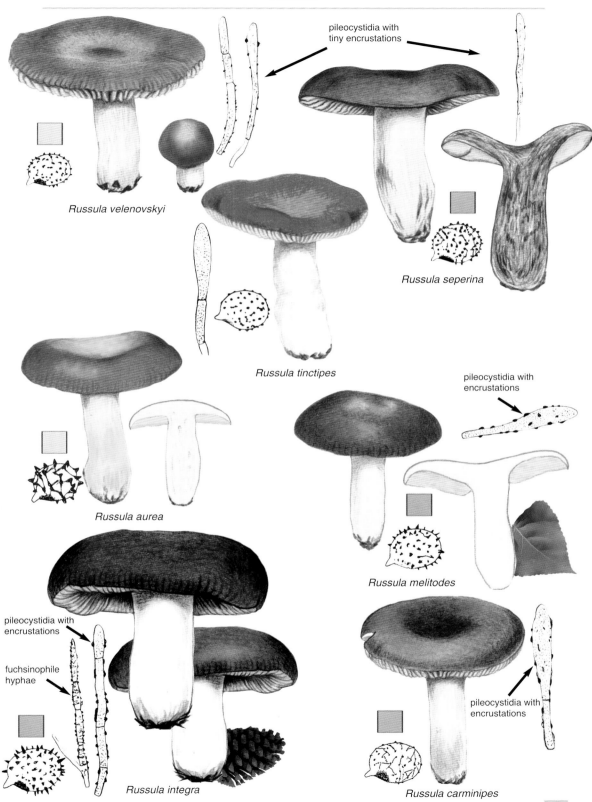

Russula cremeoavellanea Singer
[cremeoavellanea = cream-nut brown]
Cap 3–8 cm, convex to depressed, smooth, margin slightly sulcate, colour very variable but usually pale yellowish ivory to buff, with margin pale rose, to reddish brown. **Cuticle** peels 1/2. **Gills** medium spaced broad, obtuse at cap margin, cream to yellowish cream when old. **Stem** cylindric, white with surface often clearly rugulose-ridged, staining greyish brown at base. **Taste** mild. **Odour** nil to slightly fruity. **Macrochemicals**: $FeSO_4$ pale pink, guaiac rapidly intense blue. **Pileocystidia** slender clavate-capitate, mostly non-septate, encrusted. **Fuchsinophile hyphae** also present, encrusted after staining with Acid Fuchsin. **Spores** 6.5–8.5 x 6–7 μm, with isolated warts up to 1 μm. **Spore deposit** clear yellow (IVb).

Under broadleaved trees, especially *Betula*, usually at higher altitudes or in far north, but also known in southern England.

Subsection *Lepidinae*
taste bitter-acrid; spores white; flesh very firm

Russula lepida Fr.
[lepida = charming]
Syn. *R. rosea* ss. some authors
Cap 5–12 cm, soon flattening, dry, matt, bright vermilion-red, pinkish red, rose, often paler, with age and frequently spotted with whitish areas; rarely completely white; very firm to hard. **Cuticle** not peeling. **Gills** fairly crowded, ivory-white to cream, forking near the stem, hard and brittle. **Stem** hard like a stick of chalk, very variable in colour from completely white to flushed pinkish to entirely carmine-red. **Taste** mild with odd taste, of menthol, often of cedarwood pencils, sometimes slightly bitter. **Odour** pleasant, fruity-menthol. **Macrochemicals**: $FeSO_4$ slowly pinkish., guaiac from rapidly positive to almost nil. **Pileocystidia** pointed-clavate poorly staining. **Fuchsinophile hyphae** also present, but poorly defined. **Spores** subspherical 7–9 x 6.4–7.4 μm, with an almost complete reticulum and warts up to 0.5 μm. **Spore deposit** pale cream (IIa).

Not uncommon under broadleaf trees, especially *Fagus*, occasionally under conifers, widespread.

Russula amarissima Romagn. & E.-J. Gilbert
[amarissima = very bitter]
Cap 3–8 cm, rounded, matt, extraordinarily hard and brittle and a dull purple to purplish red. **Cuticle** hardly peeling. **Gills** cream showing a strong tendency to discolour yellow or brown along the edges with age. **Stem** white flushed purplish, extremely hard and brittle. **Taste** the flesh is extremely bitter and unpleasant to taste (compare with the cedarwood taste of the closely related *R. lepida*). **Odour** nil. **Macrochemicals**: $FeSO_4$ greyish rust, guaiac slowly blue. **Pileocystidia** are present, mostly cylindric to clavate, 0–2 septate. **Spores** are 7–9 x 6–7 μm with low warts and an incomplete reticulum. **Spore deposit** cream (IIa).

Usually found with *Quercus* or *Castanea* on sandy or gravelly soils. A rare species but easy to recognise.

Section *Tenellae*
Subsection *Puellarinae*
spores cream to yellow, taste mild, stem yellowing

Russula puellaris Fr.
[puellaris = resembling a girl]
Cap 3–6 cm, soon expanded and depressed, thin-fleshed, fragile with sulcate margin, viscid, shiny when wet, vinaceous-red, garnet-red, violaceous-purple, soon flushing yellowish with age. **Cuticle** peels 1/2–2/3. **Gills** broad, fragile, pale ochre-yellow when mature. **Stem** cylindric, soft and fragile, white but soon flushing dull ochre with age or on bruising. **Taste** mild. **Odour** nil. **Macrochemicals**: $FeSO_4$ pale pink, guaiac rapidly blue. **Pileocystidia** clavate, 0–2 septate, 4–7 μm across. **Spores** 7–9 x 5.7–7.2 μm, with warts from 0.6–1 μm, isolated or with very few connectives. **Spore deposit** cream (IIc-d).

Common and widespread under broadleaf and coniferous trees, usually in damper areas.

Russula versicolor Jul. Schäff.
[versicolor = of various colours]
Often difficult to distinguish from *R. puellaris* in the field, it tends to be larger and has a more diversely coloured cap, often with greenish tints and with the colours frequently concentrically zoned. **Taste** slightly acrid. **Odour** nil. **Pileocystidia** clavate or subclavate, 4–7 μm across, 4–5 septate. **Spores** 6–7.6 x 4.8–5.6 μm, warts up to 0.5 μm, with numerous short crests or connectives but no true reticulum. **Spore deposit** ochre (IIIa-c).

Frequent and widespread associated with *Betula*. Best distinguished from **R. puellaris** by the more robust stature, more acrid taste, spores with numerous short connectives and its multi-septate dermatocystidia.

Russula unicolor Romagn.
[unicolor = of one colour]
Cap 3–5 cm, matt, pruinose, violet, lilac-violet, blackish violet at centre. **Cuticle** peels 1/3–1/2. **Gills** fairly widely spaced, brittle, cream then pale ochre-yellow. **Stem** firm to spongy-brittle, white, staining ochre when bruised or with age. **Taste** strongly acrid. **Odour** pleasant, fruity. **Macrochemicals**: $FeSO_4$ pinkish rust, guaiac intensely and rapidly blue. **Pileocystidia** clavate-fusiform, broad, (5)-8–13 μm across, 0–1 septate, strongly staining in SV. **Spores** (6.5)7–8 x 5–6(6.5) μm, warts up to 0.6–0.7 μm, with numerous connectives forming a more or less complete reticulum. **Spore deposit** medium ochre (IIIa).

Very rare, only one British record from West Kent, associated with *Betula*.

Russula odorata Romagn.
[odorata = fragrant]
Cap 2–5 cm, thin-fleshed, margin sulcate, vinaceous, rosy, red-brown, buff with centre browner, fulvous-ochre, often yellowish where bruised. **Cuticle** peels almost completely. **Gills** broad, fragile, strongly interviened, saffron-yellow-ochre. **Stem** slender, fragile, white, sometimes yellowing when bruised or aged. **Taste** mild to slightly acrid in the gills. **Odour** aromatic or slight of *Pelargonium* when old. **Macrochemicals**: $FeSO_4$ pale salmon, guaiac rapidly dark blue. **Pileocystidia** numerous, cylindric to clavate or fusiform, 1–5 septate, strongly staining in SV. **Spores** 7–8.5 x 5.5–7 μm, with warts up to 1 μm high, numerous connectives or ridges sometimes forming a partial reticulum. **Spore deposit** deep yellow (IVc).

Not uncommon in the southern English counties, usually associated with *Quercus*. Very similar is **R. arpalices**, found under poplars and with a *Pelargonium* odour but with reticulate spores.

Russula parodorata ** Sarnari
[parodorata = next to odorata]
Cap 2–7 cm, very similar to *R. odorata*, cap varying from vinaceous to lilac-violet, dry, matt. **Odour** fruity. **Spores** with larger warts than *R. odorata*. **Pileocystidia** with 0–1 septa. Associated with *Quercus ilex* and *Q. suber* in the Mediterranean. Not recorded in Britain.

Russula

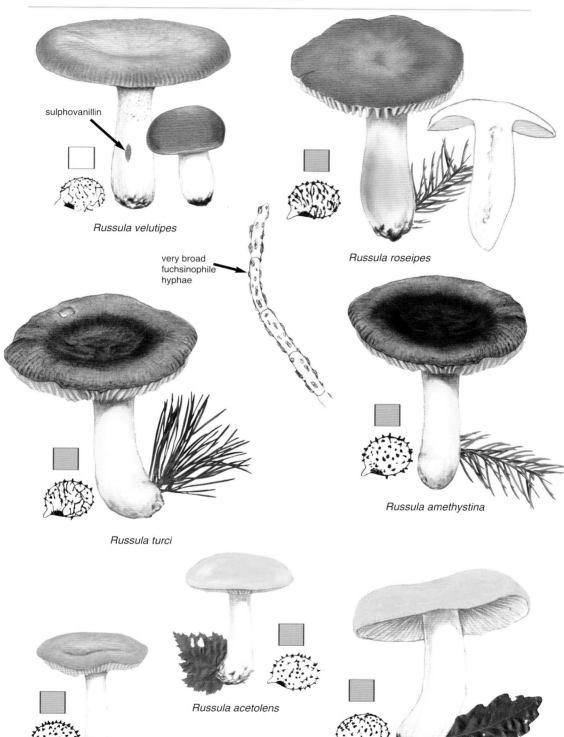

Russula postiana Romell
[postiana = after H.A. von Post, 1822–1911, mycologist]
Syn. *R. olivascens* ss. Sarnari
Cap 4–7 cm, broadly rounded, smooth and shining, pale herbage green, sage green, yellowish green. **Cuticle** peels 1/4–1/2. **Gills** crowded, very broadly rounded at front, soft, soon bright ochre-yellow. **Stem** cylindric-clavate, robust, white often stained ochre-brown. **Taste** mild. **Odour** almost nil to slightly fruity **Macrochemicals**: $FeSO_4$ pale pinkish, guaiac rather slowly blue. **Pileocystidia** absent, **fuchsinophile hyphae** numerous, 4–5 μm broad, narrow with clavate end cells, heavily encrusted. **Spores** 8–10.2 x 7–8 μm, spiny, warts isolated and up to 1 μm. **Spore deposit** medium yellow (IVc).

With conifers in subalpine and alpine areas, rare in the UK, some collections have been made in Scotland.

Sarnari uses the older, Friesian name of **R. olivascens** but this is such a confused and poorly understood name, used in many different senses by different authors, that I feel it is best treated as a *nomen confusum*.

Subsection *Integroidinae*
colours varied, red, yellow, purple, etc; with dark spores, flesh often reddening or blackening

Russula vinosa Lindblad
[vinosa = wine-red in colour]
Syn. *R. obscura*
Cap 4–12 cm, rounded then soon expanded and depressed, very firm and fleshy, smooth to matt and pruinose, deep purple, vinaceous, purplish red. **Cuticle** peels 1/3. **Gills** crowded, slightly arched, strongly interveined, pale ochre to straw-yellow, outer margin often tinted purplish, grey-black with age or bruising. **Stem** cylindric-clavate, robust, firm and fleshy, white to rarely flushed greyish rose, flushing greyish black with age or upon bruising. **Taste** mild. **Odour** nil. **Macrochemicals**: $FeSO_4$ greyish rose or even slightly greenish, guaiac slowly blue-green. **Pileocystidia** absent, **fuchsinophile hyphae** numerous, elongate, filiform, 4–7 μm across, although some cells swell to almost 12 μm, 3–5 septate, scattered uniformly across the cap surface. **Spores** large, 8.4–11.2 x 7–9 μm, with large isolated warts, cylindric to conical, up to 0.8 (1) μm high. **Spore deposit** deep ochre (IIIa-b).

Widespread in Scotland, not known with certainty in England, associated with *Betula* and *Pinus*.

Russula pubescens A. Blytt
[pubescens = with pubescent or pruinose cap]
Cap 7–13 cm, broadly rounded, robust and firm, dry and pruinose-matt, vinaceous, vinaceous-brown, extensively rust-brown to ochre-brown or yellow-ochre especially at centre with age. The margin as seen from below is obtuse and whitish. **Cuticle** peels 1/4–1/2. **Gills** rather crowded, broad, cream then soon pale yellow-ochre. **Stem** stout, cylindric-clavate, white stained brownish at base. **Taste** mild. **Odour** slightly fruity, a little like chanterelles. **Macrochemicals**: $FeSO_4$ pink, guaiac almost nil. **Pileocystidia** absent, **fuchsinophile hyphae** 4.5–7 μm across, in irregular, twisted tufts or clumps. **Spores** 8.4–12 x 6.7–9.3 μm, warts up to 0.8 μm, with a few distinct crests and connectives. **Spore deposit** medium ochre-yellow (IIIc-IVa).

Associated with *Betula* in Scandinavia, recently recorded in Scotland.

The very similar **R. vinosa** differs in its spores with entirely isolated warts, and its broader (8–12 μm) fuchsinophile hyphae which are scattered irregularly on the cap. The cap margin is not whitish and obtuse as it is in **R. pubescens**.

Russula claroflava Grove
[claroflava = clear yellow]
Cap 5–12 cm, subglobose, then broadly rounded for some time before flattening, smooth and shiny to slightly matt, bright, clear yellow, lemon-yellow, gold-yellow without olivaceous tones. **Cuticle** peeling 1/3-1/2. **Gills** moderately crowded, broad, ivory white then cream-ochre, bruising and aging grey-black especially on the edges. **Stem** cylindric-clavate, white, often briefly turning red when scratched before turning slowly grey-black. **Taste** mild. **Odour** nil. **Macrochemicals**: $FeSO_4$ pale pink, guaiac very slowly and weakly blue. **Pileocystidia** absent, **fuchsinophile hyphae** numerous, long, slender, 4.5–6 μm, with prominent external encrustations. **Spores** 8–9.6 x 6.4–7.8 μm, warts to 0.6 μm, mostly isolated with the occasional connective. **Spore deposit** medium ochre yellow (IIIc).

Common throughout Britain in wet areas under *Betula*. Most likely to be confused with **R. ochroleuca** but that species has duller, more olive-yellow tones, an often slightly yellowish, encrusted stem base, spores with reticulum and dermatocystidia present.

Russula sericatula Romagn.
[sericatula = less than silky]
Cap 5–9 cm, soon expanded and depressed, fleshy, smooth to slightly rugulose or matt, dull vinaceous brown, brown, brick-red, greyish fawn or buff sometimes recalling *R. vesca* (i.e. colour of 'old ham'). **Cuticle** peels 1/2. **Gills** fairly widely spaced, broad, interveined, pale cream to pale ochre-yellow when mature. **Stem** cylindric-clavate, fleshy, firm, white with rust-brown stains at base. **Taste** mild. **Odour** nil to fruity-acidic or even slight of shellfish. **Macrochemicals**: $FeSO_4$ greyish rose, guaiac more or less intensely blue. **Pileocystidia** absent, **fuchsinophile hyphae** long, slender, 3.5–6.5 μm across, multi-septate. Cuticular hyphae frequently with swollen end cells. **Spores** 7.2–9 x 6–7 μm, with isolated sharp warts up to 1.25 μm high. **Spore deposit** deep yellow (IVb).

Very rare, associated with broadleaf trees, one record from southern England but it may well be overlooked.

A rather nondescript species macroscopically but with very striking microscopic characters: prominent, isolated spore warts combined with fuchsinophile hyphae, plus the often swollen cuticular hyphal endings.

Russula caerulea (Pers.) Fr.
[caerulea = bluish]
Syn. *R. amara*
Cap 4–8 cm, dark violet, vinaceous-violet to brownish violet, very smooth and with a low umbo at the centre. **Cuticle** peels 1/3. **Gills** deep ochre when mature. **Stem** white and unchanging. **Taste** bitter in the cap cuticle while the flesh and lamellae are mild. **Odour** nil. **Macrochemicals**: $FeSO_4$ pink-rust, guaiac weakly blue. **Pileocystidia** absent, **fuchsinophile hyphae** – although present – are sparse and difficult to differentiate from the ordinary cuticular hyphae. **Spores** 8–9 x 6.5–7.5 μm with tall warts up to 1.25 μm high and some connectives or crests. **Spore deposit** dark yellow (IVa-b).

One of the commoner species under *Pinus* on acid soils. Usually easy to recognise with its glossy, deep violet to vinaceous, umbonate cap, dark spores and association with *Pinus*.

Lentinellus

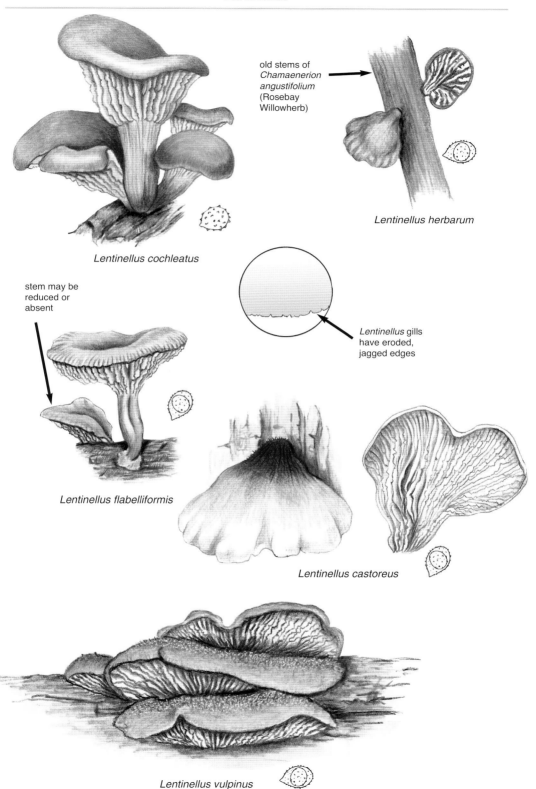

Genera & Species Index

Genera are in bold text
Species and common names are in normal text
Synonyms are in italics
Names in red are described in the text but not illustrated

abietina, Ramaria	56
abietinum, Trichaptum	80
abietinus, Hirschioporus	80
Abortiporus	72
acerrimus, Lactarius	154
acetolens, Russula	214
acrifolia, Russula	172
acris, Lactarius	168
acuta, Clavaria	50
adelphus, Paxillus	140
adulterina, Russula	196
adusta, Bjerkandera	72
adusta, Russula	172
aemilei, Boletus	118
aereus, Boletus	108
aeruginascens, Suillus	136
aeruginea, Russula	184
aerugineum, Leccinum	132
aestivalis, Boletus	108
Albatrellus	82
albida, Antrodia	88
albidum, Hydnum	36
albidus, Boletus	110
albonigra, Russula	172
alborufescens, Cantharellus	30
albostipitatum, Leccinum	134
albus, Pisolithus	20
Aleurodiscus	90
alnetorum, Russula	190
alpestre, Hericium	44
alutacea, Russula	218
amara, Russula	216
amarellus, Boletus	122
amarellus, Chalciporus	122
amarissima, Russula	208
amethysteus, Cantharellus	30
amethystina, Russula	214
ammoniavirescens, Paxillus	140
amoena, Russula	186
amoenicolor, Russula	186
amoenoides, Clavaria	52
amoenoides, Russula	200
amoenolens, Russula	178
amorphus, Aleurodiscus	90
anatina, Russula	184
angustifolius, Lentinellus	220
annosum, Heterobasidion	72
anthocephala, Thelephora	44
anthracina, Russula	172
Antrodia	88
Aphroditeola	142
appendiculatus, Boletus	116
appendiculatus, Butyriboletus	116
applanatum, Ganoderma	62
aquosa, Russula	186
Arched Earthstar	6
archeri, Clathrus	24
areolatum, Scleroderma	18
argillacea, Clavaria	52
var. sphagnicola	52
arhizus, Pisolithus	20
armeniacus, Rheubarbariboletus	124
armeniacus, Xerocomus	124
arpalices, Russula	210
artesiana, Russula	198
Artist's Bracket	62
Artomyces	56
Aseröe	24
asiaticus, Suillus	136
asperulispora, Clavaria	50
aspideus, Lactarius	152
asterospora, Clavaria	50
Astraeus	4
atlanticus, Lactarius	166
atrofusca, Clavaria	50
atroglauca, Russula	182
atropurpurea, Russula	186
atropurpureum, Lycoperdon	14
atrorubens, Russula	186
atrostipitatum, Leccinum	132
atrotomentosa, Paxillus	142
atrotomentosa, Tapinella	142
atroumbrina, Clavaria	50
aurantia, Tremella	102
aurantiaca, Hygrophoropsis	140
aurantiaca, Peniophora	92
aurantiacum, Hydnellum	38
aurantiacum, Leccinum	134
also *aurantiacum, Leccinum*	134
aurantiacus, Lactarius	162
aurata, Russula	206
aurea, Ramaria	56
aurea, Russula	206
Aureoboletus	120
aureum, Botryobasidium	90
auricula-judae, Auricularia	100
Auricularia	100
Auriscalpium	42
aurora, Cantharellus	32
aurora, Russula	214
australe, Ganoderma	62
avellaneum, Leccinum	130
azonites, Lactarius	168
azurea, Russula	212
badia, Imleria	118
badia, Russula	194
badiosanguineus, Lactarius	162
badius, Boletus	118
badius, Polyporus	68
Bankera	38
Baorangia	118
Barometer Earthstar	4
Basket Fungus	24
Battarrea	10
Beaked Earthstar	6
Bear's Head Fungus	44
bellini, Suillus	138
benzoinum, Ischnoderma	64
Berkeley's Earthstar	6
berkeleyi, Geastrum	6
bertillonii, Lactarius	170
bertillonii, Lactifluus	170
betularum, Russula	188
betulicola, Boletus	108
betulina, Fomitopsis	62
betulinus, Piptoporus	62
biennis, Abortiporus	72
Bjerkandera	72
Blackish Puffball	12
blennius, Lactarius	158
Blushing Bracket	70
Boletopsis	82
Boletus	108
Botryobasidium	90
botrytis, Ramaria	56
bovinus, Suillus	138
Bovista	14
bovista, Scleroderma	18
bresadolae, Russula	186
bresadolae, Suillus	134
var. flavogriseus	134
brevipes, Sparassis	82
britannicum, Geastrum	6
britannicus, Chroogomphus	144
britannicus, Lactarius	164
British Earthstar	6
brumale, Tulostoma	10
brumalis, Polyporus	68
brunneogriseolum, Leccinum	132
brunneoviolacea, Russula	210
brunneoviolascens, Lactifluus	170
bryantii, Geastrum	6
bubalinus, Hortiboletus	124
buccina, Guepiniopsis	104
Buchwaldoboletus	120
Buglossoporus	66
Butyriboletus	116
Byssomerulius	90
caerulea, Terana	88
caerulea, Russula	216
caeruleum, Hydnellum	38
caeruleum, Pulcherricium	88
caesia, Postia	64
caesius, Oligoporus	64
callitrichum, Leccinum	132
Caloboletus	110
Calocera	104
calocera, Favolaschia	82
calopus, Boletus	110
calopus, Caloboletus	110
Calvatia	10
camarophylla, Russula	172
campestre, Geastrum	6
camphoratus, Lactarius	166
caninus, Mutinus	26
Cantharellus	30
capsulifer, Pisolithus	20
carminipes, Russula	206
carneogrisea, Skeletocutis	86
carnosum, Ganoderma	62
carpini, Leccinum	130
carpini, Russula	204
castaneus, Gyroporus	138
castoreus, Lentinellus	220
caudatum, Lycoperdon	12
cavipes, Boletinus	136
cavipes, Russula	190
cavipes, Suillus	136
cepa, Scleroderma	18

Ceriporia	88	Coral Spine Fungus	44	duplicatus, Phallus	26
Ceriporiopsis	86	corallina, Typhula	48	duriusculum, Leccinum	132
Cerrena	72	coralloides, Hericium	44	*durus, Polyporus*	68
cessans, Russula	210	corium, Byssomerulius	90	Dwarf Earthstar	6
Chalciporus	122	corium, Mycenastrum	10	Dyer's Mazegill	68
chamaeleontina, Russula	214	cornea, Calocera	104	echinatum, Lycoperdon	12
chamiteae, Russula	200	corniculata, Clavulinopsis	52	edulis, Boletus	108
Chanterelle	30	cornucopioides, Craterellus	32	var. fuscoruber	108
Chicken of the Woods	66	corollinum, Geastrum	4	*elaeodes, Russula*	202
chloroides, Russula	172	coronatum, Geastrum	4	*elegans, Suillus*	134
Chondrostereum	90	Collared Earthstar	4	ellipsosporum, Hydnum	36
Chroogomphus	144	coralloides, Clavulina	58	emetica, Russula	188
chrysenteron, Boletus	126	corrugata, Hymenochaete	80	emilei, Baorangia	118
chrysenteron, Xerocomellus	126	*cramesinus, Aureoboletus*	120	*emeticella, Russula*	188
chrysonemus, Xerocomus	124	**Craterellus**	32	emeticicolor, Russula	212
chrysorrheus, Lactarius	154	cremeoavellanea, Russula	208	encephala, Tremella	102
cibarium, Ileodictyon	24	*cremor, Lactarius* ss. Bon	166	*engelii, Hortiboletus*	124
cibarius, Cantharellus	30	crinalis, Tomentella	96	ericaeum, Lycoperdon	14
cicatricata, Russula	202	crispa, Sparassis	82	eriksonii, Peniophora	92
ciliatus, Polyporus	68	*cristata, Clavulina*	58	erinaceus, Hericium	44
cimicarius, Lactarius	166	var. *incarnata*	58	*erythropus, Boletus*	110
cinerea, Clavulina	58	crocea, Ramariopsis	54	var. *discolor*	110
var. *gracilis*	58	crocipodium, Leccinum	130	erythropus, Typhula	48
cinerea, Peniophora	92	crosslandii, Clavaria	50	evosmus, Lactarius	154
cinereoides, Clavulinopsis	54	Crowned Earthstar	4	exalbicans, Russula	194
cinereus, Craterellus	32	cruciatus, Lysurus	24	excipuliforme, Lycoperdon	12
cinnabarinus, Pycnoporus	70	**Crucibulum**	16	*excipuliformis, Calvatia*	12
cinnamomea, Coltricia	78	cumulatum, Hydnellum	40	*excipuliformis, Handkea*	12
circellatus, Lactarius	158	cuprea, Russula	196	**Exidia**	100
cirrhatum, Hericium	44	cuprinus, Paxillus	140	*falcata, Clavaria*	50
cirrhatus, Creolophus	44	curtipes, Russula	204	*fallax, Boletus*	110
cisalpinus, Xerocomellus	126	cuticularis, Inonotus	74	*fageticola, Russula*	188
citriolens, Lactarius	154	cutifracta, Russula	180	faginea, Russula	202
citrinovirens, Boletus	122	cyaneobasileucum, Leccinum	132	False Truffles	20
citrinum, Scleroderma	18	cyanescens, Gyroporus	138	*farcta, Nidularium*	16
clariana, Russula	190	**Cyanoboletus**	120	farinipes, Russula	178
claroflava, Russula	216	cyanoxantha, Russula	180	faustiana, Russula	182
clathroides, Hericium	44	cyathuliformis, Lactarius	166	**Favolaschia**	82
Clathrus	24	**Cyathus**	16	favrei, Russula	200
Clavaria	50	**Daedalea**	64	*fechtneri, Boletus*	116
Clavariadelphus	58	**Daedaleopsis**	70	fechtneri, Butyriboletus	116
clavatus, Gomphus	32	Daisy Earthstar	4	fellea, Russula	178
clavipes, Russula	202	Dark-spored Puffball	14	felleus, Tylopilus	128
Clavulina	58	**Datronia**	72	var. *alutarius*	128
Clavulinopsis	52	debilis, Pterula	48	ferrea, Fuscoporia	88
cochleatus, Lentinellus	220	decipiens, Lactarius	164	ferruginascens, Cantharellus	30
var. *inolens*	220	decipiens, Lycoperdon	14	ferrugineum, Hydnellum	40
coliforme, Myriostoma	4	decipiens, Russula	196	ferruginosus, Fuscoporia	92
collinitus, Suillus	136	*declivitatum, Boletus*	124	Field Bird's Nest	16
Coltricia	78	decolorans, Russula	202	Field Earthstar	6
Common Earthball	18	decurrens, Ramaria	56	ferrugineus, Xerocomus	122
Common Puffball	12	deformis, Nidularium	16	fimbriatum, Geastrum	4
Common Stinkhorn	26	delica, Russula	172	fimbriatum, Steccherinum	96
communis, Boletus	124	deliciosus, Lactarius	150	firmula, Russula	198
conchata, Phellinopsis	78	densifolia, Russula	172	**Fistulina**	66
concrescens, Hydnellum	40	depilatum, Hemileccinum	118	*fistulosa, Macrotyphula*	44
confluens, Albatrellus	82	*depilatum, Leccinum*	118	var. *contorta*	44
confluens, Coltricia	78	*depilatus, Boletus*	118	fistulosa, Typhula	48
confluens, Nidularium	16	dermoxantha, Bovista	14	flabelliformis, Lentinellus	220
confluens, Phellodon	38	deterrimus, Lactarius	150	var. *herbarum*	220
confragosa, Daedaleopsis	70	Devil's Fingers	24	flaccida, Ramaria	56
confusus, Strobilomyces	128	dryadeus, Pseudoinonotus	74	Flaky Puffball	12
Coniophora	88	Dryad's Saddle	68	flava, Ramaria	56
consobrina, Russula	190	Dune Stinkhorn	26	flavispora, Russula	172
contorta, Typhula	48	Dung Bird's Nest	16	flavidus, Lactarius	152
controversus, Lactarius	156	dupainii, Suillellus	112	flavidus, Suillus	136

flavipes, Clavaria	52	**Gomphidius**	144	*impolitus, Boletus*	118
flavostellifera, Clavaria	52	**Gomphus**	32	impudicus, Phallus	26
flexuosus, Lactarius	158	gracile, Ileodictyon	24	var. togatus	26
floccopus, Strobilomyces	128	gracilis, Gomphidius	144	incarnata, Clavaria	52
floriforme, Geastrum	4	gracilis, Ramaria	56	incarnata, Clavulina	58
fluens, Lactarius	158	gracilipes, Hydnellum	40	incarnata, Peniophora	92
fluryi, Suillus	136	gracillima, Russula	194	incarnata, Typhula	48
Fluted Bird's Nest	16	granulatus, Suillus	136	*infundibuliformis, Craterellus*	32
foetens, Russula	176	Grassland Puffball	14	**Inonotus**	74
foetidum, Lycoperdon	12	*grata, Russula*	176	insignis, Russula	178
foliacea, Tremella	102	graveolens, Russula	200	*intactior, Russula*	194
fomentarius, Fomes	66	greletii, Clavaria	50	integra, Russula	206
Fomes	66	grevillei, Suillus	134	intermedia, Russula	196
Fomitopsis	62	var. badius	134	involutus, Paxillus	140
font-queri, Russula	210	**Grifola**	66	ionochlora, Russula	182
fornicatum, Geastrum	6	grisea, Russula	182	**Ischnoderma**	64
Four-rayed Earthstar	6	grisescens, Russula	188	joeides, Sarcodon	42
fragilis, Clavaria	50	**Guepinia**	100	*juncea, Macrotyphula*	48
fragilis, Russula	186	**Guepiniopsis**	104	juncea, Typhula	48
var. gilva	186	guilleminii, Clavaria	50	**Junghuhnia**	86
fragrans, Boletus	118	**Gyrodon**	138	*junquilleus, Boletus*	110
fragrans, Lanmaoa	118	**Gyroporus**	138	junquilleus, Neoboletus	110
fragrantissima, Russula	176	hadriani, Phallus	26	kluzakii, Caloboletus	110
fraxinea, Perenniporia	68	**Hapalopilus**	70	*kreiglsteineri, Clavaria*	50
friesii, Cantharellus	30	Hedgehog Puffball	12	*kunzei, Clavaria*	54
frondosa, Grifola	66	helios, Russula	214	kunzei, Ramariopsis	54
fuligineoalba, Bankera	38	helodes, Russula	192	laccata, Russula	188
fuligineoviolaceus, Sarcodon	42	helvelloides, Guepinia	100	lacera, Junghuhnia	94
fuliginosus, Lactarius	168	*helvelloides, Tremiscus*	100	lacrymans, Serpula	92
fulmineus, Chroogomphus	144	helvola, Clavulinopsis	52	Lactarius	150
fulvissimus, Lactarius	164	*hemichrysus, Buchwaldoboletus*		lacteus, Gyroporus	138
fumosa, Bjerkandera	72		120	*lacteus, Postia*	64
fumosa, Clavaria	50	**Hemileccinum**	118	lacunarum, Lactarius	164
furcata, Calocera	104	Hen of the Woods	66	laeticolor, Clavulinopsis	54
fuscoalbum, Leccinum	132	*henrici, Cantharellus*	30	**Laetiporus**	66
fuscoatra, Mycoacia	96	hepatica, Fistulina	66	laeve, Crucibulum	16
fuscosquamula, Hygrophoropsis	142	hepaticus, Lactarius	164	lakei, Suillus var. landkammeri	146
Fuscoporia	88	herbarum, Lentinellus	220	*laminosa, Sparassis*	82
fuscoroseus, Butyriboletus	116	**Hericium**	44	*lanatus, Boletus*	122
fusiformis, Clavulinopsis	52	*herinkii, Ceriporia*	88	**Lanmaoa**	118
galochroa, Russula	182	**Heterobasidion**	72	largentii, Ramaria	56
galochroides, Russula	182	heteroderma, Imleria	118	laricina, Russula	210
Ganoderma	62	heterophylla, Russula	180	laurocerasi, Russula	176
gausapatum, Stereum	90	hippophaeicola, Phellinus	76	**Leccinum**	130
geaster, Scleroderma	18	hirsuta, Trametes	70	*legaliae, Boletus*	114
Geastrum	4	hirsutum, Stereum	90	legaliae, Rubroboletus	114
gelatinosum, Pseudohydnum	102	hispidus, Inonotus	74	**Lentinellus**	220
gemmatum, Lycoperdon	12	holopus, Leccinum	132	**Lenzites**	72
Giant Polypore	66	Horn of Plenty	32	*leonis, Boletus*	120
Giant Puffball	10	**Hortiboletus**	124	Leopard-spotted Earthball	18
gibbosa, Pseudotrametes	70	**Hydnellum**	38	lepida, Russula	208
gibbosa, Trametes	70	**Hydnum**	36	lepidum, Leccinum	130
gigantea, Calvatia	10	hygrometricus, Astraeus	4	leporinus, Inonotus	74
gigantea, Langermannia	10	**Hygrophoropsis**	140	leptocephalus, Polyporus	68
gigantea, Phlebiopsis	90	**Hymenochaete**	80	**Leucogyrophana**	88
giganteus, Meripilus	66	**Hyphodontia**	90	lignicola, Buchwaldoboletus	120
gigasperma, Russula	196	hysginus, Lactarius	160	*lignicola, Pulveroboletus*	120
glandulosa, Exidia	100	ianthinoxanthus, Craterellus	32	lignyotus, Lactarius	168
glaucescens, Lactarius	170	igniarius, Phellinus	76	ligula, Clavariadelphus	58
glaucescens, Lactifluus	170	**Ileodictyon**	24	lilacea, Russula	212
globispora, Russula	196	*ilicis, Cantharellus*	30	*lilacinopruinosus, Cantharellus*	30
globispora, Tremella	102	illota, Russula	176	lilacinus, Lactarius	160
Gloeophyllum	80	imbricatus, Sarcodon	42	Lion's Mane	44
Gloeoporus	86	**Imleria**	118	*livescens, Russula*	178
glossoides, Calocera	104	**Imperator**	114	lividum, Lycoperdon	14
glutinosus, Gomphidius	144	impolitum, Hemileccinum	118	lividus, Gyroporus	138
glyciosmus, Lactarius	160	*impolitum, Leccinum*	118	lucidum, Ganoderma	62

Lumpy Bracket	70	
lundellii, Russula	196	
lupinus, Boletus	114	
lupinus, Rubroboletus	114	
luridiformis, Boletus	110	
luridiformis, Neoboletus	110	
ver. *immutatus*	110	
luridus, Boletus	112	
luridus, Lactarius	152	
luridus, Suillellus	112	
forma primulicolor	112	
var. erythroteron	112	
var. rubriceps	112	
lutea, Ramaria	56	
lutea, Russula	214	
luteoalba, Clavulinopsis	54	
luteocupreus, Imperator	114	
luteolus, Lactarius	170	
luteolus, Rhizopogon	20	
luteonana, Ramariopsis	54	
luteo-ochracea, Clavulinopsis	54	
luteotacta, Russula	194	
lutescens, Craterellus	32	
lutescens, Tremella	102	
luteus, Suillus	138	
Lycoperdon	12	
Lysurus	24	
macrospora, Hygrophoropsis	142	
maculata, Russula	196	
maculatus, Gomphidius	144	
mairei, Lactarius	156	
mairei, Russula	188	
mammiforme, Lycoperdon	12	
mammosus, Lactarius	162	
marginatum, Geastrum	4	
Meadow Puffball	14	
medullata, Russula	184	
melaleucus, Phellodon	38	
melaneum, Leccinum	130	
melanocyclum, Tulostoma	10	
melanocyclum, Polyporus	68	
melanoxeros, Craterellus	32	
melitodes, Russula	206	
melliolens, Russula	202	
mellita, Ceriporia	88	
melzeri, Russula	210	
mendax, Boletus	112	
mendax, Suillellus	112	
Mensularia	76	
meridionale, Scleroderma	18	
Meripilus	66	
mesenterica, Auricularia	100	
mesenterica, Tremella	102	
messapica, Clavaria	52	
micans, Typhula	48	
micheneri, Lentinellus	220	
microcarpus, Pisolithus	20	
microspora, Clavulinopsis	54	
minimum, Geastrum	4	
minutula, Ramariopsis	54	
minutula, Russula	212	
mirabilis, Queletia	10	
mokusin, Lysurus	24	
molare, Cerocorticium	96	
molaris, Radulomyces	96	
molle, Lycoperdon	14	
mollis, Datronia	72	
mollusca, Leucogyrophana	88	
moravicus, Aureoboletus	120	
moravicus, Boletus	120	
moravicus, Xerocomus	120	
Mosaic Puffball	12	
multifida, Pterula	48	
multizonata, Podoscypha	61	
mustelina, Russula	180	
musteus, Lactarius	160	
Mutinus	26	
Mycenastrum	10	
Mycoacia	96	
Myriostoma	4	
nauseosa, Russula	210	
Neoboletus	110	
nidulans, Hapalopilus	70	
Nidularia	16	
niger, Phellodon	36	
nigrescens, Bovista	14	
nigrescens, Leccinum	130	
nigrescens, Lycoperdon	12	
nigricans, Phellinus	76	
nigricans, Russula	172	
nitida, Junghuhnia	86	
nitida, Russula	212	
niveum, Tulostoma	10	
nobilis, Russula	188	
nodulosus, Inonotus	74	
nucatum, Leccinum	132	
nucleata, Exidia	100	
nympharum, Russula	196	
obliquus, Inonotus	74	
obscura, Russula	216	
obscuratus, Lactarius	166	
obscurisporus, Paxillus	140	
ochraceoflavum, Stereum	90	
ochraceovirens, Ramaria	56	
ochraceum, Steccherinum	96	
ochroleuca, Russula	200	
ochrospora, Russula	184	
occidentalis, Rhizopogon	20	
odorata, Russula	208	
odoratum, Gloeophyllum	80	
oedematopus, Lactifluus	170	
olida, Aphroditeola	142	
olida, Hygrophoropsis	142	
olivacea, Russula	218	
olivaceosum, Leccinum	132	
olivascens, Russula	216	
olivellus, Paxillus	170	
olivina, Russula	210	
olla, Cyathus	16	
omphaliformis, Lactarius	166	
omphalodes, Lentinellus	220	
Onion Earthball	18	
Onnia	80	
ovinus, Albatrellus	82	
ovoideisporum, Hydnum	36	
pachypus, Boletus	110	
pallens, Cantharellus	30	
pallescens, Russula	178	
pallidospathulata, Calocera	104	
pallidospora, Russula	172	
pallidus, Cantharellus	30	
pallidus, Lactarius	160	
palmata, Thelephora	44	
paludosa, Russula	204	
palustre, Leccinum	132	
pannocincta, Ceriporiopsis	86	
panuoides, Paxillus	142	
panuoides, Tapinella	142	
var. ionipes	142	
var. rubrosquamulosus	142	
paradoxa, Schizopora	86	
parasiticus, Pseudoboletus	120	
parasiticus, Xerocomus	120	
parazurea, Russula	180	
parodorata, Russula	208	
pascua, Russula	202	
Paxillus	140	
Pea-shaped Bird's Nest	16	
peckii, Hydnellum	38	
pectinata, Russula	176	
pectinatoides, Russula	176, 178	
pectinatum, Geastrum	6	
Pedicelled Puffball	12	
pelargonia, Russula	190	
pelletieri, Phylloporus	128	
Peniophora	92	
Peniophorella	90	
percandidum, Leccinum	132	
perennis, Coltricia	78	
pergamenus, Lactarius	170	
perlatum, Lycoperdon	12	
Perenniporia	68	
perplexa, Boletopsis	82	
persicina, Russula	194	
persoonii, Boletus	108	
persicolor, Rheubarbariboletus	126	
persicolor, Xerocomus	126	
pes-caprae, Albatrellus	82	
pfeifferi, Ganoderma	62	
phacorrhiza, Typhula	48	
Phaeolus	68	
phalloides, Battarrea	10	
Phallus	26	
Phellinopsis	78	
Phellinus	76	
Phellodon	38	
Phlebia	72, 90	
Phlebiella	90	
Phlebiopsis	90	
Phylloporia	78	
Phylloporus	128	
Physisporinus	86	
picinus, Lactarius	168	
picipes, Polyporus	68	
pierrhuguesii, Chalciporus	122	
pilatii, Lactarius	160	
pinetorum, Boletus	108	
pini, Peniophora	92	
pini, Porodaedalea	78	
pinicola, Boletus	108	
pinicola, Fomitopsis	62	
pinophilus, Boletus	108	
piperatus, Boletus	122	
piperatus, Chalciporus	122	
piperatus, Lactarius	170	
piperatus, Lactifluus	170	
Pisolithus	20	
pistillaris, Clavariadelphus	58	
placidus, Suillus	136	
plana, Exidia	100	
plumbea, Bovista	14	

plumbeobrunnea, Russula	180	*queletii, Boletus*	112	romellii, Leucogyrophana	92
Podoscypha	61	queletii, Russula	192	*rosea, Russula*	208
poikilochroa, Russula	186	queletii, Suillellus	112	roseipes, Russula	214
poikilochromus, Boletus	112	var. lateritius	112	roseus, Gomphidius	144
poikilochromus, Suillellus	112	var. rubicundus	112	rostratus, Lactarius	166
polychroma, Russula	206	quercilicis, Russula	198	Rosy Earthstar	4
polygonia, Peniophora	92	quercina, Daedalea	64	rubellus, Hortiboletus	124
Polyporus	68	quercina, Peniophora	92	ruber, Clathrus	24
polyrhizum, Scleroderma	18	*quercinum, Leccinum*	132	rubicundulus, Paxillus	140
pomaceus, Phellinus	76	quercinus, Buglossoporus	66	rubicundus, Phallus	26
porninsis, Lactarius	154	*quercinus, Piptoporus*	66	rubiginosa, Hymenochaete	80
Porodaedalea	78	quieticolor, Lactarius	150	rubinus, Chalciporus	122
porosporus, Xerocomellus	128	quietus, Lactarius	162	*rubinus, Rubinoboletus*	122
Porphyrellus	128	quisquiliaris, Typhula	48	rubra, Aseröe	24
porphyrosporus, Porphyrellus	128	radiata, Phlebia	90	rubra, Russula	198
Postia	64	*radicans, Boletus*	110	rubroalba, Russula	204
postiana, Russula	216	radicans, Caloboletus	110	**Rubroboletus**	112
Potato Earthball	18	radula, Schizopora	94	rubrocinctus, Lactarius	166
praestigiator, Neoboletus	110	**Radulomyces**	96	rubrosanguineus, Rubroboletus	114
praetermissa, Peniophorella	90	rameale, Stereum	90	rufa, Hygrophoropsis	142
praetervisa, Russula	176	**Ramaria**	56	rufa, Phlebia	90
pratense, Lycoperdon	14	**Ramariopsis**	54	rufescens, Geastrum	4
pruinatus, Boletus	126	*ramosum, Hericium*	44	rufescens, Hydnum	36
pruinatus, Xerocomellus	126	raoultii, Russula	188	rufipes, Clavulinopsis	54
pseudoaeruginea, Russula	182	ravenelii, Mutinus	26	rufomarginata, Peniophora	88
pseudoaffinis, Russula	178	reae, Clavulina	58	rufus, Lactarius	162
Pseudoboletus	120	recisa, Exidia	100	rugatus, Lactifluus	170
pseudodelica, Russula	172	*recisa, Tremella*	100	ruginosus, Lactarius	168
Pseudohydnum	102	Red Cage Fungus	20	rugosa, Clavulina	58
Pseudoinonotus	74	regalis, Sarcodon	42	rugosum, Stereum	90
pseudointegra, Russula	218	*regius, Boletus*	116	**Russula**	172
pseudoregius, Boletus	116	regius, Butyriboletus	116	rutila, Russula	198
pseudoscaber, Porphyrellus	128	renidens, Russula	194	rutilus, Chroogomphus	144
pseudoscabrum, Leccinum	130	repandum, Hydnum	36	salentina, Clavaria	52
pseudostriatum, Geastrum	6	repraesentaneus, Lactarius	152	*salicola, Leccinum*	134
pseudosulphureus, Boletus	110	resimus, Lactarius	152	salmonicolor, Lactarius	150
pterosporus, Lactarius	168	resinaceum, Ganoderma	62	sambuci, Hyphodontia	90
Pterula	48	resinosum, Ischnoderma	64	Sandy Stiltball	10
pubescens, Lactarius	156	reticulatus, Boletus	108	sanguifluus, Lactarius	150
pubescens, Russula	216	*reticulatus, Rhizopogon*	20	sanguinaria, Russula	192
pubescens, Trametes	70	**Rhizopogon**	20	sanguinolentus, Stereum	90
puellaris, Russula	208	**Rheubarbariboletus**	124	sanguinolentus, Physisporinus	86
puellula, Russula	210	rhodomelanea, Russula	190	sapinea, Russula	210
pulchella, Clavaria	54	*rhodopurpureus, Boletus*	114	**Sarcodon**	42
pulchella, Ramariopsis	54	rhodopurpureus, Imperator	114	sardonia, Russula	192
pulchra, Russula	194	var. *gallicus*	114	forma viridis	192
pulchrae-uxoris, Russula	190	forma *polypurpureus*	114	forma mellina	192
pulchrotinctus, Boletus	114	var. xanthopurpureus	114	sarnarii, Xerocomellus	126
pulchrotinctus, Rubroboletus	114	rhodopus, Russula	192	*satanas, Boletus*	112
pullei, Clavaria	50	*rhodoxanthus, Phylloporus*	128	satanas, Rubroboletus	112
pulverulentus, Boletus	120	ribis, Phylloporia	78	forma crataegi	112
pulverulentus, Cyanoboletus	120	**Rigidoporus**	70	sat*anoides, Boletus*	114
pumila, Russula	190	ripariellus, Xerocomellus	126	scabrosus, Sarcodon	42
punctatus, Phellinus	76	romagnesianus, Cantharellus	30	scabrum, Leccinum	130
pungens, Russula	198	romellii, Russula	204	Scaly Earthball	18
purpurascens, Chroogomphus	144	rosea, Clavaria	52	schistophilum, Leccinum	132
purpurea, Ceriporia	88	*rosea, Russula*	214	**Schizopora**	86
purpureum, Chondrostereum	90	roseofractum, Leccinum	130	schmidelii, Geastrum	6
pusillus, Cantharellus	32	roseolus, Rhizopogon	20	schweinitzii, Phaeolus	68
puteana, Coniophora	88	roseotinctum, Leccinum	132	**Scleroderma**	18
Pycnoporus	70	*rhodoxanthus, Boletus*	*114*	scoticus, Lactarius	156
pyriforme, Lycoperdon	14	rhodoxanthus, Rubroboletus	114	scrobiculatum, Hydnellum	40
pyrogalus, Lactarius	158	rigidipes, Leccinum	130	scrobiculatus, Lactarius	152
pyxidata, Clavicorona	56	risigallina, Russula	214	semisanguifluus, Lactarius	150
pyxidatus, Artomyces	56	rivulicola, Russula	202	seperina, Russula	206
quadrifidum, Geastrum	6	robertii, Russula	212	sepiarium, Gloeophyllum	80
Queletia	10	romagnesianus, Lactarius	168	sericatula, Russula	216

serifluus, Lactarius	166	
Sessile Earthstar	4	
silvestris, Russula	188	
silwoodensis, Xerocomus	124	
simplex, Sparassis	82	
simplex, Tremella	102	
sinuosus, Craterellus	32	
sinuosus, Pseudocraterellus	32	
Skeletocutis	86	
Soft-spined Puffball	14	
solaris, Russula	188	
sororia, Russula	178	
Southern Bracket	62	
Sparassis	82	
spathulata, Sparassis	82	
speciosus, Boletus	116	
Sphaerobolus	16	
sphagneti, Lactarius	162	
sphagnophila, Russula	212	
spiculosa, Exidia	100	
spinosulus, Lactarius	156	
spretus, Boletus	118	
squamosus, Polyporus	68	
squamosus, Sarcodon	42	
Starfish Fungus	24	
Steccherinum	96	
stellatus, Spaerobolus	16	
stercoreus, Cyathus	16	
Stereum	90	
stevenii, Battarrea	10	
stillatus, Dacrymyces	104	
stiptica, Postia	64	
stipticus, Oligoporus	64	
straminea, Clavaria	52	
Striated Earthstar	6	
striatum, Geastrum	6	
striatus, Cyathus	16	
stricta, Ramaria	56	
strobilaceus, Strobilomyces	128	
Strobilomyces	128	
Stump Puffball	14	
suaveolens, Hydnellum	38	
subappendiculatus, Boletus	116	
subappendiculatus, Butyriboletus	116	
subbotrytis, Ramaria	56	
subcaesia, Postia	64	
subcaesius, Oligoporus	64	
subcinnamomeum, Leccinum	130	
subdulcis, Lactarius	164	
subfoetens, Russula	176	
sublevispora, Russula	184	
subochracea, Phlebia	90	
subpruinosus, Cantharellus	30	
subrubens, Russula	200	
subrubescens, Albatrellus	82	
subterfurcata, Russula	182	
subtilis, Clavariopsis	54	
subtilis, Ramariopsis	54	
subtomentosus, Xerocomus	122	
subumbonatus, Lactarius	166	
subvolemus, Lactifluus	170	
Suillellus	112	
Suillus	134	
Sulphur Polypore	66	
sulphureus, Laetiporus	66	
syringinus, Lactarius	158	
tabidus, Lactarius	166
Tapinella	142
taxicola, Gloeoporus	86
tenuipes, Clavaria	50
tenuiramosa, Ramariopsis	54
tephroleuca, Postia	64
Terana	92
terrestris, Thelephora	44
Thelephora	44
tinctipes, Russula	206
tinctorius, Pisolithus	20
Tiny Earthstar	4
Tomentella	96
tomentosa, Onnia	80
tomentosus, Phellodon	36
torminosus, Lactarius	156
torulosa, Russula	192
torulosus, Phellinus	76
torosus, Imperator	116
torosus, Boletus	116
trabeum, Gloeophyllum	80
Trametes	70
Tremella	102
tremellosa, Phlebia	72
tremellosus, Merulius	72
Tremiscus	100
tremulae, Phellinus	78
Trichaptum	80
tricolor, Daedaleopsis	70
tridentinus, Lentinellus	220
tridentinus, Suillus	136
triplex, Geastrum	4
trivialis, Lactarius	160
truncata, Exidia	100
truncatus, Clavariadelphus	58
tubaeformis, Craterellus	32
Tulostoma	10
tumidus, Boletus	120
turci, Russula	214
turpis, Lactarius	156
Tylopilus	128
Typhula	48
uda, Mycoacia	96
uda, Phlebia	96
ulmarius, Rigidoporus	70
umbellatus, Polyporus	68
umbilicatum, Hydnum	36
umbrinella, Clavulinopsis	54
umbrinum, Lycoperdon	12
undulatus, Pseudocraterellus	32
unicolor, Cerrena	72
unicolor, Russula	208
urens, Russula	196
ursinus, Lentinellus	220
utilis, Lactarius	160
utriforme, Lycoperdon	12
utriformis, Clavaria	12
utriformis, Handkea	12
uvidus, Lactarius	152
vaga, Phlebiella	90
validus, Paxillus	140
variegatus, Suillus	138
variicolor, Leccinum	130
variisporus, Dacrymyces	104
velenovskyi, Russula	206
vellereus, *Lactarius*	170
vellereus, Lactifluus	170
velutipes, Russula	214
var. cretacea	214
venturii, Boletus	108
vermicularis, Clavaria	50
verrucosum, Scleroderma	18
versicolor, Boletus	124
versicolor, Russula	208
versicolor, Trametes	70
versipelle, Leccinum	132
var. flavescens	132
vesca, Russula	180
vesterholtii, Hydnum	36
veternosa, Russula	198
vietus, Lactarius	158
villosulus, Rhizopogon	20
vinosa, Russula	216
vinosobrunnea, Russula	218
var. perplexa	218
vinosopurpurea, Russula	198
vinososordida, Russula	204
violacea, Russula	190
violascens, Bankera	38
violeipes, Russula	184
virescens, Russula	184
viscida, Russula	198
viscidus, Suillus	136
viscosa, Calocera	104
vitellina, Russula	214
volemus, Lactarius	170
var. *oedematopus*	170
volemus, Lactifluus	170
vulgare, Auriscalpium	42
vulgaris, Rhizopogon	20
vulpinum, Leccinum	134
vulpinus, Lentinellus	220
wakefieldiae, Aleurodiscus	90
Weather Earthstar	4
White-egg Bird's Nest	16
xanthocyaneus, Boletus	116
xanthopus, Cantharellus	32
xanthopus, Neoboletus	110
xerampelina, Russula	200
Xerocomellus	126
Xerocomus	122
zollingeri, Clavaria	52
zonarius, Lactarius	154
zvarae, Russula	212
var. pusilla	212